Plant variation and evolution

Plant variation and evolution

D. BRIGGS

S.M. WALTERS

SECOND EDITION

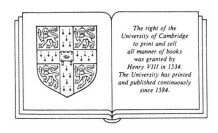

The right of the
University of Cambridge
to print and sell
all manner of books
was granted by
Henry VIII in 1534.
The University has printed
and published continuously
since 1584.

CAMBRIDGE UNIVERSITY PRESS

Cambridge
New York   Port Chester
Melbourne   Sydney

Published by the Press Syndicate of the University of Cambridge
The Pitt Building, Trumpington Street, Cambridge CB2 1RP
40 West 20th Street, New York, NY 10011, USA
10 Stamford Road, Oakleigh, Melbourne 3166, Australia

First published by Weidenfeld and Nicolson 1969
Second edition published by Cambridge University Press 1984
Reprinted 1986 (with corrections), 1988, 1990

Printed in Great Britain by University Press, Cambridge

Library of Congress catalogue card number: 83 – 14310

British Library cataloguing in publication data
Briggs, D.
Plant variation and evolution – 2nd ed.
1. Plant genetics
I. Title II. Walters, S.M.
581.1'5   QH433

ISBN 0 521 25706 9 hardback
ISBN 0 521 27665 9 paperback

UP

'The standing objection to botany has always been, that it is a pursuit that amuses the fancy and exercises the memory, without improving the mind, or advancing any real knowledge: and, where the science is carried no farther than a mere systematic classification, the charge is but too true. But the botanist that is desirous of wiping off this aspersion should be by no means content with a list of names; he should study plants philosophically, should investigate the laws of vegetation, should examine the powers and virtues of efficacious herbs, should promote their cultivation; and graft the gardener, the planter, and the husbandman, on the phytologist. Not that system is by any means to be thrown aside; without system the field of Nature would be a pathless wilderness; but system should be subservient to, not the main object of, pursuit.'

GILBERT WHITE

from *The natural history of Selborne* 1789, letter written on 2 June 1778

'... there is no better method for scientists of one period to bring to light their own unconscious, or at least undiscussed, presuppositions (which may insidiously undermine all their work) than to study their own subject in a different period. And ... when the writings of an earlier author have apparently been taken as the basis of subsequent work, constant scrutiny is necessary to prevent his presuppositions becoming fossilized, so to speak, in the subject and producing unnoticed inconsistencies when modifications have been made as a result of subsequent work.'

A.J. CAIN, 1958

To Daphne and Lorna

Contents

Preface to the Second Edition

When it was first proposed to establish laboratories at Cambridge, Todhunter, the mathematician, objected that it was unnecessary for students to see experiments performed, since the results could be vouched for by their teachers, all of them men of the highest character, and many of them clergymen of the Church of England.
BERTRAND RUSSELL, 1931

Although experimental science is firmly established, introductory books in biology are often influenced by the 'Todhunter' attitude – the teacher has all the answers. In many text books a judicious mixture of concepts, mathematical ideas and selected results of laboratory and field experiments is combined in an elaborate pastiche to provide a more or less complete edifice. Perhaps one or two areas of uncertainty may be indicated, but the general impression is of a house well built but awaiting the placing of the last few roof-tiles. Conversations with research biologists, however, quickly reveal a different picture. Almost nothing is settled: current views represent a provisional framework, and even some parts of the subject long held to be clarified are suddenly overturned by new discoveries.

Teaching experience reinforces our view that students of science should be shown the way in which, slowly and painstakingly, our present partial pictures have been built up, how and to what extent they are testable by experiment and observation, and in what way they remain vague or defective. A healthy scepticism in the face of the complexity of organic evolution is the best guarantee of real progress in understanding its patterns and processes.

The reception given to the first edition of our book was sufficiently encouraging to persuade us to venture on this new, largely rewritten, edition. We were helped in this decision by two fortunate circumstances: the return of one of us (D.B.) to rejoin the other in Cambridge in 1974, and the enterprise of our present publishers, Cambridge University Press, in acquiring from our first publishers, Weidenfeld and Nicolson, the remaining (American) stock of the first edition which has to some extent bridged the gap until this new book could be prepared. The general shape and intention of the book remain as they were: it aims to provide an authoritative, introductory text for students of botany, whilst at the same time satisfying a general reader with real interest in the subject. We have retained, with slight additions and modifications, the historical material in the earlier chapters, but have largely recast most of the second half of the book.

Our aim, as in the first edition, is to show how one branch of biology – namely the study of variation and evolution in plants – has developed in the last 300 years. This development has been increasingly scientific, leading to a realisation of the crucial roles of hypothesis and experiment. Throughout the book, in which we are uncommitted and even sceptical about neat explanations and simple formulations, we have tried to provide a critical but concise account of current excitements and advances in the subject, while paying full attention to the difficulties and uncertainties. Moreover, we have tried to engender a critical attitude in the mind of the reader.

We regret two changes in our second edition, one inevitable, the other voluntary. To keep the book within the reach of students in the 1980s, coloured illustrations which embellished the first edition have been sacrificed, although we have been encouraged by our publishers to use as much black and white illustration as is useful. The other cause for regret is of a quite different nature. In the late 1960s, when we drafted the first text, it did seem possible that a unified 'deme' terminology might become widely used, and we accordingly used it quite freely ourselves, believing that we thereby achieved extra clarity and consistency. Regretfully we have to admit that the terminology has not achieved wide acceptance and we have therefore reduced (though not entirely abandoned) our use of these terms.

In our account we have intentionally introduced and shown the connections between many complex subjects, and have provided references to important books and research papers in order that the student may build on our framework. Our survey concentrates on four main themes:

    1. We have examined the logical and historical framework – a fascinating story of early observation and experiment to do

justice to which would require a book to itself but which is normally, in our experience, almost wholly neglected in school and university courses.

2. While readily admitting the importance of laboratory-based observations and experiments, we believe that biologists who wish to understand evolution must study in detail the variation found in plants out of doors, whether in wild, cultivated or weedy habitats. Accordingly our book concentrates attention on the patterns of variation discovered in natural and in disturbed or artificial habitats, as well as examining the various processes which determine the patterns.

3. Taxonomists were the first biologists to study plant variation. An important theme in our book is, therefore, the impact that experimental studies have had on the theory and practice of taxonomy.

4. Our subject has a peculiar attraction in that it is developing very rapidly and at many levels, and experiments with sophisticated equipment have yielded important insights. As our examples show, however, knowledge of wild populations is so incomplete that a keen student appreciating the extent and limitations of our present understanding may make a very real contribution. We believe that the biologist interested in field observations and experiments has a very great advantage over other scientists, to whom the possibility of significant discovery is greatly restricted, or even denied, by the nature of their material.

A final word about the relevance of our subject in the twentieth century. As we shall see, our insights into microevolution owe as much to the study of cultivated as to that of wild plants. There should be no schism between the pure and applied aspects of biology. Moreover, it is to be hoped that in the next decade professional scientists will concentrate more and more on examining the living plant in its outdoor environment, for we believe that increased food production, enlightened nature conservation and a proper use of biological resources can only come if people appreciate and understand the patterns of plant and animal variation found all around us. We hope that this book, in its new edition, will continue to make a small but significant contribution to filling what we have found in our experience to be a very real gap in the available biological literature.

D. Briggs
S.M. Walters

Acknowledgments

In the preparation of the first edition of this book we received assistance from many people. It is not easy to select from these but we must record our special obligation to the following colleagues for careful reading and criticism of part or all of the manuscript and proofs: Dr A.M.M. Berrie, Mr A.O. Chater, Dr C.G. Elliott, Dr D. MacColl, Dr H. McAllister and Dr D.J. Ockendon.

For help with the second edition, we are especially grateful to Dr J.R. Akeroyd, Dr D.E. Coombe, Dr J.W. Kadereit, Professor A.J. Pollard, Professor D.M. Porter, Mr P.D. Sell, Dr S.I. Warwick, Dr H.L.K. Whitehouse and Professor D.H. Valentine. In thanking them we should make it clear that, although we have not in every case acted upon their advice, we have always valued it.

In the first edition we used photographs taken to our specification by Dr M.C. Lewis and Professor R. Sibson, and diagrams drawn by Mr K.G. Farrell, Mr J. Messenger and Design Practitioners Ltd. To reproduce some of these and many other new black and white illustrations for the second edition we used photographs taken by Mr C.B. Chalk and Miss J.A. Bulmer in the Botany School, Cambridge. The complete revision of the text needed several new drafts, which were carefully typed for us by Mrs A. Hill and Mrs A. James, and a new index was prepared for the second edition by Mrs L.M. Walters. We are particularly indebted to Dr M. Block, who spent many hours checking the typescripts and preparing the bibliography. We gratefully acknowledge all this help.

Finally, we wish to acknowledge our appreciation of the support of successive Professors of Botany in Cambridge. To Professor Sir Harry Godwin, F.R.S., in particular, we owe the original stimulus to write a book on this subject; and from his successors, the late Professor P.W. Brian, F.R.S. and the present holder of the Chair, Professor R.G. West, F.R.S., we have received sympathetic encouragement throughout.

Note on names of plants

Scientific names of plants are generally in accordance with Clapham, Tutin & Warburg: *Excursion Flora of the British Isles* (3rd edn, 1981) for British plants, and Tutin *et al.*: *Flora Europaea* (1964–1980) for European plants not in the British flora. In the few other cases not covered by either work we have used the name we believe to be correct. No authorities for names are given, since this would unduly overload the text and would be of little value to the general reader. Vernacular names, with capital letters, are used where they are familiar and unambiguous. Normally both scientific and vernacular names are given at the first mention.

1

Looking at variation

The endless variety of organisms, in their beauty, complexity and diversity, gives to the biological sciences a fascination which is unrivalled by the physical world. Some recognition of these different 'kinds' of organisms is a feature of all primitive societies, for the very good reason that man had to know, and to distinguish, the edible from the poisonous plant, or the harmless from the dangerous animal. We now distinguish more than a quarter of a million species of plants and many more than a million species of animals. What do we mean by 'species'? This is one of the questions we intend to explore in the chapters that follow.

'Kinds', species and natural classification

Suppose we assemble and examine an array of living organisms. We find breaks in the pattern of variation and can recognise a number of different 'kinds'. It is not, however, a question of simple discontinuity in particular features of the form of the organism. We can appreciate this if we think of a concrete example. In an Oak–Ash wood, the trees with simple, lobed leaves are Oak (*Quercus robur*) and others with pinnately compound leaves are Ash (*Fraxinus excelsior*). Looking more closely at these two 'kinds' of trees, we can find differences in bark, twig, flower or fruit – indeed in the characters of any part of the tree. It is clearly not only discontinuity which characterises the gross variation of organisms, but also the fact that the discontinuously varying characters are highly significantly correlated. Thus we see in the case of Oak that lobed leaves, alternately arranged on the twig, go together with an acorn fruit, whilst pinnate leaves are correlated with opposite arrangement and a winged 'key' fruit in Ash. Of course, this simple distinction may not hold in every case. It is possible, for example, to see in Botanic Gardens a peculiar Ash tree with simple leaves (*Fraxinus excelsior* forma *diversifolia*) and we know it is an Ash because it conforms in every other character. Nevertheless, the correlations of characters are strong enough to make broad agreement on the delimitations of 'kinds' a reasonably satisfactory aim for plant (or animal) taxonomy. Moreover, if we look at the 'folk taxonomies' developed independently by primitive peoples, we find a fair measure of agreement between the 'kinds' recognised and the genera and species scientifically named and classified by modern man.

There is a further dimension to the variation pattern, however. 'Kinds' of organisms can be arranged in a hierarchy, each higher group containing one or more members of a lower group. A study of primitive biological classifications reveals that this hierarchical classification is also a widespread feature of languages in general, and the particular hierarchy of genus and species which modern biology uses is really only a special case of a general linguistic phenomenon. The inescapable conclusion from such comparative studies is that hierarchical taxonomic classifications arise in human societies wherever those societies develop and that the detail of the treatment which we find in folk taxonomies reflects the importance of the organisms concerned in the life of the particular tribe or culture. This view of taxonomy as a product of man's need to understand, describe and use the plants and animals around him can in fact be applied to the Latin biological classification which we use at the present day.

Continuing our example, we find that the common European Ash tree, *Fraxinus excelsior*, is now grouped with some 60 other North American, East Asiatic and European species in the genus *Fraxinus*. This genus is in the Oleaceae, a family of 21 genera with 400 species, amongst which is also the Olive (*Olea europaea*). Likewise, *Quercus robur* belongs to a genus of some 300 species in the family Fagaceae. Other notable plants in the Fagaceae include the Beech (*Fagus*), the Southern Beech (*Nothofagus*) and the Sweet Chestnut (*Castanea*).

Although we may talk of associating species into genera, and genera into families, this is not what happened in the early days of biological classification. It is indeed arguable that the ordinary man's idea of a 'kind' of plant or animal corresponds in the history of classification more closely to a modern genus than it does to a species. The reason for this is that the classical and mediaeval ideas of kinds of plants were available in the eighteenth century to Linnaeus, who stabilised the scientific names in the 'binomial' form in which we still use them; and so the Linnaean genera (*Quercus* = Oak, *Fraxinus* = Ash, etc.) indicate the level of recognition of 'kinds' in the botany of mediaeval Europe. This is beautifully illustrated by the Carrot family (Umbelliferae), many of which were familiar plants in classical times in Europe, mainly because they were cultivated for food or flavouring (for example, *Daucus* = Carrot, *Pastinaca* = Parsnip) or because they were poisonous (for example, *Conium* = Hemlock). All these familiar European plants were accurately described and given what were later to become their generic names, long before Linnaeus. Fig. 1.1 shows a page of illustrations of the fruits of Umbelliferae from the earliest monograph of a family of plants, published by Morison in Oxford in 1672. With closer examination, and particularly as the exploration of the plants of the world proceeded, Linnaeus and his successors then distinguished other 'kinds' of Oak, Ash, Carrot, etc., retaining the name of the genus for all the species so distinguished (see Walters, 1961, 1962).

One other important element in our understanding of the history of biological classification must now be introduced: the idea of a 'natural' classification. Looking back as we can now do, equipped with our modern evolutionary picture of the diversity of living organisms, we are apt to feel that a natural classification must be one which accurately reflects the evolutionary history of the plants or animals concerned. As we shall see in later chapters, this was a view held by Charles Darwin and indeed others before him, although not widely expressed until after Darwin's ideas were generally accepted. Yet it is abundantly clear in the history of botanical and zoological thought that so-called 'natural classifications' were used and discussed long before evolution was an accepted picture. What did a 'natural' classification mean to John Ray in the seventeenth century or to Linnaeus in the eighteenth? This is a complex subject, but two things can be said. First, Ray, Linnaeus and all earlier biologists believed that, in describing and naming different 'kinds' of organisms, they were discovering a divine pattern in the created world, a pattern in which the different 'kinds' (modern genera or species) showed patterns of 'affinities' or relationships with each other. This was the 'natural order' awaiting Man's discovery and it had been fixed by God, who had separately created all the different 'kinds' of plants and animals. Secondly, the idea of natural classification should be logically contrasted with artificial classification; in the former, the sum total of 'affinity' or resemblance is taken into account in classifying groups together, whereas in artificial classifications a single character (or a small set of characters) defines the group. Thus, Linnaeus developed a so-called 'sexual system' to classify his genera of plants into what we would now call 'families', but in spite of the initial success of this system, which was frankly artificial, it was superseded by the 'natural system' of families which we use today.

Fig. 1.1. The fruits of the Umbelliferae, from Morison's monograph of the family published in 1672.

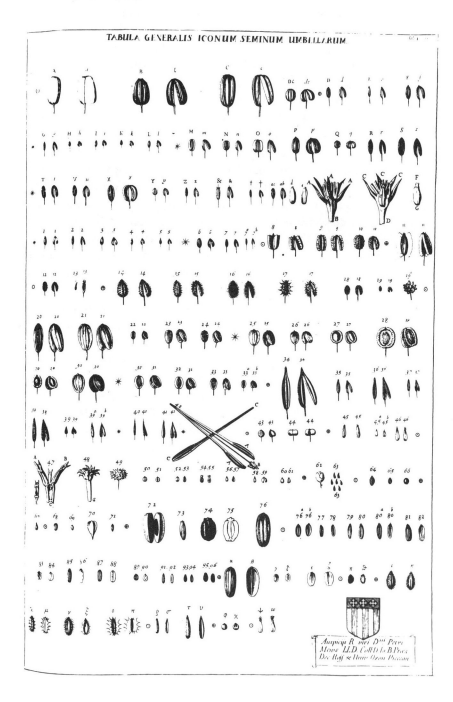

TAB. PRIMÆ GENERALIS.
ICONUM SEMINUM EXPLICATIO.

Nota prima genera novem intermedia contineri inter ⊖ & ⊕ genera intermedia secunda distingui ⁕ eorumque varietates distingui *

Nota 2do notas majusculas tam in Alphabeto, quam figuris, designare semina adhuc viridia conjunctim adhaerentia, ut videantur quasi unum; minusculas autem, eadem arida & exsiccata, hiulca, & à fibris pendentia exhibere.

A. a Indicant semina cachryos foliis ferulæ, sem. fungoso, pericarpio lævi incluso.
B. b. Ind. semina cachryos, foliis peucedani, sulcata, aspera.
C. c. Ind. sem. cachryos foliis peucedani, fungosa, sulcata, plana, majora.
D C. d c. Ind. sem. cachryos foliis peucedani, fungosa, sulcata, plana, minora.

D. d. Ind. semin. fœniculi vulgaris.
E. e. Ind. fem. cumini.
F. f. Ind. fem. Mei Athamantici.

G. g. Ind. fem. Mei spurii.
H. h. Ind. fem. Bulbocastani.
I. i. Ind. fem. selini montan. Pumili.
K. k. Ind. fem. visnagæ.
L. l. Ind. fem. saxifragæ Pannonicæ.

M. m. Ind. fem. levistici.
N. n. Ind. fem. Angelicæ sativæ.
O. o. Ind. fem. Angelicæ Canadensis.
P. p. Ind. fem. Smyrnii majoris.
Q. q. Ind. fem. Smyrnii Cretici.
R. r. Ind. fem. aftrantiæ, seu imperatoriæ nigræ.
S. s. Ind. fem. fileris, seu libanotidis fol. aquilegiæ.

Quoniam sequentis schematis, seu tabulæ in doctrina nostra contentæ supra, semina quoad formam & figuram omnino sibi invicem sunt similia, ideo nec delineanda nec sculpenda ipse curavimus; suasque figurarum pimpinellæ saxifragæ, sison, & sium. Omnia haec semina sese habent ut apis quasi, cum facili inter se distinguantur foliis, radice, odore, sapore, duratione, caeterisque, ut observare licet in cap. 2. doctrina istæs tradita de umbellis labatis minoribus, sem. striato laevi, minore praeditis. ⁕

T. t. Ind. fem. seselios Æthiopici fruticis.
V. u. Ind. fem. seselios pratensis herbæ.
Y. y. Ind. fem. cicutæ maximæ fœtidissimæ.
Z. z. Ind. fem. cicutæ minoris fatuæ.
Z. z. Ind. fem. cicutæ palustris.
&.&. Ind. fem. œnanthes odore viroso, cicutæ facie.
†. †. Ind. fem. œnanthes millefol. palustr. folio.

a c. a b. Ind. fem. crithmi marini vulgaris.
d & e. Ind. crithmi spinosi marini part. convexam, & cavam.
A A A B. Ind. capsulam cartilagineam crithmi spinosi.
C C C D. Ind. capsulam scissam crithmi spinosi.
e. locum vacuum indicat unde semen est exemptum.
F. G. Ind. bina semina conjuncta & exempta crith. spinosi.

†.1. Ind. fem. ammeos vulgaris.
2. 2. Ind. fem. ammeos perennis, nobis.
3. 3. Ind. fem. apii hortensis maximi B. in Prod.
4. 4. Ind. fem. apii Macedonici.
5. 5. Ind. fem. apii peregrini.

6. 6. Ind. fem. perfoliatæ longifoliæ , J. B.
7. 7. Ind. fem. perfoliatæ vulgatissimæ.
7b 7b Ind. fem. Bupleuri angustifolii.

8. 8. Ind. fem. laserpitii foliis saturate viridibus.
9. 9. Ind. fem. laserpitii fol. dilute virentibus.
10. 10. Ind. fem. laserpitii lob. latioribus, sem. crispo.

11. 11. Ind. fem. Thapsiæ latifol. villosæ 1. Clus.

12. 12. Ind. fem. Dauci seu carotæ lucidæ.
13. 13. Ind. ejusdem sem. reticulo suo involutum , & tectum.
14. 14. Ind. fem. caucalidis maj. purpur. Col.
15. 15. Ind. fem. caucalidis Monsp. seu lappæ Boariæ.
16. 16. Ind. fem. caucalidis tenuifol. mag. fl. albo.
17. 17. Ind. fem. caucalidis maj. Tingitanæ , nobis.
18. 18. Ind. fem. caucalidis parvo fl. & fructu, C. B. P.
19. 19. Ind. fem. caucalidis minoris sem. nodoso.
19a ... Ind. semina prioris in capitulum congesta.

20. 20. Ind. fem. ferulæ Matth.
21. 21. Ind. fem. panacis Asclepii.

---

22. 22. Ind. fem. anethi.
23. 23. Ind. fem. peucedani Italici.
24. 24. Ind. fem. peucedani fol. conjugatim positis.

25. 25. Ind. fem. sphondylii.
26. 26. Ind. fem. pastinacæ latifoliæ sativæ.
27. 27. Ind. fem. Tordylii maj. vulgaris, nobis.
28. 28. Ind. fem. Tordylii Syriaci limbo granul.
29. 29. Ind. fem. Tordylii Apuli minimi, Col.
30. 30. Ind. fem. panacis peregrini , Dod.

31. 31. Ind. fem. oreoselini , Clus.
32. 32. Ind. fem. Thysselii , Dod.
33. 33. Ind. fem. seselios palustris lactescentis , C B P.
33a 33b Ind. fem. libanotidis nigræ, Dod.

Nota harum omnium umbellarum (ab ⊙ supra, bucusque ad ⊕) semina membranacea , compressa , rotunda , aut subrotunda ichnographice depingi , & sculpi, utamque tantum visui offeri dorso ad dorsum adhaerens, cum autem exsiccata sunt , biulca dependent ex fibrillis binatim, & sciagraphice pinguntur & sculpuntur.

34. 34. Ind. fem. myrrhidis maj. albæ , odoratæ.
35. 35. Ind. fem. myrrhidis feminæ striat. aureo.
36. 36. Ind. fem. myrrhidis Daucoidis luteæ.
37. 37. Ind. fem. myrrhidis annuæ , fem. striat. lævi.
38. 38. Ind. fem. myrrhidis nodosæ , fem. aspero longo.
39. 39. Ind. fem. myrrhidis fem. asp. brevi.
39a 39a Ind. fem. myrrhidis fem. villosæ , seu incano.

40. 40. Ind. fem. cerefolii sativi.
41. 41. Ind. fem. cerefolii sylvestr. fem. lævibus.

42. Ind. scandicis , seu pectinis veneris rostrum in mucronem pungentem definens.
A A Indicant bina semina striata oblonga ad basin fibrillæ mediæ aperta , &hiulca, visui lateraliter apparentia.
B. Indicat tres radiolos quæ fibrillas sustinebant , quibus semina adhuc viridia nutriebantur.
C C Indicant semen & rostrum seminis inversa & repanda , in convexam partem resupinata; atque in ventricositate cava apparet fibrilla quæ fuit seminis adhuc viridis nutritiva.

43. 43. Ind. fem. coriandri sativi.
44. 44. Ind. fem. coriandri sylvestris , testicul.

UMBELLÆ
improprie dictæ.

45. 45. Ind. fem. valerianæ hortens. cum pappo.
45a 45b Ind. fem. valerianæ palustris cum pappo.
46. 46. Ind. fem. valerianæ marinæ maj. cum pappo.

47. Semen valerianellæ Indicæ monstrat fitum inter radios A & B divaricatos.
A B C Indigitant substantiam fungosam , seu membranaceam, continentem semina valerianæ Indicæ.
E   Ostendit semen inde exemptum.
48. Exhibet valerianellæ cornucopoidis , Col. semina membranacea, spinosa , seu echinata, in capitulum congesta , habita ratione feminum singularium. A summitate in spinas sursum versus desinunt , & ad basin lata sunt, crassa , unde cornucopoides ei nomen est.
G Indicat partem caulis , prædicta semina adhuc viridia sustinentis, & nutrientis.
49. indic. valerianellæ stellatæ feminæ maj. femina in capitulum congesta scabiosæ ritu.
50 Ind fem. valerianellæ fem. umbilic. nudo , rotund. partem convexam.
51. ind. ejusdem fem. umbilic. part. cavam.
52. ind. valerianellæ fem. umbilic. nudo longo part. convexam.
53. Ind. feminis ejusdem part. concavam.
54. Ind. fem. valerianellæ fem. hirsut. maj. umbilic. in medio & extremo part. convexam.
55. Ejusdem fem. part. concavam indicat.
56. Ind. valerianellæ fem. umbilic. minore hirsuto seminis partem convex.
57. Ejusdem feminis part. concavam indicat.
58. Valerianellæ Arvensis, præcocis humilioris bina fem. conjuncta indicat.
59. Ejusdem bina femina hiulca offert.
60. Ird. valerianellæ serotinæ , altioris fem. turgid. præditæ , partem convexam.
61. Seminis ejusdem partem concavam indicat.

62. Ind. capsulam valerianæ Græc.
63. 63. Ind. femina , quæ ibidem fuêre contenta.

---

64. Pimpi. sanguisorb. nostratis fem. quadratum indic.
65. Ind. fem. striatum pimp. agrimonoidis.
66. Ind. fem. pimp. spinolæ semper virentis.

67. Semen exhibet filipendulæ.

68. Involucrum striatum ostendit Thalictri vulgar. fl. luteis.
69. Ejusdem semen denotat inde exemptum cylindraceum , planum.
70. Involucrum triquetrum Thalictri Canaden. Corn. indicat.
71. Ejusdem semen ex involucro exemp. indicat oblongum, ad utrumque extremum graciles. *Icones seminum quæ sequuntur nunc partem convexam, nunc concavam exhibent.*

72. Cachryos sem. fungosæ substantiæ lævi inclusa, partem fungosam in qua semen latitat scissam , & apertam offert; atque ibidem semen in binas partes & visum ostenditur.
73. Seminis singularis ejusdem cachryos ex pericarpio fungoso exempti , partem gibbam ostendit.
74. Seminis cachryos pericarpio fungoso, sulcato, adhuc inclusi , partem convexam denotat.
75. Ejusdem feminis partem denotat concavam , qua jungebantur bina femina adhuc viridia.
76. Indicat feminis cachryos pericarpio fungoso, plano tecti , partem convexam seu exteriorem.

76a Ind. fem. fœniculi Azorici partem convexam.
76b Ind. ejusdem fem. part. concavam.
77. Ind. fœniculi vulgaris part. convexam seu gibbam.
78. Ejusdem seminis indic. part. concavam.
79. Sem. Mei Athamantici part. convex. indica.
80. Ejusdem fem. part. concavam exhibet.
81. Ind. fileris montani maj. part. striatam gibbam.
82. Ejusdem fem. part. concavam indicat.
83. Ind. fileris (non libanotidis ex festucis thorum) aquilegiæ fol. partem striatam convexam.
84. Ejusdem seminis partem concavam indicat.
85. Ind. Seminis imperatoriæ part. gibbam striatam offert.
86. Sem. aftrantiæ , seu imperatoriæ nigræ part. gibbam striat. rugosam offert.
87. Ejusdem seminis partem concavam ostendit.
88. Sem. sefelios Æthiopici , fruticis part. convex. ostendit.
89. Ind. ejusdem partem concavam exhibet.
90. Ejusdem sem. partem striat. convexam.
91. Ejusdem fem. partem offert concavam.
92. Ejusdem feminis part. concavam exhibet.
93. Ejusdem perfoliatæ vulgaris annuæ convexam striatam orthographice.
94. Indic. ejusdem feminis part. concavam orthographice.
95. Indic. feminis perfoliatæ annuæ longioribus J. B. part. convexam & rugosam orthographice.
96. Ejusdem fem. partem concavam, rugosam indic.

A Ind. fem. ferulæ Matth. partem gibbam , striatam.
B Ind. ejusdem fem. partem concavam.
C Ind. fem. anethi partem gibbam.
D Ind. ejusdem feminis part. concavam.
E Ind. fem. peucedani Italici part. gibbam.
F Ind. ejusdem feminis part. concavam.

G Ind. seminis carotæ lucidæ , part. convexam.
H Ind. ejusdem fem. partem concavam.

I Ind. fem. caucalidis purp. latifol. col. part. seu convexam echinatam.
K Ejusdem fem. partem concavam offert.
L Seminis caucalidis lappæ Boariæ Plinii Hist. part. partem convexam echinatam offert.
M Ejusdem feminis ostendit part. cavam.
N Sem. caucalidis magno fl. albo partem convexam gibbam echinatam.
O Ind. ejusdem fem. part. concavam.
P Sem. caucalidis maj. Tingitanæ Daucoidæ convexam echinatam indigitat.
Q Ejusdem fem. partem concavam offert.

R Sem. myrrhidis minoris sem. aureo , part. convex. striatam exhibet.
S Ejusdem feminis part. concavam offert.
T Indigitat fem. myrrhidis ann. nodosæ part. gibbam.
V Ejusdem indicat feminis part. concavam.
X Sem. coriandri syl. fœtidissimi part. exhibet gibbam seu convexam.
Y Ejusdem feminis part. concavam duobus foraminibus perviam ostendit.
Z Sem. valerianæ maj. hortens. partem convexam cum pappo in fem. superiore parte ostendit.
W Ejusdem feminis partem concavam cum pappo offert.

---

13. Sem. fem. singulare cumini repletus glob. C. B. P. dicunt vacuum.

13. Sem. ejusd. simen reticulo involutum & tectum.

Individual variation

To introduce the main themes of this book let us first look at individual variation. If we examine carefully a group of plants of any one species it is soon clear that not all the individuals are alike. For instance, in a seedling Ash the seedling leaves are simple, quite unlike the pinnate leaves of mature individuals. Here we have to consider developmental variation. Another source of variation between individuals can be plausibly attributed to factors of the external environment, such as coppicing by man or grazing by animals. Further, we may note that adjacent specimens of ordinary Ash and 'Weeping' Ash remain distinct in cultivation. Thus we can distinguish three main types of differences between individuals, which we might call 'developmental', 'environmentally induced' and 'intrinsic'. For many purposes we may very usefully distinguish, in a study of variation, a component which is fixed and heritable, which is the intrinsic genetic character of the 'kind' (ordinary growth habit *versus* 'weeping' habit in the case of Ash), from a component which is environmentally induced, non-heritable and imposed, as it were, from outside. In addition to these we have the phenomenon of developmental variation, by which the adult differs, often strikingly, from the immature individual.

In Oak and Ash, seedlings and all stages of development to dying and dead trees can be studied to provide information about individual variation. An interesting problem arises in many plants, however. How can one delimit individuals in grassland turf, for example? Many plants spread freely, in some cases exclusively, by vegetative means, establishing uniform clones, which may at least temporarily remain in physical contact by means of a common rhizome system, but whose ramets or branches are able to spread the plant and continue a quite independent existence. In such cases the concept of the individual is inapplicable. Looked at from this viewpoint, there is more in common between Man and the Ash tree than between, say, the Ash tree and the Daisy on the lawn. We are, perhaps naturally, inclined to think that Man and Ash trees are 'normal', for in these species individuals are born, mature, reproduce sexually and die, while the Daisy, apparently immortal, is 'odd'; but we ought to see this for what it is – an anthropocentric distortion – and be prepared to look at the facts objectively. In later chapters we shall examine recent work on the nature of individual variation and realise some of the importance of this difference.

The nature of species

The main problems to be examined in this book concern the nature of species. Anyone familiar with the vegetation of an area has to face a number of questions which have puzzled biologists increasingly in the last hundred years. For instance, why are certain species clearly distinct, while in other cases we find a galaxy of closely similar species, often difficult to distinguish from each other? Is there, in fact, any objective way of delimiting species? How can one account for the different degrees of intraspecific variation found in species? Further, why is it that hybridisation occurs in certain groups of plants and not in others?

In the early nineteenth century an examination of these questions, as we shall see in Chapter 2, produced a static picture of variation. Since 1859, however, with the publication of *On the origin of species*, all such studies have been made in the light of Darwin's profound generalisation of evolution by natural selection. Even though this theory has not always been accepted by biologists, it could never be ignored. It is too easy for this generation to forget the tremendous impact made upon biology by Darwin's work. The fact of evolution is taken for granted, in part because of the wealth of evidence assembled by Darwin and other scientists. There is often at the same time an uncritical acceptance of the theory – a tendency to say 'it must be true, for it is in all the books'. Implicit in Darwin's ideas is the assumption that evolution is still taking place. Thus in this book we shall not only look at the problems of species and patterns of variation but also indicate evidence for evolution, particularly experimental evidence for evolution on a small scale, often called 'microevolution'.

In discussing variation and microevolution it is essential to realise that the basic raw material for our studies exists in every country of the world. Even though we use mainly European and North American examples, because in these regions variation has been most carefully examined, similar examples can be found in countries where the flora is comparatively unknown. There is a further point of importance. It is not only 'natural', unspoiled vegetation which we can usefully study; equally illuminating results may be obtained from the study of communities radically altered by man, and in fact some of the most important insights into microevolution have come from studies of introduced plants, agricultural crops, weeds, and the vegetation of areas subject to pollution.

2

From Ray to Darwin

In 1660 Robert Sharrock, Fellow of New College, Oxford, wrote a book entitled '*History of the propagation and improvement of vegetables by the concurrence of art and nature*'. He was concerned in its early pages to debate a live issue of the day (Bateson, 1913):

> It is indeed growen to be a great question, whether the transmutation of a species be possible either in the vegetable, Animal or Minerall Kingdome. For the possibility of it in the vegetable; I have heard Mr Bobart and his Son often report it, and proffer to make oath that the Crocus and Gladiolus, as likewise the Leucoium, and Hyacinths by a long standing without replanting have in his garden changed from one kind to the other.

The Bobarts were both professional botanists. Sharrock investigated their claim, and found '... diverse bulbs growing as it were on the same stoole, close together, but no bulb half of the one kind, and the other half of the other'. In this age we find it hard to understand a belief in the possibility of transformation of Crocus into Gladiolus. Our reason for disbelief is partly concerned with the nature of evidence; we are not satisfied with the test for the alleged transmutation and would not have been content merely to examine the crowded underground parts. Another reason relates to current ideas of the nature of species. We have a different notion of species from that of the seventeenth century.

Ray and the definition of species

It was the English naturalist John Ray (1628–1705) who was probably the first man to seek a scientific definition of species. In his definition is an implied rejection of the sort of transmutation of species claimed by the Bobarts of Oxford, although in other passages in Ray's work he does not wholly dismiss the possibility of transmutation. For instance, he cites as reliable the case of cauliflower seed supplied by a London dealer, which on germination produced cabbage. Richard Baal, who sold the seed, was tried for fraud and ordered by the court at Westminster to refund the purchase money and pay compensation (De Beer, 1964).

Ray's views on species were published in 1686 in *Historia plantarum*. He wrote (trans. Silk in Beddall, 1957):

> In order that an inventory of plants may be begun and a classification of them correctly established, we must try to discover criteria of some sort for distinguishing what are called 'species'. After a long and considerable investigation, no surer criterion for determining species has occurred to me than distinguishing features that perpetuate themselves in propagation from seed.

He is concerned to define species as groups of plants which breed true within their limits of variation. This definition of species, based as it is partly upon details of the breeding of the plant, was a great advance upon older ideas, which relied entirely upon consideration of the external form.

Ray was also very interested in intraspecific variation. In his letters to various friends (collected by Lankester, 1848), he noted several striking variants of common plants discovered on his journeys around Britain. For example, at Malham in Yorkshire he noticed white-flowered as well as the normal blue-flowered Jacob's Ladder (*Polemonium caeruleum*), and from other localities he reported white-flowered Foxglove (*Digitalis purpurea*), double-flowered specimens of Water Avens (*Geum rivale*) and white-flowered Red-rattle (*Pedicularis palustris*). Ray also made observations on a prostrate variant of Bloody Cranesbill (*Geranium sanguineum* var. *prostratum*). He wrote to a friend: 'Thousands hereof I found in the Isle [Walney] and have sent roots to Edinburgh, York, London, Oxford, where they keep their distinction'. This report on the constancy of this distinct variant of *Geranium sanguineum* in cultivation is of particular interest and is referred to again in Chapter 4.

We may learn more of Ray's ideas on the nature of species and intraspecific variation by examining a discourse given to the Royal Society on 17 December 1674 (Gunther, 1928). In this, he expresses his concern that great care should be taken in deciding what constitutes a species and what variation is insufficient for specific distinction. He shows, for instance, that within a species there might occur individuals different from the normal in one or more of the following characters: height, scent, flower colour, multiplicity of leaves, variegation, doubleness of flower, etc. Plants differing by such 'accidents', as Ray calls them, should not be given specific status. He records the origin of one notable variant in his own garden: 'I found in my own garden, in yellow-flowered Moth-Mullein (*Verbascum*), the seed whereof sowing itself, gave me some plants with a white flower'. Concerning other variants Ray suggests that they are caused by growing plants under unnatural conditions, for example, a rich or a poor soil, extreme heat, and so on.

He concludes his analysis of specific differences and the problem of intraspecific variation as follows:

> By this way of sowing [rich soil, etc.] may new varieties of flowers and fruits be still produced in infinitum, which affords me another argument to prove them not specifically distinct; the number of species being in nature certain and determinate, as is generally acknowledged by philosophers, and might be proved also by divine authority, God having finished his work of creation, that is, consummated the number of species, in six days.

Ray's views on the origin of specific and intraspecific variation are here laid bare. Given sufficient regard for the variation patterns of a particular group of plants, a botanist should be able to avoid elevating 'accidental' variants to the level of the species. Species themselves were, for Ray, all created at the same time, and all therefore of the same age. That new species can come into existence, Ray denies, as this is inconsistent with the account of the Creation given in Genesis. This idea is again expressed in a passage written towards the end of his life: 'Plants which differ as species preserve their species for all time, the members of each species having all descended from seed of the same original plant' (Stearn, 1957).

Ray, an ordained minister himself, firmly upholds the doctrine of special creation. This view was almost universally accepted in the seventeenth century, Protestants being particularly influenced by the works of Milton. Indeed, a fundamentalist approach to the Biblical account of the Creation was characteristic of most biologists until the middle of the last century.

Linnaeus

In our examination of historical aspects of the subject, we must next study Linnaeus (Carl von Linné, 1707–78), the great Swedish systematist, who made extremely important contributions. He too, in *Critica botanica* (1737), championed the idea of the fixity of species:

> All species reckon the origin of their stock in the first instance from the veritable hand of the Almighty Creator: for the Author of Nature, when He created species, imposed on his Creations an eternal law of reproduction and multiplication within the limits of their proper kinds. He did indeed in many instances allow them the power of sporting in their outward appearance, but never that of passing from one species to

another. Hence today there are two kinds of difference between plants: one a true difference, the diversity produced by the all-wise hand of the Almighty, but the other, variation in the outside shell, the work of Nature in a sportive mood. Let a garden be sown with a thousand different seeds, let to these be given the incessant care of the Gardener in producing abnormal forms, and in a few years it will contain six thousand varieties, which the common herd of Botanists calls species. And so I distinguish the species of the Almighty Creator which are true from the abnormal varieties of the Gardener: the former I reckon of the highest importance because of their Author, the latter I reject because of their authors. The former persist and have persisted from the beginning of the world, the latter, being monstrosities, can boast of but a brief life. (trans. Hort, 1938)

The approaches of both Ray and Linnaeus were typological; they upheld the Greek philosophical view that beneath natural intraspecific variation there existed a fixed, unchangeable type of each species. It was the job of botanists to see these 'elemental species': 'natural variation' was in a sense an illusion.

We see also in the passage quoted above that Linnaeus had a very similar attitude to intraspecific variation to that of Ray. Stearn (1957), in an interesting analysis of the origin of Linnaeus' views, draws attention to his love for gardening and his experience as personal physician and superintendent of gardens to George Clifford, a banker and director of the East India Company. For some years Linnaeus, working on his great illustrated book on the plants in Clifford's gardens – the *Hortus Cliffortianus* – lived at Hartekamp, near Haarlem, in the centre of the Dutch bulb-growing area. Here thousands of varieties of Tulips and Hyacinths were grown. Linnaeus wrote the *Critica botanica* during this period, and no doubt his personal observations at the time prompted the following outburst: 'Such monstrosities, variegated, multiplied, double, proliferous, gigantic, wax fat and charm the eye of the beholder with protean variety so long as gardeners perform daily sacrifice to their idol: if they are neglected these elusive ghosts glide away and are gone'.

Other observations of Linnaeus in the *Critica botanica* show his familiarity with variation in wild plants and his experimental approach to problems. For instance, he studied flower colour, noting that purple flowers tend to fade after a few days, turning to a bluish colour; but '... sprinkle these fading flowers with any acid, and you will recover the pristine red hue'. Concerning aquatic plants he notes: 'Many plants which are purely aquatic put forth under water only multifid leaves with capillary segments, but above the surface of the water later produce broad and relatively entire leaves. Further, if these are planted carefully in a shady garden, they lose almost all these capillary leaves, and are furnished only with the upper ones, which are more entire.' As an example, Linnaeus gives *Ranunculus aquaticus folio rotundo et capillaceo*, the aquatic species of *Ranunculus* to which we refer later.

Linnaeus was particularly interested in cultivation and its effect upon plants:

> *Martagon sylvaticum* is hairy all over, but loses its hairiness under cultivation.
>
> Hence plants kept a long while in dry positions become narrow-leaved, as *Sphondylium, Persicaria* ... Hence broad-leaved plants, when grown for a long while in spongy, fertile, rich soil have been known to produce curly leaves, and have been distinguished as varieties, ... the following have been distinguished as '*crispum*': *Lactuca, Sphondylium, Matricaria*, etc.

The early botanical work of Linnaeus is extremely important in the history of ideas about species and variation. He championed firmly the reality, constancy and sharp delimitation of species. He was also concerned to refute the Ancient Greek idea of transmutation of species, which was still widely believed in his day. In *Critica botanica* he wrote:

> No sensible person nowadays believes in the opinion of the Ancients, who were convinced that plants 'degenerate' in barren soil, for instance, that in barren soil Wheat is transformed into Barley, Barley into Oats, etc. He who considers the marvellous structure of plants, who has seen flowers and fruits produced with such skill and in such diversity, and who has given more credence to experiments of his own, verified by his own eyes, than to credulous authority, will think otherwise.

Linnaeus is immortalised for botanists by his great work *Species plantarum* (1753), in which are described in a concise and methodical fashion all the approximately 5900 species of plants then known to man. In earlier works, and most explicitly in the theoretical *Philosophia botanica* (1751), he stresses the clear distinction between *species*, which were constituted as such by the Creator from the beginning, and mere *varieties*, which may be induced by changed environmental conditions, or raised by the art of gardeners. Nevertheless, there are not infrequently appended to particular species described in *Species plantarum* comments which show that Linnaeus did not always find specific distinctions clear: for example, under *Rosa indica* we find that 'the species of *Rosa* are with difficulty to be distinguished, with even greater difficulty to be defined; nature seems to me to have blended several or by way of sport to have formed several from one' (Stearn, 1957). It is even true that Linnaeus speculates, in a few cases, on the possible evolutionary derivation of one species from another in the pages of *Species plantarum*. Thus, under *Beta*

Fig. 2.1. Treatment of *Beta* in Linnaeus' *Species plantarum*, 2nd edn, 1763. (Now called *Beta vulgaris* with ssp. *maritima* and ssp. *vulgaris*.)

322   PENTANDRIA DIGYNIA.

Caulis *spithamæus, inæqualis, ramoso-patulus.* Folia *carnosa, lanceolata, obtusa.* Corymbi *axillares, dichotomi, folio longiores, terminati spinis inermibus.* Flores *sessiles in divaricationibus.*

BETA.

maritima.   1. BETA caulibus decumbentibus.
Beta caulibus decumbentibus, foliis triangularibus petiolatis. *Mill. dict.*
Beta sylvestris maritima. *Bauh. pin.* 118. *Raj. angl.* 4. *p.* 127.
Habitat in Angliæ, Belgii *littoribus maris.*

vulgaris.   2. BETA caule erecto.
Beta. *Hort. cliff.* 83. *Hort. upf.* 56. *Mat. med.* 113. *Roy. lugdb.* 220.

rubra.   α. Beta rubra vulgaris. *Bauh. pin.* 118.
β. Beta rubra major. *Bauh. pin.* 118.
γ. Beta rubra, radice rapæ. *Bauh. pin.* 118.
δ. Beta lutea major. *Bauh. pin.* 118.
ε. Beta pallide virens major. *Bauh. pin.* 118.

Cicla.   ζ. Beta alba vel pallescens, quæ Cicla officinarum. *Bauh. pin.* 118.
η. Beta communis viridis. *Bauh. pin.* 118.
Habitat - - - - - , ♂, *forte a maritima, in exoticis, prognata.*

vulgaris, we find, (Fig. 2.1), after a list of seven agricultural crop varieties, the fascinating statement: 'Probably born of *B. maritima* in a foreign country'. *B. maritima*, the wild Beet, is given separate treatment as a distinct species! This and several other cases are interestingly discussed by Greene (1909), who points out that there is good evidence to support the view that the dogmatic 'special creation' statements of *Philosophia botanica* and similar writings of Linnaeus did not, even in his earlier days, represent Linnaeus' real views, but were diplomatic writings to satisfy the 'orthodox ecclesiastics who, in his day, ruled the destinies of all seats of learning in Sweden'.

If he was orthodox on these matters in the main works, which established his academic and scientific reputations, Linnaeus allowed

Fig. 2.2. *Linaria vulgaris* (*a*) and its *Peloria* variant (*b*). (Illustration by Sowerby in Boswell Syme, 1866.)

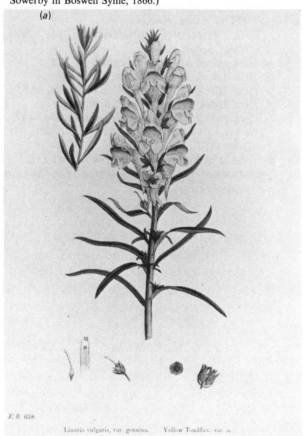

himself much more freedom in several of the 186 'dissertations' which his research students, following the mediaeval rules of disputation, had to 'defend' in Latin. It is clear from these writings that Linnaeus came to believe less rigidly in the fixity of species. For instance, in 1742 a student brought to him, from near Uppsala, an unusual specimen of Toad-flax (*Linaria vulgaris*). The flower was not of the usual structure but had five uniform petals and five spurs. Experiments showed that the plant bred true and Linnaeus called it *Peloria* (Fig. 2.2). After close study Linnaeus decided that *Peloria* was a new species which had arisen from *L. vulgaris* (Linnaeus, 1744). He also considered that certain other species might have arisen as a result of hybridisation. In *Plantae hybridae* (1751, see Linnaeus (1749–90)) records are given of 100 plants which might be regarded as hybrids. In *Somnus Plantarum* (1755, see Linnaeus (1749–90)) we read:

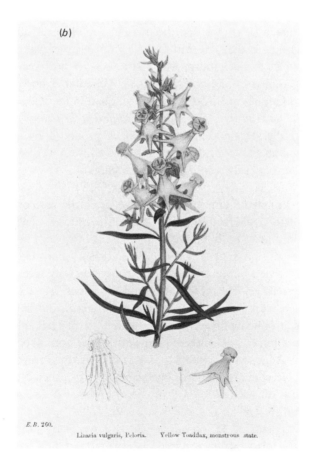

(b)

E.B. 260.

Linaria vulgaris, Peloria.   Yellow Toadflax, monstrous state.

'The flowers of some species are impregnated by the farina (pollen) of different genera, and species, inasmuch that hybridous or mongrel plants are frequently produced, which if not admitted as new species, are at least permanent varieties'. Later, in the summer of 1757, Linnaeus made what might be considered to be the first scientifically produced interspecific hybrid, between the Goatsbeards *Tragopogon pratensis* (yellow flowers) and *T. porrifolius* (violet flowers). Ownbey (1950), who studied *Tragopogon* in America, gives the following details of Linnaeus' experiment. After rubbing the pollen from the flower-heads of *T. pratensis* early in the morning, Linnaeus sprinkled the stigmas with pollen of *T. porrifolius* at about 8 a.m. The flower-heads were marked, the seed eventually harvested and subsequently planted. The first generation hybrid plants flowered in 1759, producing purple flowers yellow at the base. Seed of the cross, together with an account of the experiment and its bearings upon the problems of the sexuality of plants, formed the basis for a contribution to a competition arranged by the Imperial Academy of Sciences at St Petersburg. Linnaeus was awarded the prize in September 1760. It is of great historical interest that the seed sent by Linnaeus was planted in the Botanic Garden in St Petersburg, where the progeny flowered in 1761. Here it was examined by the great hybridist Kölreuter, who concluded that 'the hybrid Goatsbeard ... is not a hybrid plant in the real sense, but at most only a half hybrid, *and indeed in different degrees*'. It is also interesting that the second generation progeny produced by the intercrossings of Linnaeus' hybrid plants clearly showed segregation of different types, a very early record of genetic segregation which we discuss in Chapter 4.

We see how Linnaeus came to believe that, as in the case of *Peloria*, certain species had arisen from others in the course of time, and also that new species could arise by hybridisation. There is, however, contemporary evidence against Linnaeus' views (Glass, 1959). Adanson, an eighteenth-century French botanist whose originality has only recently been appreciated, tested *Peloria* more fully than Linnaeus. He found that *Peloria* specimens supplied by Linnaeus to the Paris Jardin des Plantes were not stable, producing flowering stems with both 'peloric' and normal flowers. Germination of seed of these plants often gave normal progeny as well as 'peloric'. Adanson concluded that the plant was a monstrosity, not a new species. He came to similar conclusions in two other cases, after experiments with an entire-leaved Strawberry (*Fragaria*) discovered by the horticulturalists Duchesnes and son at Versailles in 1766, and the famous laciniate plant of *Mercurialis annua* discovered by Marchant in 1715. There was also evidence against the origin of new species by hybridisation. Kölreuter made a large number of crosses in Tobacco (*Nicotiana*) and other genera. True-breeding new species were not produced by hybridisation; indeed the hybrids were often almost completely sterile, and even when they were fertile there was great variation in the progeny.

Returning to the writings of Linnaeus, we find that in later life he also gave further thought to the origins of the patterns of variation in plant groups. He speculated on the Creation as follows (*Fundamenta fructifica-tionis*, 1762, trans., quoted from Ramsbottom, 1938):

> We imagine that the Creator at the actual time of creation made only one single species for each natural order of plants, this species being different in habit and fructification from all the rest. That he made these mutually fertile, whence out of their progeny, fructification having been somewhat changed, Genera of natural classes have arisen as many in number as the different parents, and since this is not carried further, we regard this also as having been done by His Omnipotent hand directly in the beginning; thus all Genera were primeval and constituted a single Species. That as many Genera having arisen as there were individuals in the beginning, these plants in course of time became fertilised by others of different sort and thus arose Species until so many were produced as now exist ... these Species were sometimes fertilised out of congeners, that is other Species of the same Genus, whence have arisen Varieties (see Fig. 2.3).

Linnaeus ascribes here almost an evolutionary origin to present-day species, genera having been formed at the Creation, species formation being a more recent process. This most important change in Linnaeus' views relates to his hybridisation studies. He appears to have been convinced in later life that species can arise by hybridisation and moved away from the idea of a fixed number of species all created at the same moment in time. Linnaeus' early views on the fixity of species received wide circulation in Europe in his main works, *Critica botanica*, *Systema Naturae*, *Species plantarum*, while his more mature views, presented in the dissertations, did not have such a wide readership. So it is not surprising that even today he is often credited with rigid views on the question.

Buffon and Lamarck

In the mid-eighteenth century zoologists, too, were considering special creation. Linnaeus' contemporary, the French zoologist Buffon (1707–88), had also started his career with orthodox beliefs: 'We see him, the Creator, dictating his simple but beautiful laws and impressing upon each species its immutable characters'. Later, in 1761, however, he speculated on the mutability of species: 'How many species, being

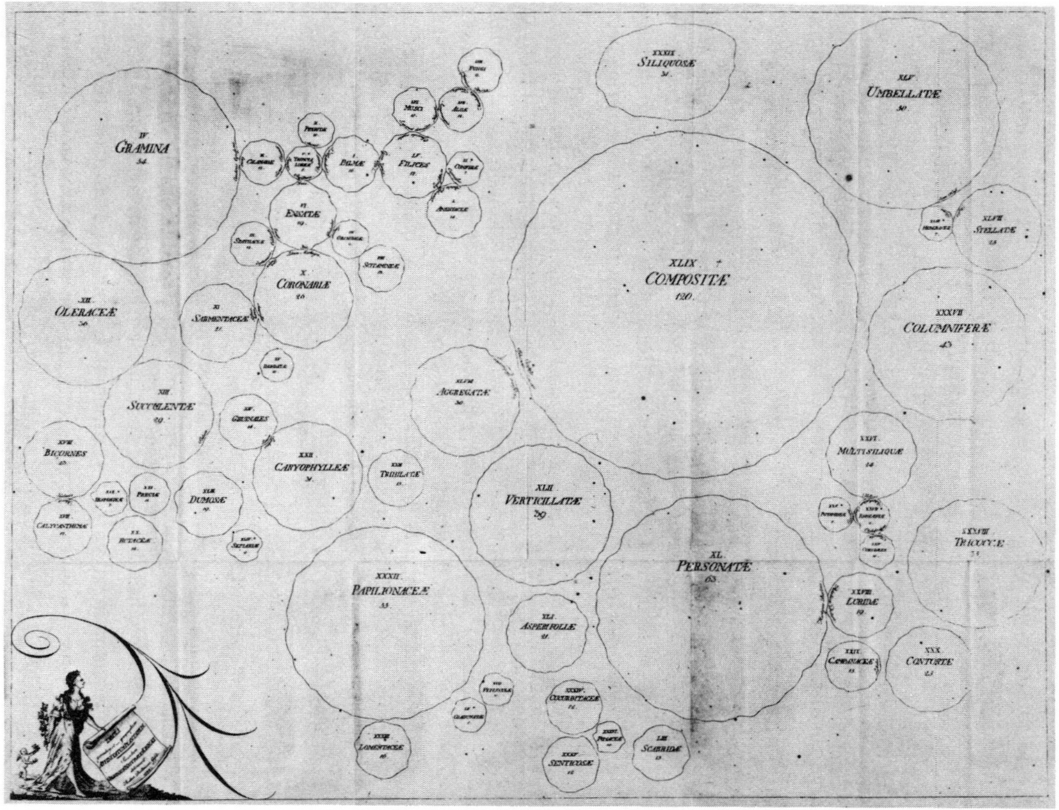

18

Fig. 2.3. Representation of Linnaeus' ideas on the relationships between plant families, by his pupil Giseke in 1792. The Roman numerals given for each family – are as in Linnaeus' *Genera plantarum* (6th edn, 1764), whilst Arabic numerals give the number of genera. The size of each circle is roughly proportional to the size of the family expressed as number of genera. At the edge of certain circles appear the names of genera which bridge the gap between groups. Such groups were intended to be a well-hidden secret, hence the more or less illegible names. Jonsell (1978) sees this as yet another example of Linnaeus' view that new forms could arise by hybridisation.

perfected or degenerated by the great changes in land and sea, ... by the prolonged influences of climate, contrary or favourable, *are no longer what they formerly were?*' (Osborn, 1894).

The speculative ideas of Buffon and others remained untested by experiment; the majority of botanists and zoologists, engaged as they were in the late eighteenth century on the naming and classification of the world's flora and fauna, believed in the fixity of species. This belief was indeed so firmly held by naturalists that Cuvier (1769–1832), who had studied many fossil animals, accounted for extinct species by postulating a series of great natural catastrophies, which wiped out certain intermediate species. Cuvier believed that there had been only one Creation, and that after each disaster the earth was repopulated by the offspring of the survivors. The last catastrophe was the Great Flood recorded in Genesis.

The doctrine of fixity of species was not without its critics in the nineteenth century. Lamarck (1744–1829), in his *Philosophie zoologique* (1809), attacked the belief that all species were of the same age, created at the beginning of time in a special act of Creation. He believed, much as Ray and Linnaeus did, that species could be changed by growth in different environments, but he also believed that modifications in plant structure brought about by environmental change were inherited (Elliot, 1914):

> In plants, ... great changes of environment ... lead to great differences in the development of their parts ... and these acquired modifications are preserved by reproduction among the individuals in question, and finally give rise to a race quite distinct from that in which the individuals have been continuously in an environment favourable to their development ... Suppose, for instance, that a seed of one of the meadow grasses ... is transported to an elevated place on a dry, barren and stony plot much exposed to the winds, and is there left to germinate; if the plant can live in such a place, it will always be badly nourished, and if the individuals reproduced from it continue to exist in this bad environment, there will result a race fundamentally different from that which lives in the meadows and from which it originated.

Thus Lamarck believed that a normally tall plant, dwarfed by growth at high altitude, would produce dwarf offspring. His belief in such an inheritance of acquired characters, which is closely paralleled in the writings of Erasmus Darwin (1731–1802), formed the basis of his evolutionary speculation – one species evolved into another as hereditary changes arose in a plant under the impact of environmental variation. Lamarck, who suffered ill-health at the end of his life and was totally blind for the last 10 years, did not make any experimental investigations in search of evidence for his hypothesis. He did, however, cite a number of possible cases of apparent change of species brought about by environmental agency. For example:

> So long as *Ranunculus aquatilis* is submerged in the water, all its leaves
> are finely divided into minute segments; but when the stem of this plant
> reaches the surface of the water, the leaves which develop in the air are
> large, round and simply lobed. If several feet of the same plant succeed
> in growing in a soil that is merely damp without any immersion, their
> stems are then short, and none of their leaves are broken up into minute
> divisions, so that we get *Ranunculus hederaceus*, which botanists regard
> as a separate species.

In this interesting quotation we see that Lamarck puts quite a different interpretation upon variation exhibited by aquatic *Ranunculus* species, from that of Linnaeus, who considered such changes in leaf characters part of intraspecific variation. We consider modern interpretations of this variation in Chapter 6.

Darwin

The views of Lamarck, at least in their simple form, are now rejected by most biologists. Our ideas on evolution are based on the work of Charles Darwin (1809–82). In an attempt to understand observations made on the voyage of the *Beagle*, Darwin began a series of note books on transmutation, the first dated July 1837 (De Beer, 1960), and he also wrote a sketch of his views in 1842 and a longer 'essay' in 1844 (Francis Darwin, 1909a, b). Many historians have investigated the development of Darwin's ideas, as revealed by his writings and annotations in his books (see, for example, Smith, 1960; Schweber, 1977), researches which suggest the crucial role of such writers as Lyell, Comte, Adam Smith, Quetelet and Malthus. Darwin delayed publication of his work. There is evidence that he wished to collect further information relevant to a theory of evolution (De Beer, 1963). It has also been suggested that he was anxious about the materialistic aspects of the theory and did not wish to offend the deeply religious views of his wife. In 1844 *Vestiges of the natural history of Creation*, a famous book advocating an evolutionary interpretation of nature, was published anonymously. It received strong condemnation in reviews and this may have contributed to the delay in the publication of Darwin's ideas (Schweber, 1977). In June 1858, Darwin received an essay from the naturalist Wallace (1823–1913), which set out a hypothesis almost identical to his own. Darwin's friends helped to resolve the delicate question of priority (Darwin, F., 1888, vol. 2, Chapters I–III) and, at a meeting of the Linnean Society on 1 July 1858, at which Wallace's paper was read, Darwin's views were represented by unpublished extracts from his papers and a letter he wrote to Professor Asa Gray in 1857. (The contributions of Darwin and of Wallace, which were prefaced by a letter from Lyell and Hooker, explaining the historical background, have been reprinted in Jameson, 1977.)

The main strands of the hypothesis of Darwin and Wallace may be summarised as follows:

1. Plants and animals vary. Darwin recognised two sorts of intra-specific variation: discontinuous variants (sports, monstrosities, jumps, saltations) and continuous variations (small, slight or individual differences, deviations or modifications). In his letter to Gray, Darwin wrote '*Natura non facit saltum*', making clear his view that it was individual differences and not saltations which were important in evolution.

2. Because of the fecundity of organisms there would be a geometri-cal increase in numbers unless checked. Such natural checks occur. Darwin and Wallace both acknowledged a debt to Mal-thus in their understanding of natural checks to population increase. (See Habakkuk, 1960; Pantin, 1960.)

3. As a consequence of these checks, only those individuals survive which have an inherent advantage over others in the population.

4. These better-fitted organisms, surviving this 'natural selection', pass on their 'advantage' to a proportion of their offspring.

5. Selection continues over thousands of generations and in a rapidly changing environment new variants take the place of the original organisms.

The principal ideas of Darwin's hypothesis are set out in the following quotations from the extract read at the Linnean Society meeting.

> De Candolle, in an eloquent passage, has declared that all nature is at war, one organism with another, or with external nature ... It is the doctrine of Malthus applied in most cases with tenfold force ... Reflect on the enormous multiplying power *inherent and annually in action* in all animals; reflect on the countless seeds scattered by a hundred ingenious contrivances, year after year, over the whole face of the land; and yet we have every reason to suppose that the average percentage of each of the inhabitants of a country usually remains constant. Finally, let it be borne in mind that this average number of individuals (the external conditions remaining the same) in each country is kept up by recurrent struggles against other species or against external nature (as on the borders of the Arctic regions, where the cold checks life), and that ordinarily each individual of every species holds its place, either by its own struggle and capacity of acquiring nourishment in some period of its life, from the egg upwards; or by the struggle of its parents ... with other individuals of the *same* or *different* species. But let the external conditions of a country alter ... Now, can it be doubted, from the struggle each individual has to

> obtain subsistence, that any minute variation in structure, habits or instincts, adapting that individual better to the new conditions, would tell upon its vigour and health? In the struggle it would have a better *chance* of surviving; and those of its offspring which inherited the variation, be it ever so slight, would also have a better *chance*. Yearly more are bred than can survive; the smallest grain in the balance, in the long run, must tell on which death shall fall, and which shall survive. Let this work of selection ... go on for a thousand generations, who will pretend to affirm that it would produce no effect ...

Darwin then goes on to give an example:

> If the number of individuals of a species with plumed seeds could be increased by greater powers of dissemination within its own area (that is, if the check to increase fell chiefly on the seeds), those seeds which were provided with ever so little more down, would in the long run be most disseminated; hence a greater number of seeds thus formed would germinate, and would tend to produce plants inheriting the slightly better-adapted down. (Darwin & Wallace, 1859)

After the meeting of the Linnean Society, Darwin spent the next few months writing the text which was eventually published in 1859 under the title *On the origin of species by means of natural selection*. Darwin saw the *Origin* as an introduction to a series of works. As Vorzimmer (1972) has shown, the book, which provided insights into so many facets of biology, provoked very considerable controversy and, reworking the material to accommodate the views of critics and the development of his own ideas, Darwin produced six editions in all. Sometimes substantial changes were made, as can be seen in the *variorum* text of the *Origin* produced by Peckham (1959).

We rightly give credit to Darwin for establishing, against considerable opposition, a plausible mechanism in natural selection to explain organic evolution. As is usually the case in the history of ideas, however, a careful reading of the literature reveals a number of statements of 'selectionist' ideas long before Darwin and Wallace. Indeed Zirkle (1941), in a remarkably interesting and little-quoted paper, provides abundant evidence of such ideas, tracing them back to the writings of Empedocles (495–435 B.C.) and Lucretius (99–55 B.C.).

In the sixth edition of the *Origin* Darwin himself provides a short historical review in which he makes it clear that the idea of natural selection had occurred to others and indeed had been published, for example by Dr W.C. Wells (1818) in 'An account of a white female, part of whose skin resembles that of a Negro' and by Patrick Matthew (1831) in his book *On naval timber and arboriculture*.

Many of the difficulties raised by contemporary critics, almost all of which are of interest to modern biologists, are very involved. Mivart (1871) produced a fascinating book which reviews all the contemporary difficulties in accepting Darwin's views, and Vorzimmer (1972) may be consulted for full details of many problems; here is a brief treatment of some of the most important.

The role of saltations in evolution

As Darwin pointed out, discontinuous variants – sports, monstrosities, etc. – are often sterile and therefore of little or no consequence in evolution. However, various biologists including Harvey (who studied a mutant *Begonia*, 1860), Huxley (in his review of the first edition of the *Origin*), and Asa Gray all suggested that saltations may be important in evolution.

> When you suppose one species to pass, by insensible degrees into another, so many facts of variation support your view that it does not seem very improbable; but where a generic limit has to be passed, bearing in mind how *persistent* generic differences are, I think we require a *saltus* (it may be a small one) or a real break in the chain, namely, a sudden divarication. (Unpublished letter from Harvey to Darwin, quoted by Vorzimmer, 1972.)

The mechanisms of heredity

In many ways this was the most important problem raised by Darwin's critics. For evolution to take place there must be selection of favoured varieties, such variants on crossing leaving better-adapted offspring. Darwin was unable to understand the mechanism of heredity, believing in a type of blending inheritance (see Chapter 4). Fleeming Jenkin, Professor of Engineering at Edinburgh University, writing anonymously in *The North British Review*, June 1867, showed a very serious weakness in Darwin's argument. He pointed out that if a rare variant favoured by natural selection appeared in a population, it would cross with the more abundant less-favoured plants in the population. Its hereditary advantage would then be lost in blending inheritance. How did favoured genetic variants ever become abundant? Darwin was never able satisfactorily to answer this criticism. Apparently Jenkin was not alone in appreciating this particular difficulty, for Vorzimmer (1972) argues that before Jenkin's review several biologists, including Darwin himself, were aware of the implication of blending inheritance.

The effect of chance

Jenkin also drew attention to another important problem – the effect of chance on the survival of favoured variants. He discusses his ideas in relation to hares:

... let us here consider whether a few hares in a century saving themselves by this process [burrowing] could, in some indefinite time, make a burrowing species of hare. It is very difficult to see how this can be accomplished, even when the sport is very eminently favourable indeed; and still more difficult when the advantage gained is very slight, as must generally be the case. The advantage, whatever it may be, is utterly out-balanced by numerical inferiority. A million creatures are born; ten thousand survive to produce offspring. One of the million has twice as good a chance as any other of surviving; but the chances are fifty to one against the gifted individuals being one of the hundred survivors. No doubt, the chances are twice as great against any one other individual, but this does not prevent their being enormously in favour of *some* average individual. However slight the advantage may be, if it is shared by half the individuals produced, it will probably be present in at least fifty-one of the survivors, and in a larger proportion of their offspring; but the chances are against the preservation of any one 'sport' in a numerous tribe.

The limits of variation

While not the first to raise this problem, Jenkin, in the following quotations, points to what many biologists saw as an important difficulty in Darwin's theory.

If we could admit the principle of a gradual accumulation of improvements, natural selection would gradually improve the breed of everything, making the hare of the present generation run faster, hear better, digest better, than his ancestors; ... Opinions may differ as to the evidence of this gradual perfectibility of all things, but it is beside the question to argue this point, as the origin of species requires not the gradual improvement of animals retaining the same habits and structure, but such modification of those habits and structure as will actually lead to the appearance of new organs. We freely admit, that if an accumulation of slight improvements be possible, natural selection might improve hares as hares, and weasels as weasels, that is to say, it might produce animals having every useful faculty and every useful organ of their ancestors developed to a higher degree; more than this, it may obliterate some once useful organs when circumstances have so changed that they are no longer useful, for since that organ will weigh for nothing in the struggle of life, the average animal must be calculated as though it did not exist.

We will even go further: if, owing to a change of circumstances some organ becomes pre-eminently useful, natural selection will undoubtedly produce a gradual improvement in that organ, precisely as man's selection can improve a special organ ... Thus, it must apparently be conceded that natural selection is a true cause or agency whereby in some cases variations of special organs may be perpetuated and accumulated, but the importance of this admission is much limited by a consideration of the cases to which it applies: ... Such a process of improvement as is described could certainly never give organs of sight, smell, or hearing to organisms which had never possessed them. It could not add a few legs to a hare, or produce a new organ, or even cultivate any rudimentary organ which was not immediately useful to an enormous majority of hares ... Admitting, therefore, that natural selection may improve organs already useful to great numbers of a species, does not imply an admission that it can create or develop new organs, and so originate species.

The origin of complex organs and structures

Many biologists, assuming a useless incipient stage in the development of complex organs (for example, the eye) could not see how such structures could have evolved as a consequence of natural selection either of small individual differences or of saltations. As can be seen by the example in Fig. 2.4, similar difficulties are encountered in complex floral structures.

What has been called 'the creative power of selection' remains a difficult concept, providing material for arguments essentially similar to those which Darwin faced but often, of course, in the more recent literature, based on more sophisticated examples of claimed 'adaptations'. Williams (1966) provides a particularly interesting and well-argued treatment of these themes.

The role of isolation

Whilst it may be argued from Darwin's writings (Mayr, 1963) that he saw isolation as providing a 'favourable circumstance' for speciation, it was the German naturalist Wagner (1868), pointing to the probable loss of favourable variants in 'blending' inheritance, who argued that spatial isolation was a necessary condition of speciation. This important difference of opinion is discussed in some detail later.

The age of the earth

The modern biologist may find it difficult to realise the extent to which controversies about the age of the earth and its rocks preceded and eventually made possible the Darwinian revolution. The study of fossils, or 'figured stones' as they were first called, undoubtedly produced questioning about the age of the earth as early as the middle of the sixteenth century, when, for example, Conrad Gesner wrote an illustrated

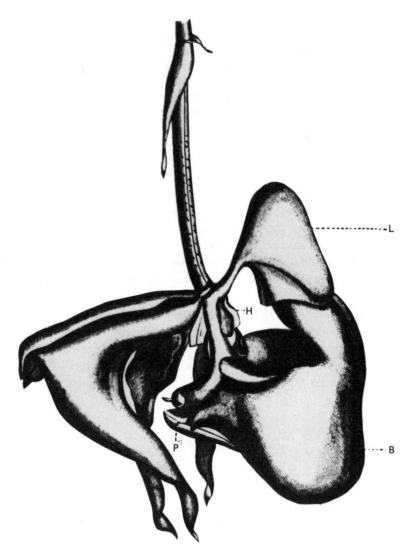

Coryanthes speciosa.   (Copied from Lindley's 'Vegetable Kingdom.')

L. labellum.
B. bucket of the labellum.
H. fluid-secreting appendages.

P. spout of bucket, over-arched by
    the end of the column, bearing
    the anther and stigma.

work on fossils. As the botanist H. Hamshaw Thomas wrote in an excellent short paper on 'The rise of geology ...' published in 1947: 'fossils provided the first challenge to the orthodox cosmogony of the day'. By the end of the eighteenth century, when Hutton published his '*Theory of the Earth*', geology was recognised as a separate and quite well-based science, and the Genesis account of the origin of the earth was already under great strain. After overcoming naïve Church opposition, which offered a date of 4004 B.C. for the Biblical Creation, Darwin's ideas were, however, faced with much more sophisticated criticism from the eminent physicist, Lord Kelvin, who estimated the probable age of the earth from calculations of average rate of heat loss based upon measurable temperature increases down bore-holes and mines. In his famous paper (Kelvin, 1871) 'On geological time', read at a meeting of the Geological Society of Glasgow in 1868, Kelvin calculated that the consolidation of the crust of the earth took place at a maximum of 400 million years ago, and expressed the view that this would have allowed insufficient time for the slow evolutionary processes postulated by Darwin (see Burchfield, 1975). (We now know that Kelvin's estimate was far too small; using evidence from the rate of decay of radioactive minerals, and in other ways, modern estimates of the age of the earth are of the order of 4000 million years.) Full details of this interesting controversy are given by Mivart (1871).

Fig. 2.4. Floral structure of the Orchid *Coryanthes*: the illustration used by Darwin (1877*b*). The extraordinary adaptations shown by many Orchids, by which cross-pollination is brought about by particular insect visitors to the flower, were used by Darwin himself as evidence for the variety and complexity of adaptation to be explained by natural selection. Writers such as Mivart (1871), however, who were opposed to any completely selectionist interpretation of evolution, were inclined to turn Darwin's 'evidence' against himself. Thus Darwin had mentioned the remarkable Orchid *Coryanthes* in which the labellum, modified to a bucket with a lip, is filled with water secreted by special glands; he accurately described the process of pollination, which involves the visiting bees in an involuntary bath from which they can only rescue themselves by crawling through the narrow passage at the lip and thus effecting pollination. As Mivart observes (p. 62): 'Mr Darwin gives a series of the most wonderful and minute contrivances ... structures so wonderful that nothing could well be more so, except the attribution of their origin to minute, fortuitous and indefinite variations'.

A century of intervening research has done little to remove this particular difficulty: the evolution of highly complex structures, dependent for their function upon the joint operation of many different structures, still presents a very real problem of interpretation. Yet, Mivart's offering – the role of a Creator God, manifest as innate forces, in meticulous detailed design – is quite unacceptable to modern science.

The nature of specific difference

Darwin's view of species, based as it was on a thorough study of living organisms as well as on the pertinent literature, is very different indeed from that of Ray and Linnaeus.

In chapter two of the sixth edition of the *Origin* we read:

> Hence, in determining whether a form should be ranked as a species or a variety, the opinion of naturalists having sound judgment and wide experience seems the only guide to follow. We must, however, in many cases, decide by a majority of naturalists, for few well-marked and well-known varieties can be named which have not been ranked as species by at least some competent judges.

He was impressed in his study of the variability of plants and animals by how difficult it was, in many groups, to delimit species, and gives many examples. In polymorphic groups he notes: 'With respect to many of these forms, hardly two naturalists agree whether to rank them as species or as varieties. We may instance *Rubus*, *Rosa* and *Hieracium* amongst plants.' He also considered the opinion of such great taxonomists as de Candolle, who, completing his monograph of the Oaks of the world, wrote:

> They are mistaken, who repeat that the greater part of our species are clearly limited, and that the doubtful species are in a feeble minority. This seemed to be true, so long as a genus was imperfectly known, and its species were founded upon a few specimens, that is to say, were provisional. Just as we come to know them better, intermediate forms flow in, and doubts as to specific limits augment.

Darwin concludes:

> Certainly no clear line of demarcation has as yet been drawn between species and sub-species – that is, the forms which in the opinion of some naturalists come very near to, but do not quite arrive at, the rank of species: or, again, between sub-species and well-marked varieties, or between lesser varieties and individual differences. These differences blend into each other by an insensible series; ...

In chapter nine, on hybridism, he notes, after examining the extensive writings of Gärtner, Kölreuter, etc., that even though exceptions are known: 'First crosses between forms, sufficiently distinct to be ranked as species, and their hybrids, are very generally, but not universally, sterile'. In the section on intraspecific crosses he notes: 'It may be urged, as an overwhelming argument, that there must be some essential distinction between species and varieties, inasmuch as the latter, however much they may differ from each other in external appearance, cross with perfect facility, and yield perfectly fertile offspring'. Notwithstanding these views on the crossing of different groups, one is impressed on reading the *Origin* by the absence of any definition of species incorporating both morphological and crossing information. Why Darwin provides no definition of species is very clear from the discussion of species in the concluding chapter:

> When the views advanced by me in this volume, and by Mr Wallace, or when analogous views on the origin of species are generally admitted, we can dimly foresee that there will be a considerable revolution in natural history. Systematists will be able to pursue their labours as at present; but they will not be incessantly haunted by the shadowy doubt whether this or that form be a true species. This, I feel sure and I speak after experience, will be no slight relief. The endless disputes whether or not some fifty species of British brambles are good species will cease. Systematists will have only to decide (not that this will be easy) whether any form be sufficiently constant and distinct from other forms, to be capable of definition; and if definable, whether the differences be sufficiently important to deserve a specific name. This latter point will become a far more essential consideration than it is at present; for differences, however slight, between any two forms, if not blended by intermediate gradations, are looked at by most naturalists as sufficient to raise both forms to the rank of species. Hereafter we shall be compelled to acknowledge that the only distinction between species and well-marked varieties is, that the latter are known, or believed, to be connected at the present day by intermediate gradations whereas species were formerly thus connected. Hence, without rejecting the consideration of the present existence of intermediate gradations between any two forms, we shall be led to weigh more carefully and to value higher the actual amount of difference between them. It is quite possible that forms now generally acknowledged to be merely varieties may hereafter be thought worthy of specific names; and in this case scientific and common language will come into accordance. In short, we shall have to treat species in the same manner as those naturalists treat genera, who admit that genera are merely artificial combinations made for convenience. This may not be a cheering prospect; but we shall at least be freed from the vain search for the undiscovered and undiscoverable essence of the term species.

From this passage it is abundantly clear that Darwin considered the taxonomist's task in recognising and naming species as a severely practical one, to be decided on criteria of degree of discontinuity and clarity of descriptive diagnosis. Undoubtedly he was influenced by his close study of domesticated plants and animals, and his views were not very acceptable to most practising taxonomists working with wild plants and animals, who continued to be impressed by the high proportion of apparently clear-cut entities to be described and named in the natural world. To them, species were 'real' in a way that genera were not.

How does one reconcile distinct species with Darwin's ideas of evolution? Bateson (1913) expressed the most extreme viewpoint on the arbitrariness of species when he wrote that systematists 'will serve science best by giving names freely and by describing everything to which their successors may possibly want to refer, and generally by subdividing their material into as many species as they can induce any responsible society or journal to publish'. We shall return to this question in Chapter 9.

Tests of specific difference

We have examined so far in this chapter a number of ideas about what constitutes a species. Linnaeus stressed the morphological difference between species whilst Darwin, considering both external morphology and the results of hybridisation experiments, found the species difficult to define. Many other botanists were interested in the species problem in the mid-nineteenth century, and tests of specific rank were devised.

At first, it seemed that hybridisation experiments might provide an objective guide as to whether a plant was a species or a variety. A number of scientists supported this view, in particular Professor Godron of the University of Nancy. In 1863 he published the following opinion (*fide* Roberts, 1929). If two given plants could be crossed without difficulty giving fertile offspring, they were to be called varieties of one species. If, on the other hand, two plants crossed with difficulty, if sterility barriers existed between different plants, then such plants were to be considered different species. Further, crossing between plants of different genera was impossible. The categories of variety, species and genus were therefore to be determined by crossing experiments. Godron's rigid ideas, which were based upon his own work as much as on the extensive publications of earlier hybridists, contrast sharply with the cautious views of Darwin. Of other botanists interested in the problems of defining species experimentally one must mention Professor von Nägeli of Vienna, with whom Mendel fruitlessly corresponded (see Chapter 4). He published a massive review of hybridisation in 1865, noting in particular the difficulties in the sort of ideas published by Godron.

A second test of species is associated with the name of Alexis Jordan of Lyons in France. He considered that cultivation experiments, with progeny testing, provided an objective means of distinguishing species. In 1864 he published a great many of his results. He is perhaps best remembered for his work on *Erophila verna*, in which he described 53 'elementary species', each retaining its distinctive characters in cultivation and coming true from seed. An even greater number could easily have been described, for he indicates that he had more than 200 distinct lines of *E. verna* in cultivation (Fig. 2.5). His experiments were not confined to this taxon, and it is of considerable interest to note the large number of 'elementary species' he described in several common genera, as shown in Table 2.1.

Jordan was followed by others in his practice of describing 'elementary species' within Linnaean species; for instance, Wittrock in Sweden working with *Viola tricolor*, and later De Vries (1905) experimenting with *Oenothera* species. The practice was condemned by many botanists as it led to an inordinate number of new plant names.

Table 2.1. *Numbers of 'elementary species' in various taxa published by Jordan in 1864*

Arabis	23	*Iberis*	23
Biscutella	21	*Ranunculus*	25
Erophila	53	*Thalictrum*	47
Erysimum	26	*Thlaspi*	21

Fig. 2.5. 'Elementary species' in *Erophila verna*. (*a*) Enlargements of flowers showing petal variation ($\times 1.6$). (*b*) Habit variation ($\times 0.75$). (From Rosen, 1889)

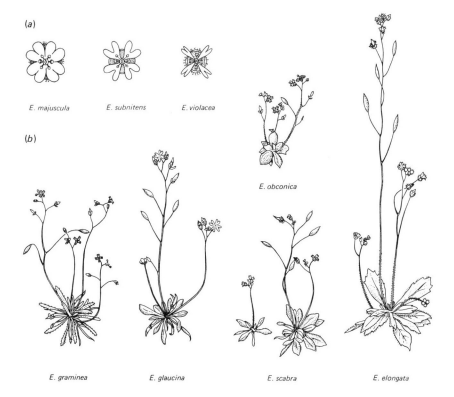

(*a*)

E. majuscula   E. subnitens   E. violacea

(*b*)

E. obconica

E. graminea   E. glaucina   E. scabra   E. elongata

The idea of using only a single line of experimental evidence as a test of specific rank did not meet with universal approval. For example, Hoffmann, Professor of Botany in the University of Giessen, carried out a large number of experiments with many taxa. He observed plants closely in the wild and also carried out cultivation and crossing experiments. In a review of his researches (1881) he considered the many different lines of evidence which could be used in judging specific rank. Not only did he study the performance of plants in cultivation and the results of crossing plants, but he also took into account geographical distribution and the extent of hybridisation *in the wild.*

This historical review of species and their variation has brought us almost to the end of the nineteenth century, and it is in this period that there emerged two new aspects of the study of variation. First, the statistical examination of biological variation: some of the results of this work are the subject of the next chapter. Secondly, following the epoch-making rediscovery in 1900 of Mendel's work on heredity, the science of genetics made its appearance.

3

Early work on biometry

In the second half of the nineteenth century, as Darwinism was making its impact upon biology, an interesting new approach to biological variation, especially intraspecific variation, was being examined. Instead of trying to describe variation patterns in words, the investigators, examining large samples of organisms, collected numerical data and subjected them to statistical analysis. In the account which follows we discuss selected themes. For more detail on the history of the subject see Pearson & Kendall (1970) and Kendall & Plackett (1977).

The first worker to study natural variation statistically – a science which became known as biometry – was probably the Belgian Quetelet (1796–1874). He wrote a famous series of letters on the subject to his pupil, the Grand Duke of Saxe-Coburg and Gotha. Later in the century, Darwin's cousin, Francis Galton (1822–1911), made notable contributions to the statistical investigation of variation and inheritance. Like Quetelet he was particularly fascinated by human variation. Very important statistical work, some on botanical material, was also carried out by Pearson (1857–1936).

It is important to examine first the general characteristics of this new approach. Instead of contenting themselves with the study of a few herbarium specimens or cultivated plants, the early biometricians took large samples, often using living material of common species. These samples were then carefully scrutinised and measured, as we read in Davenport (1904): 'Having settled upon the general conditions, of race, sex, locality, age, which the individuals to be measured must fulfil, take the individuals methodically at random and without possible selection of individuals on the basis of the magnitude of the character to be measured'. Finally, having collected the samples and obtained the numerical data, the worker performed the analysis, observing Quetelet's precept that statistics must be collected without any preconceived ideas and without neglecting any numbers (Quetelet, 1846).

These early studies of biological material established the important point that there are two main kinds of intraspecific variation. Firstly, much of the variation is discontinuous. If, for instance, one is examining the number of chambers in a capsule, the number of seeds in a fruit, the number of leaves on a plant – in fact any variation in the number of parts – then the numbers found must be integers. One never discovers 14.5 undamaged peas in a pod. Often, in considering variation in the number of parts – so-called meristic variation – a more or less complete series of numbers is found. For instance, Pearson noticed, in a cornfield in the Chiltern Hills, England, that Poppies (*Papaver rhoeas*) had different numbers of stigmatic bands on the capsule (Fig. 3.1). He collected a very large sample of 2268 capsules: the frequency of different numbers of stigmatic bands is given in Table 3.1. In other instances of discontinuous variation, however, only two or a few strikingly different variants are found. For instance, the Opium Poppy (*Papaver somniferum*) may or may not have a dark spot at the base of the petal; Groundsel (*Senecio vulgaris*) has either radiate or non-radiate flowers; and Foxglove (*Digitalis purpurea*) may have white or red flowers, and hairy or glabrous stems. On the other hand, a second type of variation – continuous variation – is also common in plants. In considering variation in such characters as height, weight, leaf length and root spread, any value is possible within a given range. There are no breaks in the variation for particular characters.

Fig. 3.1. Young and old capsules of the Poppy, *Papaver rhoeas* (× 0.5).

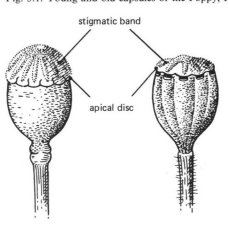

Young capsule   Old capsule

Strikingly discontinuous variation patterns, as in white or red-purple flower colour in *Digitalis*, presented little difficulty in examination or classification. The analysis of arrays of data, however, whether of discontinuous or continuous variates, posed somewhat more complex problems. With an array of data, how was it possible to show numerically where the bulk of the variation lay; how could a numerical estimate or spread of the data within the sample be obtained and, further, how could the variability of two samples be compared? Using Pearson's data for variation in stigmatic band number in *Papaver rhoeas* (Table 3.1), we may now briefly examine some of the statistics employed by the early biometricians.

Commonest occurring variation in an array

Sometimes a knowledge of the mode, or most frequent class, and the median, or middle value of an array, is a useful indication of where the bulk of the variation lies in a sample. These are, however, less useful than

Table 3.1. *Calculation of mean, variance, standard deviation and coefficient of variation in number of stigmatic bands in capsules of Poppy,* Papaver rhoeas. *(Data from Pearson, 1900)*

Number of bands x	Frequency, f	fx	Difference from mean $x - \bar{x}$	Square of difference $(x - \bar{x})^2$	$f \times$ square of difference $f(x - \bar{x})^2$
5	1	5	−4.8	23.04	23.04
6	12	72	−3.8	14.44	173.28
7	91	637	−2.8	7.84	713.44
8	295	2360	−1.8	3.24	955.80
9	550	4950	−0.8	0.64	352.00
10	619	6190	+0.2	0.04	24.76
11	418	4598	+1.2	1.44	601.92
12	195	2340	+2.2	4.84	943.80
13	54	702	+3.2	10.24	552.96
14	25	350	+4.2	17.64	441.00
15	5	75	+5.2	27.04	135.20
16	3	48	+6.2	38.44	115.32
	2268	22 327			5032.52

$$\text{Mean} = \frac{22\,327}{2268} = 9.8$$

$$s^2 = \frac{\sum f(x - \bar{x})^2}{n-1} = \frac{\sum d^2}{n-1} = \frac{5032.52}{2267} = 2.2 \quad s = 1.49$$

$$\text{Coefficient of variation} = \frac{s}{\bar{x}} \times 100 = \frac{1.49}{9.8} \times 100 = 15.2\%$$

the arithmetic mean, $\bar{x}$. This is calculated quite simply by summing ($\sum$) the observed values (x) and dividing by the number of observations, n.

$$\text{Mean} = \bar{x} = \frac{\sum x}{n} \tag{1}$$

Estimates of dispersion of the data

Values of the mean give no indication of the variation within a sample. Identical means may be obtained if the data are all clustered very closely to the mean or if the data are markedly above and below the mean (Fig. 3.2). It is clearly very important to have an estimate of the degree of dispersion of the data within a particular sample.

There are several possible ways of examining dispersion. Early biometricians often noted the extreme values of the array, or alternatively they calculated how much each value of the array differed from the mean, and after summing the differences, calculated an *average deviation from the mean*. They also used a statistic, now seldom if ever calculated, called the *probable error*. Details of this calculation may be found in statistics books. In more recent times, however, dispersion has been estimated by calculating the *variance, s^2*, and *standard deviation, s*.

Fig. 3.2. Hypothetical frequency distribution for two population samples with the same mean. Sample B is much more variable than sample A. (From Srb & Owen, 1958)

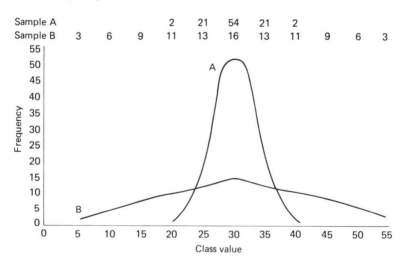

The variance, s^2, is calculated by summing the squares of the deviations of all the observations from their mean (d^2) and dividing by $n-1$.

$$s^2 = \frac{\sum d^2}{n-1} \tag{2}$$

Except where samples are small, d^2 is more readily calculated by employing the equations:

$$\sum d^2 = \sum (x - \bar{x})^2 \tag{3}$$

$$\sum (x - \bar{x})^2 = \sum x^2 - \frac{\left(\sum x\right)^2}{n} \tag{4}$$

In calculating the variance, s^2, it is important to note (and statistics books should be consulted for justification) that the divisor is $n-1$. The standard deviation, s, is found by obtaining the square root of the right-hand side of eqn 2. The calculation of variance and standard deviation for Pearson's Poppy data is given in Table 3.1.

The variance is a valuable statistic, giving a measure of the dispersion of the data about the mean. It is used a good deal in more complex statistics, where different populations are being compared. The standard deviation too is a useful measure of dispersion, especially as the 'spread' of the data is here expressed in the same units as the mean. (The probable error – the statistic estimating dispersion which was often calculated by early biometricians – is 0.6745 times the standard deviation.) Now that the variance and standard deviation values have been calculated, how are they to be interpreted? Before we examine this point, let us look at early work on the visual representation of arrays of data.

Histograms, frequency diagrams and the normal distribution curve

Most people find it easier to comprehend the significance of data expressed visually rather than numerically. The variation in *Papaver* may be expressed as: $\bar{x} = 9.8$, $s^2 = 2.22$, $s = 1.49$, or it may be represented in the form of a diagram. Histograms and plotted curves were frequently employed in early biometrical studies. The distribution of the values for stigmatic band number in *Papaver* has been plotted as a histogram in Fig. 3.3; the distribution is roughly bell-shaped, being almost symmetrical about the mean value. Small irregularities in the distribution are the result of small sample size; a closer fit to a bell-shaped curve would result from an even larger set of data for stigmatic band number. The results for *Papaver* are an example of a very common frequency distribution in biological material – the 'normal' or Gaussian distribution, the latter after Gauss (1777–1855), one of the investigators of this type of distribution.

In the last decades of the nineteenth century, approximately normally distributed variation was demonstrated in a great range of biological materials. Davenport (1904), in the second edition of his book *Statistical methods with special reference to biological variation*, first published in 1889, provides a very valuable survey of early biometrical results, giving

Fig. 3.3. Histogram of Pearson's data (Table 3.1) for the variation in the number of stigmatic bands in a sample of capsules of *Papaver rhoeas*. Such histograms were often used in early biometrical studies. Campbell (1967), discussing the use of histograms, suggests that they should be used only in cases of continuous variation. For examples of meristic or other discontinuous variation, frequency diagrams are preferable; in these, the frequency of each class is indicated by a vertical line on the graph.

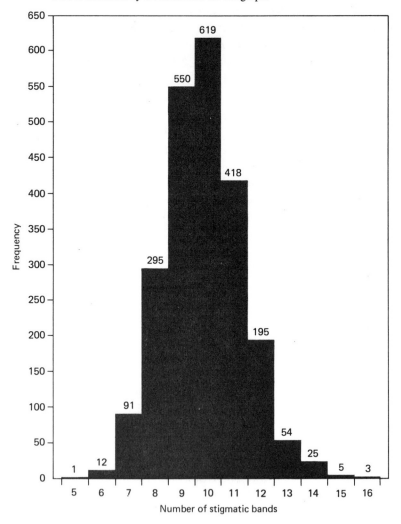

Fig. 3.4. Early botanical results showing approximately normal distributions.
(*a*) Fruit length in Evening Primrose, *Oenothera erythrosepala*
(*O. lamarckiana*), for 568 plants collected in October 1893 (De Vries, 1894). (*b*)
Number of primary umbel rays in *Anethum graveolens* for 552 plants collected
in July 1893 (De Vries, 1894). (*c*) Number of ray–florets in Ox-Eye Daisy,
Leucanthemum vulgare (*Chrysanthemum leucanthemum*), collected from 1133
heads in Keswick, England, July 1895 (Pearson & Yule, 1902). (*d*) Number of
main-branch veins in leaves of Beech (*Fagus sylvatica*) from 2600 leaves
(Pearson, 1900). (*e*) Number of prickles on 2600 leaves of Holly (*Ilex
aquifolium*) from trees in Somerset, England (Pearson *et al.*, 1901). (*f*) Seta
length in the moss *Bryum cirrhatum* for 522 plants (Amann, 1896).

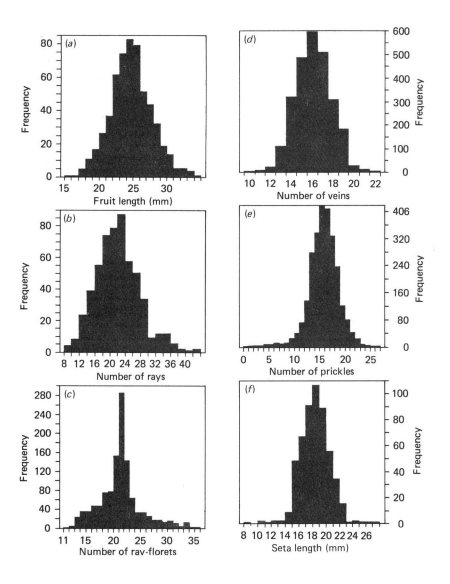

references and details of scores of botanical and zoological examples. Fig. 3.4 shows some typical cases.

As approximately normal distributions are frequently encountered in biological material, it is important to look at some of their properties. First, Fig. 3.5 shows that in a normal curve the median, mode and mean of the array fall at the same point. Secondly, and of great importance, is the relation of standard deviation to the curve. We have outlined above how to calculate variance and standard deviation. Now, how precisely does knowledge of the standard deviation help us to understand the dispersion of the data within the sample? Examining Fig. 3.5 we see that about two-thirds (68.26% to be exact) of the total variation under a normal curve falls within the range 'mean ± one standard deviation'. Twice the standard deviation on each side of the mean excludes about 5% of the variation, 2.5% in each tail of the normal curve. For different sets of data which are normally distributed, different values of the standard deviation will be found. Thus, if we have a large amount of variability in a sample, a wide curve corresponding to the large standard deviation will be obtained. Whatever the width of the curve, however, the 'mean ± one standard deviation' always contains 68.26% of the variation. We can see now how the standard deviation is so useful in indicating dispersion. One

Fig. 3.5. A normal distribution, showing proportions of the distribution that are included between ±1 standard deviation or ±2 standard deviations with reference to the mean. (From Srb & Owen, 1958)

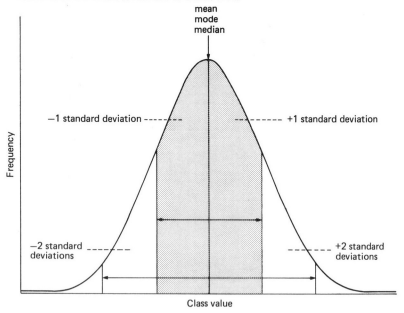

last point remains to be considered. How is the appropriate normal curve to be fitted to a histogram? It cannot, of course, be drawn 'by eye'. Recourse to a statistics book will indicate full details, and it is sufficient for our purposes to note that it involves the substitution of values of the mean and standard deviation into the equation for the normal curve.

Other types of distribution

Not all the biometrical studies of plant materials gave normally distributed variation, however. De Vries (1894) was one of the first to point to deviant distributions, calling attention to what he called 'Half-Galton' curves. Table 3.2 gives sets of data for the number of compartments in the fruit of Sycamore (*Acer pseudoplatanus*), and petal number in Marsh Marigold (*Caltha palustris*) and Silverweed (*Potentilla anserina*), which in each case approximates to half a normal curve. In *Caltha* and *Acer*, the 'right-hand half' of the curve is represented, whilst in *Potentilla* only the 'left-hand half' is found.

Other researches of this period, especially those dealing with numbers of plant parts, revealed further asymmetrical and deviant frequency distributions. For instance, examining the figures of Pledge for petal frequency in the Buttercup *Ranunculus repens* (Table 3.3), we see that the frequency distribution when plotted would have a long tail to the right: such a curve is described as 'positively skewed'. (A curve with a long

Table 3.2. *Half-Galton curves (De Vries, 1894)*

Caltha palustris	Petal number	5	6	7	8
	Frequency	300	87	25	4
Acer pseudoplatanus	Number of fruit compartments	2	3	4	
	Frequency	50	17	3	
Potentilla anserina	Petal number	3	4	5	
	Frequency	6	537	1819	

Table 3.3. Ranunculus repens *(Data of Pledge (1898) in Vernon, 1903)*

	3	4	5	6	7	8	9	10	11	12	13
Sepal frequency	1	20	959	18	2						
Petal frequency		8	706	145	72	38	15	7	7	1	1

tail to the left is said to be negatively skewed.) The data for sepal numbers, collected in the same study, also depart from a normal distribution, in this case by being too tightly bunched together. Such a distribution is said to be leptokurtic (high-peaked). Sometimes flat-topped curves (platykurtic) have been discovered (Bulmer, 1967; Rayner, 1969; David, 1971). As we shall see in Chapter 13, samples of biological specimens often exhibit non-normal distributions.

Comparison of different arrays of data

By visual inspection, it is often possible to see that a group of plants is more variable in, say, height than in flower size, and the problem of investigating this biometrically particularly fascinated Pearson. In the late 1890s he first devised a statistic known as the *coefficient of variation*. Easy to calculate, it is merely the ratio of the standard deviation to the mean. In order to have a scale of reasonable-sized numbers, the resulting coefficient is usually expressed as a percentage:

$$C \text{ (coefficient of variation)} = \frac{s}{\bar{x}} \times 100\%$$

An important property of the coefficient is that, as it is calculated as a *ratio*, direct comparison of different coefficients is possible. This even applies when the original figures were calculated in different units, as in metres, inches, grams, etc. Table 3.4 shows some data for human height

Table 3.4. *Variation in human height and weight; means* $= \bar{x}$, *coefficients of variation* $= C$ *(Data of Pearson and others in Davenport, 1904)*

		n	$\bar{x}$	C
Height				
English upper middle class	male	683	69.215 in	3.66
English criminals		3000	166.46 cm	3.88
US recruits		25878	170.94 cm	3.84
Cambridge University	male	1000	68.863 in	3.66
students	female	160	63.883 in	3.70
English newborn babies	male	1000	20.503 in	6.50
	female	1000	20.124 in	5.85
Weight				
Cambridge University	male	1000	152.783 lb	10.83
students	female	160	125.605 lb	11.17
English newborn babies	male	1000	7.301 lb	15.66
	female	1000	7.073 lb	14.23

and weight. Taking the figures for height first, a comparison of means is impossible in certain cases, as some of the measurements are in inches and others in centimetres. Direct comparison of coefficients of variation is, however, possible, and we can see that English criminals and US recruits show a similar degree of variation in height. There are small differences in height between male and female students and between males and females at birth. Considering the information for variation in weight, again we have differences between male and female at birth and at college. Finally, as high values of the coefficient of variation indicate greater variation for a particular character, we can see that there is a much greater variation in weight than in height in the samples examined.

Coefficients of variation continue to be very useful in the study of variation. Fig. 3.6 shows the coefficients calculated by Gregor (1938) for different parts of the Sea Plantain, *Plantago maritima*. These data illustrate convincingly a fact known before Linnaeus, namely that floral parts are generally less variable than vegetative parts.

Fig. 3.6. Coefficients of variation in *Plantago maritima*: typical mean values from a number of populations. (Data from Gregor, 1938)

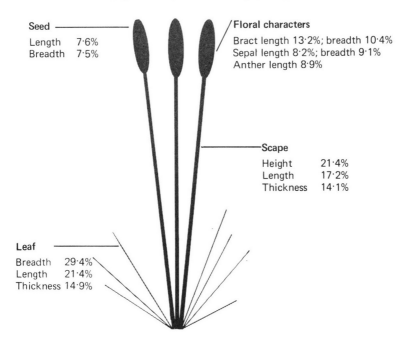

Seed
Length   7·6%
Breadth   7·5%

Floral characters
Bract length 13·2%; breadth 10·4%
Sepal length 8·2%; breadth 9·1%
Anther length 8·9%

Scape
Height   21·4%
Length   17·2%
Thickness   14·1%

Leaf
Breadth   29·4%
Length   21·4%
Thickness 14·9%

Complex distributions

Other biometrical studies in the 1890s revealed more complex frequency distributions. Some of the results of Professor Ludwig of Greiz in Germany may be used as an illustration. He counted the numbers of ray-florets in 16 800 heads of the Ox-eye Daisy, *Leucanthemum vulgare* (*Chrysanthemum leucanthemum*), collected from Greiz, Plauen, Altenberg and Leipzig between the years 1890 and 1895. The frequency distribution he obtained was not of the 'normal' type but had several peaks (Table 3.5). He obtained similar multimodal distributions for many species: for example, in number of ray-florets in Daisy (*Bellis perennis*), number of disc-florets in Yarrow (*Achillea millefolium*) and number of flowers in the umbel of Cowslip (*Primula veris*).

In collecting his *Leucanthemum* (*Chrysanthemum*) data, Ludwig records some interesting differences in ray-floret numbers in different localities. Mountain plants showed 'peaks' at 8 and 13, while lowland plants on the other hand had a 'peak' at 21 ray-florets. In fertile soil Ludwig often found a strong 'peak' at 34. He considered that these variations between plants from different areas were the result of nutritional factors. What interested Ludwig most of all about the *Chrysanthemum* was the presence (in the results) of clear peaks at 8, 13, 21 and 34 ray-florets. These numbers, he pointed out, belong to the famous Fibonacci sequence of numbers discovered by Leonardo Fibonacci of Pisa in the twelfth century. The sequence runs 0, 1, 1, 2, 3, 5, 8, 13, 21, 34, 55, 89, 144 ..., each term being the sum of the two terms which precede it. It represents a set of whole numbers which satisfy almost exactly an exponential growth curve. Not all Ludwig's results gave such clear peaks at the Fibonacci numbers, and he was hard-pressed to explain peaks at 11 and 29, which he discovered in certain plants. Nevertheless, he believed that the Fibonacci sequence of numbers was important in understanding complex patterns of variation.

Certain other biometricians, notably Weldon (1902*b*), were sceptical about Ludwig's claim for the Fibonacci sequences. They pointed to the fact that plants from different areas had been amalgamated in collecting the data. Plants from a single locality often gave a different picture of the variation; for instance, counts for *Chrysanthemum* ray-florets from Keswick, England, in 1895 gave an approximately normal distribution (Fig. 3.4). Weldon also pointed out that sampling at different times could have an important influence upon the results. This is well brought out in the results of Tower (1902). He collected, at the beginning and end of July 1901, two sets of *Chrysanthemum* plants from a locality at Yellow Springs, Ohio, USA. His results show clearly that early flowers have more ray-

Table 3.5. *Variation in the number of ray-florets in* Leucanthemum vulgare (Chrysanthemum leucanthemum)

Number	Ludwig (1895)[a]	5 July 1901 Tower (1902)[b]	30 July 1901 Tower (1902)[b]	Total Tower (1902)[b]
7	2			
8	9			
9	13			
10	36			
11	65			
12	148		1	1
13	427		8	8
14	383		3	3
15	455		6	6
16	479	1	8	9
17	525	0	9	9
18	625	0	8	8
19	856	2	12	14
20	1568	8	19	27
21	3650	17	26	43
22	1790	23	11	34
23	1147	22	10	32
24	812	21	10	31
25	602	22	8	30
26	614	19	5	24
27	375	16	4	20
28	377	14	6	20
29	294	12	4	16
30	196	10	2	12
31	183	16	4	20
32	187	18	2	20
33	307	29	1	30
34	346	20	1	21
35	186	6	0	6
36	64	6	0	6
37	28	0	0	0
38	16	0	0	0
39	16	2	0	2
40	14			
41	0			
42	3			
43	2			
Total	16 800	284	168	452

[a]Ludwig (1895): Plants from Greiz, Plauen, Altenberg and Leipzig, 1890–5
[b]Tower (1902): Plants from Yellow Springs, Ohio, USA (two collections and total from same locality)

florets than those produced later in the season. Tower went on to show that it was not a question of different plants in flower at the beginning and end of July; marked plants continued to flower throughout the summer, producing flowers with different numbers of parts at different times in the season. Different peaks for ray-floret numbers are found in early July (22, 33) compared to those found later in the month (13, 21). It is interesting to note that it is only in the amalgamated data that these peaks are found, and also that the highest peak is not at the Fibonacci number of 34, as found by Ludwig, but at 33. Such results as these cast some doubt upon the importance of the Fibonacci numbers, indicating that the location of peaks in a complex distribution, far from conforming to a mathematical sequence, was greatly influenced by the method of collecting the data. The precise results obtained would depend upon whether all the plants were collected at the same stage of maturity, a point particularly difficult to ascertain if data for plants from widely different localities and ecological conditions were amalgamated.

Local races

Close study of local variation in the species occupied the attention of many early biometricians. For instance, Ludwig (1901) made a special analysis of variation in the Lesser Celandine (*Ranunculus ficaria*). He showed that plants from different localities had different numbers of carpels and stamens. Details of the two most dissimilar populations, from Gais and Trogen, are given in Table 3.6. Clearly Gais has plants with more carpels and stamens than Trogen.

Ludwig called these local populations, characterised by different mean numbers of floral parts, '*petites espèces*' or 'local races'. Until this time the term 'local race' had been used rather loosely for plants from particular areas used for biometrical study or experiments, but Ludwig sought to demonstrate the reality of 'local races', using biometrical evidence. In his

Table 3.6. *Mean number of stamens and carpels in* Ranunculus ficaria *(Ludwig, 1901)*

		Mean number	Standard deviation
Gais	Stamens	23.8250	2.8872
(80 plants)	Carpels	18.1125	4.2885
Trogen	Stamens	20.3682	3.8234
(385 plants)	Carpels	13.2635	3.0606

view these races could be distinguished on the basis of the mean number of floral organs, amalgamation of data for a number of races giving a multimodal distribution curve, such as we described above.

Ludwig's views were again challenged by British and American biometricians, particularly by Lee (1902) who, using the data of MacLeod on *Ranunculus ficaria*, pointed to the great seasonal variation in floral parts (Table 3.7). Her criticism of Ludwig's 'local races' is particularly telling as the variation in early and late flowers from a single locality covers almost the entire range between the Gais and Trogen plants.

This criticism of Ludwig's results did not clinch the issue, however, as there was earlier work by Burkill (1895) on two dissimilar *Ranunculus ficaria* populations in which large differences in mean numbers of floral parts were maintained (although not completely) on later sampling on the same site (Table 3.8).

The reality of 'local races' was an important issue in the early volumes of the journal *Biometrika*, which was launched in 1901. In an editorial (**I:** 304–6, 1902) it was contended that the polymorphism found in most results was spurious. It was difficult to defend the notion that each peak of a complex distribution represented a 'local race', especially as peaks often

Table 3.7. *MacLeod's data for seasonal variation in floral parts in a population of* Ranunculus ficaria *(Lee, 1902)*

		Mean number	Standard deviation
Early flowers	Stamens	26.7313	3.7609
(268 plants)	Carpels	17.4478	3.8942
Late flowers	Stamens	17.8633	3.2984
(373 plants)	Carpels	12.1475	3.3878

Table 3.8. *Variation in* Ranunculus ficaria *(Burkill, 1895)*

	Date of collection	Number of flowers	Mean number of stamens	Mean number of carpels
Cambridge	3 March	32	22.87	13.41
(under trees)	16 April	75	19.49	11.95
Cayton Bay				
(open field,	31 March	100	38.24	32.32
top of cliffs)	4 May	43	30.67	25.72

disappeared as sample size was increased. Another important point concerned sampling techniques. It was stressed that random sampling was essential, a point perhaps neglected by early workers. Further, the problem of what constitutes a locality was raised, and the validity of putting together data for samples taken from different areas was questioned. Finally, the editorial stressed the difficulties of seasonal variation and environmental effects, and concluded that a species is not broken up into 'local races'.

Returning once again to variation in *Ranunculus ficaria*, we find the same conclusion is reached by Pearson *et al.* (1903) in a paper in *Biometrika*, which draws together published records, together with new results of variation in floral parts in different areas of Europe. The tables of data are too large for inclusion here, but the following conclusions were drawn from the extensive statistics. 'Local races' could not be distinguished by the number of floral parts, and the influence of the environment and seasonal variation would seem to be sufficient to mask any difference due to 'local races'. The problem of how to eliminate seasonal and environmental variables from experimental studies was not seriously investigated until later, as we shall see in Chapter 4.

Correlated variation

Many early biometricians examined closely a further aspect of variation, namely the simultaneous variation in pairs of characters. For instance, Pearson was interested in the relation of measurements of different parts of the human body.

Suppose we consider body height and its relation to forearm length. It may be that there is some relation between the two variables or they may be independent. Three different situations are possible:

1. The taller the person, the longer the forearm.
2. The taller the person, the shorter the forearm.
3. A tall person is as likely to have a long or a short forearm as is a short person.

The first situation is one of positive correlation, the second of negative correlation, whilst if the last were discovered we should conclude that there was no correlation between the traits.

In investigating correlation, a statistic called the correlation coefficient (r) is often calculated. It is not necessary for our purposes to give the formulae and details of calculation, which may be found in any statistics book. What is important is the way in which r values indicate correlation or lack of it. $r = +1$ indicates complete positive correlation; $r = -1$ signifies complete negative correlation. If $r = 0$, then correlation is absent.

In biological material, perfect correlation – either positive or negative – is very rare; the various degrees of positive and negative correlation which are often found are indicated by figures which lie between $r = +1$ and $r = -1$.

Examining the relation between stature and forearm length, Pearson demonstrated positive correlation: in one case $r = +0.37$. A number of botanical situations were also studied at this time. Among the problems investigated was the correlation in the size of leaves in the same rosette in *Bellis perennis* (Verschaffelt, 1899), correlation between pairs of measurements of leaves and fruits of various species (Harshberger, 1901) and correlation between various parts in the Desmid *Syndesmon* (Kellerman, 1901). The sort of figures obtained for correlation in the floral parts of plants may be illustrated with data, summarised from various authors, on *Ranunculus ficaria* (Table 3.9). Clearly there is a stronger correlation between numbers of stamens and carpels than between other organs.

Correlation coefficients – and a further method of studying the association of pairs of measurements known as regression analysis – were used, particularly by Galton and Pearson, for studying heredity. It is a matter of common experience that tall fathers tend to have tall sons, and that short fathers usually have short sons. The association is by no means complete, however. Galton examined the situation biometrically, analysing data from a large number of human families (Galton, 1889): 'Mr Francis Galton offers 500L in prizes to those British Subjects resident in the United Kingdom who shall furnish him, before May 15 1884, with the best Extracts from their own Family Records'. Galton sifted through particulars of 205 couples of parents with their 930 adult children of both sexes. He examined his data carefully, looking for association between the characteristics of parent and offspring. In many cases r values proved to be positive – as high as $r = +0.5$ for height of parents and offspring. We shall examine Galton's interpretation of these results in Chapter 4.

Table 3.9. *Correlation coefficients in* Ranunculus ficaria *(Davenport, 1904)*

Numbers of	Values of r
Sepals to petals	$+0.34$ to -0.18
Sepals to stamens	$+0.06$ to $+0.02$
Sepals to carpels	$+0.25$ to $+0.03$
Petals to stamens	$+0.38$ to $+0.22$
Petals to carpels	$+0.35$ to $+0.19$
Stamens to carpels	$+0.75$ to $+0.43$

Problems of biometry

In this short survey of early biometrical work a number of problems remain to be examined. In our opening remarks we indicated that there are two main types of variation found on sampling. Arrays of data may be obtained showing either discontinuous or continuous variation. Also there may be found markedly discontinuous patterns of variation, with two or more very distinct non-overlapping categories. The reality of these distinct groups is important, as they figure widely in genetic work. As we shall show in Chapter 4, Mendel's work on genetics, published in 1866 and rediscovered in 1900, involved crossing peas with different coloured cotyledons (green or yellow), or plants of different height (tall or dwarf). Early geneticists crossed glabrous and hairy plants of *Biscutella laevigata* (Saunders, 1897), and *Silene* spp., especially *S. dioica* and *S. alba* (De Vries, 1897; Bateson & Saunders, 1902). Among the biometricians it was Weldon (1902*a*), an opponent of Mendelism, who pointed out a certain ambiguity in defining discontinuities. For instance, he showed that if a large range of cultivated pea stocks was examined it was found that there was a continuous range of cotyledon colour from green to yellow. It was impossible to sort into green and yellow categories. Similarly, he also showed that there was an enormous range of hairiness in *Silene* species and that it is very hard to accept a classification into glabrous or hairy variants. The important point to bear in mind, however, is the scale of the operation; it may be that general discontinuities do not occur, but marked discontinuities in limited collections and in the progeny from carefully controlled crosses certainly exist. When we read of Mendel crossing tall and dwarf peas, yellow and green peas, it is as well to remember that he deliberately chose stocks with markedly contrasting characters and that, even though there would have been variation in, say, height in his tall and dwarf stocks – perhaps normally distributed variation – there was no overlap in the distribution curves of tall and short plants.

A further problem raised by early biometrical work is that of the significance of differences between sets of numerical data. For instance, the coefficient of variation for weight in Cambridge University students (Table 3.4) shows that females ($C = 11.17\%$) show greater weight variation than males ($C = 10.83\%$). The difference in values is, however, quite small. Now, is this result due to differences in sample size? The female student population in Cambridge in the 1890s was very small and there was difficulty in getting even 160 measurements. Or is the variation due to chance? Would further samples taken in different years give the same basic pattern of greater weight variation in female students?

This type of problem is widespread in biometry. Is there any statistically significant difference in the frequency distributions of two sets of data? Do the peaks in a multimodal distribution reveal a true polymorphism or is it the result of sample size or chance? Questions of this type are now tackled by applying statistical significance tests. In Chapter 13 we shall go further into these problems; it is sufficient at this point to note that most of these tests came into being because biometricians wrestled with the problems of interpreting data from biological material.

Finally, we must return to another problem: the vexed question of the underlying basis of variation, which fascinated and puzzled early workers. What part of the variation was due to environmental variation and what part was genetic? In the next chapter we examine this problem.

4

Early work on the basis of individual variation

In the last chapter we saw how the early biometricians found great difficulty in analysing some of their data because they were unable to decide which part of the variation had a genetic basis and which part was environmentally induced. For animal studies it was Galton (1876) who appreciated the unique value of twins in investigations of the relative roles of nature and nurture in the development of the individual. To study genetic and environmental effects in plants, specimens selected for comparison may be cultivated under a standard set of environmental conditions. Experiments, both historical and recent, have been performed on the assumption that residual differences between plants of the same species, collected from the same or different habitats and grown under such standard conditions, might be considered to have a genetic basis. What follows is a brief survey of early studies. In Chapter 13 we will discuss in some detail the assumptions, design and interpretations of garden experiments.

It is very interesting to see how cultivation techniques have developed as methods of analysing variation in plants. Experimental cultivation of plants undoubtedly arose as an adjunct to gardening and horticulture, and in Chapter 2 we saw how Ray, collecting the striking prostrate variant of *Geranium sanguineum* from Walney Island, demonstrated its constancy by cultivating plants in different gardens. The most valuable of these experimental tests were undoubtedly those of a comparative nature. For instance, Mendel cultivated two variants of the Lesser Celandine (*Ranunculus ficaria*), which he called *Ficaria calthaefolia* and *F. ranunculoides*, and reported to Dr von Niessl that each remained distinct (Bateson, 1909).

In a paper of quite remarkable scope, Langlet (1971) has reviewed the extent to which foresters in the eighteenth and nineteenth centuries were using experimental cultivation to study adaptive variation in some of the widespread forest trees of Europe. He cites, for example, the neglected (and largely unpublished) work of Duhamel du Monceau, Inspector-General of the French Navy, who, around the time when Linnaeus published his *Species plantarum* (1753), brought together an impressive collection of samples of Scots Pine, *Pinus sylvestris*, from Russia, the Baltic countries, Scotland and Central Europe, and established the first experimental provenance tests for any wild plant. This early development of what we could now call 'genecology' is understandable because of the economic and military importance of the timber supply, but the neglect by most modern writers of the further expansion of such studies in the nineteenth century is less easy to explain and probably, as Langlet suggests, is in part due to the fact that much of this forestry research was published in German. Darwin himself, of course, was greatly interested in the variation of cultivated plants; but forestry differed from agriculture and horticulture, as Langlet shrewdly observes, because its source material was almost entirely the wild species not already subject to artificial selection.

These examples show the importance of simple cultivation of carefully examined material, comparing performance in the wild with that in culture, and comparing also the behaviour of samples of the same or closely related species in the same garden. The method of comparative cultivation, whether seeds or plants are collected from the wild, permits us to investigate the basis of variation patterns. It is easy with hindsight to get a false impression of the ideas of the past and here is a case in point. Even though ideas about the balance between genetic and environmental variation are implicit in some of the writings of the nineteenth century and even discernible in the work of Linnaeus, an explicit statement came only with the researches of the Danish botanist Johannsen carried out in the years 1900–7. He worked with dwarf beans of the species *Phaseolus vulgaris*, which is naturally self-fertilising.

Phenotype and genotype

Johannsen obtained commercial seeds of the variety 'Princess' and grew 19 of them in an experimental garden. The progeny from each of these beans had a different mean seed weight, and Johannsen inferred that these differences were genetic. From each of these 19 original beans, he established a separate line by self-fertilisation, growing up to six generations of daughter beans. For each line he raised a sub-line by selecting heavy seeds at each generation and a separate sub-line in which light seeds were selected. Very great care was taken to label the plants, and in each generation the mean seed weight for a line was calculated separately for progeny from heavy and light mother beans. Table 4.1 gives the results for two lines. Johannsen found that for a particular line in any one year the mean seed weight for progenies from light and heavy beans did not differ significantly. From each of the 19 original beans a pure line was established, selection having no effect upon mean seed weight. The implication of these results may be more readily understood later, when it will be shown that habitual self-fertilising leads to genetic invariability, and thus genetically identical plants were produced from the progeny of a single bean. Even though the pure lines from the 19 beans were each genetically uniform, Johannsen found great differences in individual bean weights, approximately normally distributed, giving slightly different mean values for a line in different years. He attributed these differences to the effects of the environment.

Table 4.1. *Two pure lines of* Phaseolus vulgaris *(Johannsen, 1909)*

Mean weight (grams) of selected small seeds	Mean weight of progeny	Mean weight (grams) of selected large seeds	Year	Mean weight (grams) of selected small seeds	Mean weight of progeny	Mean weight (grams) of selected large seeds
30		40		60		70
	↘36	35↙	1902		↘63	65↙
25↗				55↙		↘80
	↘40	42↘ 41↙	1903		↘75	71↙ ↘87
31↙		43↘		50↙		
	↗31	33↗	1904		↘55	57↗
27↙		39↘		43↙		↘73
	↘38	39↗	1905		64	64↙
30↗		46↘		46↙		↘84
	↘38	40↙	1906		↘74	73↗
24↗		47↘		56↙		↘81
	↘37	37↙	1907		↘69	68↗

Pure line 'A'   Pure line 'B'

These experiments led him to define clearly the distinction between genetic and environmental effects upon an organism. Of first importance were the hereditary properties of an individual – the *genotype* – which were largely fixed at fertilisation. The appearance, or *phenotype*, of particular individuals of the same genotype might, however, be different because of environmental factors. Even though Johannsen's results were obtained for a habitually self-fertilising species, there is no reason to doubt that the concept of genotype and phenotype is of general validity.

Transplant experiments

Besides the rather simple cultivation experiments we have examined so far, nineteenth-century botanists also investigated, through transplant and transfer experiments, the degree of adaptation that a plant showed when placed in a habitat different from that in which it was collected in the wild. Not only were they interested in what we now call changes in phenotype of a plant but also in the persistence of any changes which occurred during the experiment.

As part of a general study of adaptation Bonnier studied many European plants. His experimental technique is of special interest as he used cloned material. Experimental plants were allowed to grow to a convenient size. They were then divided into pieces, and these pieces or 'ramets' were transplanted into experimental beds at different altitudes in the Alps, the Pyrenees and in Paris. His alpine sites were not gardens – ramets were planted into natural vegetation, protected sometimes by fencing. No fertiliser was added and no watering of the plants took place. In the first reports of his experiments (begun in 1882) he showed how 'alpine' ramets grew into very dwarf compact plants with very vivid flowers, in comparison with 'lowland' ramets (Fig. 4.1). In the 1890s, in a series of largely neglected papers, he published a great deal about the physiological and anatomical adaptation of these plants.

In 1920 Bonnier presented a summary of his researches and claimed that in the course of his experiments certain lowland species became modified to such a degree that they were transformed into related alpine and subalpine species or subspecies. This claim, which, Bonnier notes, supports the ideas of Lamarck, is of very great interest, and if true would have a profound effect upon the interpretation of natural variation patterns. It is worthy of note that Bonnier did not publish his conclusions in his *earlier* papers. Writing in 1890, he does not mention any transmutation of *Lotus corniculatus* into *L. alpinus*, although he had grown plants for some years both in the Pyrenees and in the Alps. He merely reported the dwarfing of the alpine clones in comparison with lowland ones.

Bonnier's claims were supported by the researches of Clements working in Colorado and California. He made a large series of clone-transplant experiments. In these experiments, too, it was asserted that lowland species had been transformed into alpine ones by growth at high altitude.

Fig. 4.1. Two examples of Bonnier's transplant experiments, showing the dwarfing effect of cultivation of ramets of the same clone at high altitudes. (*a*) & (*b*): Lowland and mountain *Leucanthemum vulgare* (*Chrysanthemum leucanthemum*). (*c*) & (*d*): Lowland and mountain *Prunella vulgaris*. (From Bonnier, 1895)

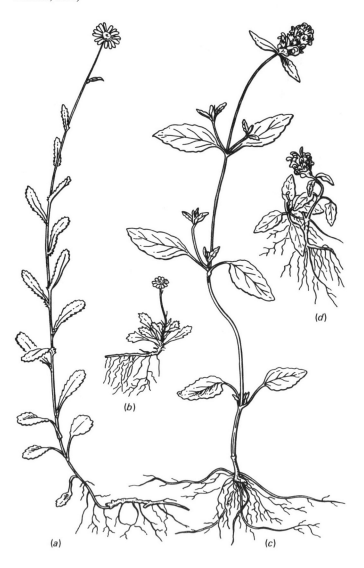

Epilobium angustifolium was considered to have been changed into *E. latifolium*, and Clements claimed that the grasses *Phleum alpinum* and *P. pratense* could be reciprocally converted (see Clausen, Keck & Hiesey, 1940).

Before examining the alleged transformations we should note that a number of central European botanists, notably von Nägeli and Kerner, had been carrying out similar experiments and had come to different conclusions. Von Nägeli was one of the first to study alpine populations in experimental gardens. He brought a wide range of alpine plants into cultivation at the Botanic Garden in Munich, and many changed their appearance greatly. This was particularly true of species of the genus *Hieracium*. Small alpine plants grown at Munich on rich soil became very large, much-branched plants. Von Nägeli was most interested to discover, however, that the acquired characters disappeared when plants were transplanted to gravelly soil within the garden, and the specimens again assumed the appearance of alpine plants.

Kerner, Professor of Botany in Vienna, carried out many transplant and reciprocal sowing experiments using an alpine garden at Blaser at 2195 m in the Tyrol, and the Botanic Gardens at Vienna and Innsbruck. He discovered that, for many species, if seeds were grown in two contrasting environments, dwarf plants with more vivid flowers were produced in alpine conditions. He noted the parallel case of more vivid colours in snails and spiders transferred to alpine conditions from the lowlands. Writing of his experiments in his famous book *The natural history of plants their forms, growth, reproduction and distribution*, (1895), Kerner noted:

> *in no instance was any permanent or hereditary modification in form or colour observed* ... They (the modifications) were also manifested by the descendants of these plants *but only as long as they grew in the same place as their parents.* As soon as the seeds formed in the Alpine region were again sown in the beds of the Innsbruck or Vienna Botanic Gardens the plants raised from them immediately resumed the form and colour usual to that position.

Kerner, therefore, came to very different conclusions from Bonnier and Clements as to the nature of the changes which had taken place in the material planted at high altitude. Since Kerner's experiments, thousands of experimental plantings have been carried out, deservedly the most famous being those of Clausen, Keck & Hiesey (1940) in California. No evidence of transformation of the kind claimed by Bonnier and Clements has been discovered. The most reasonable explanation for their anomalous results is that their experimental areas became invaded by the related alpine species which were growing naturally at these high altitudes.

From these observations it can be seen that experimental cultivation can be of very great value in investigating variation in plants. Simple cultivation tests, in which a range of material is grown under standard conditions, may reveal genetic differences between the stocks under investigation. Transfer and transplant experiments, properly carried out with special care in labelling and organisation, will give information upon the plasticity and adaptation of different plants. Especially useful are clone-transplants, as the performance of material of a single individual is investigated in different environments. In this respect Bonnier's experiments were to be preferred to those of Kerner who often used seeds. Seeds, except in special circumstances (see Chapter 7), will be genetically heterogeneous, and raise difficulties in interpretation not present in the clone-transplant method.

The work of Mendel

Let us now suppose that cultivation and transplant experiments in a particular instance have established a *prima facie* case of genetic difference between two plants. What is the nature of this difference? Our present knowledge of heredity stems from the various experiments of Mendel which he carried out over many years. Mendel, an Augustinian monk of the monastery of St Thomas at Brünn (now Brno in Czechoslovakia), reported his work on crossing Garden Peas in two papers to the Natural History Society of Brünn on 8 February and 8 March 1865, and the proceedings of these meetings were subsequently published in the *Transactions* of the Society. It seems inexplicable now that Mendel's epoch-making discoveries, which shed such great light on the mechanism of hereditary transmission, were overlooked in his lifetime. It was not in fact until 1900 that three botanists, De Vries, Correns and Von Tschermak, who had all been conducting breeding experiments at the end of the nineteenth century, rediscovered Mendel's paper. They immediately realised its significance.

Even though Mendel may be credited with the discoveries leading to the establishment of genetics, in many elementary textbooks the accounts of his work lack historical perspective. There is a wealth of pre-Mendelian experiments in hybridisation (Roberts, 1929), although it is true that early workers often had different objectives from those of Mendel. Kölreuter and Linnaeus investigated the phenomenon of sex in plants. Others, such as Laxton and the de Vilmorins, tried to improve varieties of plants of horticultural and agricultural importance. Another group of hybridists, as we saw in Chapter 2, were trying to find criteria for the experimental definition of species, using the data from experimental and natural hybridisation. Darwin was extremely interested in all aspects of hybridisation and published a book on the effects of self- and cross-fertilisation in plants.

Many findings of Mendel were in fact anticipated by earlier hybridists, although they were not connected into a coherent theory (see Zirkle, 1966, for examples). Kölreuter, for example, discovered that *Nicotiana paniculata × N. rustica* and the reciprocal cross gave identical hybrids of the first filial (F_1) generation. He also had crosses which showed dominance: *Dianthus chinensis* (normal flowers) × *D. hortensis* (double flowers) resulted in dominance in the F_1 of double flowers. The phenomenon of segregation was also known long before Mendel's day.

Turning now to discuss the main points of Mendel's contribution, we might ask what are the ways in which his approach to the problem of heredity differed from those of his predecessors? The contemporary preoccupation with species led to many interspecific crossing experiments but relatively few crosses between plants of a single species, except for those aimed at improving food-plant varieties. For the purpose of elucidating the mechanisms of heredity, species-crosses are not very helpful because species differ in innumerable characters, and in a number of generations a bewildering array of hybrid variants may appear. Before Mendel, hybridists did not in general concern themselves with the numbers of progeny of different sorts, and sometimes they did not even keep separate the progeny from different plants or different generations. In Mendel's paper we see that he is aware of the defects of past experiments (trans. Bateson, 1909):

> Those who survey the work done in this department will arrive at the conviction that among all the numerous experiments made, not one has been carried out to such an extent and in such a way as to make it possible to determine the number of different forms under which the offspring of hybrids appear, or to arrange these forms with certainty according to their separate generations, or definitely to ascertain their statistical relations.

In selecting peas for his work, Mendel knew that they are usually self-fertilising and that different cultivated varieties differ from each other in a number of respects. First he tested a selection of stocks (34 in all) and found them true-breeding. This is one of the most important facets of Mendel's work. Then, after carefully removing the unopened stamens of selected flowers, he crossed pairs of pea plants which differed in a single character. We may use as an example the cross between unpigmented plants (white seeds, white flowers, stem in axils of leaves green) and

pigmented plants (grey or brownish seeds – with or without violet spotting, flowers with violet standards and purple wings, stem in axils of leaves red). Here Mendel discovered that the first generation of hybrids, F_1, were all pigmented plants: 'pigmented' Mendel spoke of as 'dominant', the character 'unpigmented' he termed 'recessive'. He obtained the same result in the F_1 when pigmented plants were seed or pollen parent – in other words reciprocal crosses gave the same results. Mendel allowed

Fig. 4.2. One of Mendel's single-factor crosses, in which, 'pigmented' (allele C) is dominant to 'unpigmented' (allele c). Note the 3:1 ratio of dominant to recessive phenotype in the F_2 progeny.

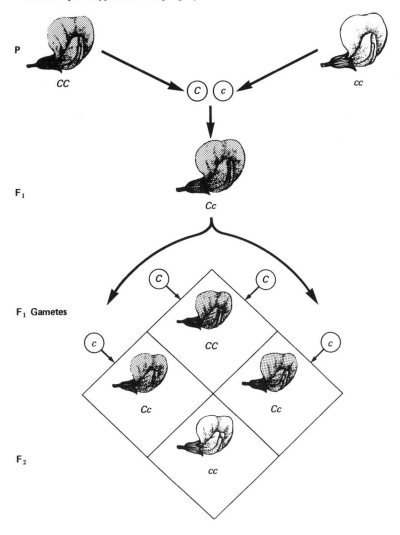

the F_1 plants to self and cross *inter se*, and in the next generation (the F_2) he discovered that the recessive character (unpigmented) reappeared along with pigmented plants in a numerical ratio of 3 pigmented:1 unpigmented (Fig. 4.2). Next Mendel studied the F_3 generation, and showed that unpigmented plants bred true, whereas only one-third of the pigmented plants did so. On selfing, the other two-thirds of pigmented plants gave pigmented to unpigmented plants in a 3:1 ratio. The 3:1 ratio in the F_2 was in reality a ratio of 1 true-breeding pigmented:2 non-true-breeding pigmented:1 unpigmented. We may note at this point that von Nägeli, in correspondence with Mendel, refused to believe that the unpigmented plants produced in the F_2 were true-breeding. He thought that the progeny of hybrids must be variable and that in the case of unpigmented F_2 peas, repeated selfing would eventually lead to segregation. (For an English translation of the letter from Mendel to von Nägeli, see Stern & Sherwood, 1966.)

Mendel obtained essentially similar results in crossing other peas differing in single characters:

Character	Dominant	Recessive
Stature	Tall	Dwarf
Seed shape	Round	Wrinkled
Cotyledon colour	Yellow	Green
Pod shape	Inflated	Constricted
Unripe pod colour	Green	Yellow
Flower position	Axillary	Terminal

Particulate inheritance

To explain his results Mendel postulated the existence of physical determinants, or 'factors' as he called them. The dominant character, in our example 'pigmented', may be denoted by a factor C, and the recessive 'unpigmented' by c. True-breeding parental stocks were CC and cc, giving C and c gametes, respectively, which at fertilisation gave an F_1 of constitution Cc. These F_1 plants, in appearance pigmented, produced in equal numbers two sorts of gametes, C and c, which (mating events being at random) gave three kinds of plants in the F_2 generation in the proportion $1CC:2Cc:1cc$ – a ratio of 3 pigmented:1 unpigmented. Mendel realised that, owing to the operation of chance, an *exact* 3:1 ratio would not be achieved in practice. His results came close to expectation: in our example his F_2 consisted of 705 pigmented:224 unpigmented plants, giving a ratio of 3.15:1.

We may note here a point of interest, first raised by Fisher (1936). Mendel's data, taken as a whole, fit expected ratios far too well, and consistently do not deviate as much as would be expected by the operation of the laws of probability. Fisher argues cogently that Mendel probably knew what his results would be before he started his experiments. Improvement of the numerical data was probably carried out to make the theoretical treatment of his results more acceptable, perhaps by an assistant. The excessive goodness of fit of Mendel's results does not seem to be in dispute, but the conclusion that deliberate falsification was involved has not been accepted by Wright (1966). For further details, Mendel's paper, with comments by Fisher (ed. Bennett, 1965) should be consulted.

Mendel's hypothesis of physical determinants or factors which can co-exist in an F_1 without blending and which segregate intact at gamete formation, was subject to a further test, that of back-crossing the F_1 (Cc) to the recessive parent (cc). As expected, the progeny were in the ratio 1 pigmented:1 unpigmented. These confirmatory results of Mendel, vindicating his theory of segregation, form the basis of modern genetics. Dominance or recessiveness of characters, however, which are so beautifully demonstrated in his experiments, are not an obligatory part of a genetic system, as we shall see later.

Mendel's two-factor crosses

We must now examine what happened when Mendel made 'two-factor' crosses. One of his experiments, incorporating his theory of determinants, may be represented by the following outline (Fig. 4.3). Mendel confirmed the genetic constitution of each category of plants by examining the progeny of selfed F_2 individuals. The important principle he discovered in these experiments was that the F_1, besides producing YR and yr gametes as did its parents, also produced gametes Yr and yR in numbers equal to those of the parental type. This equality of numbers of the four types of gametes established the *independent assortment* of the pairs of factors. Verification of independent assortment was made by Mendel when he crossed the F_1 ($YyRr$) with the double recessive parent ($yyrr$). As he predicted from his earlier results, four classes of offspring were produced in a 1:1:1:1 ratio.

The principles of independent assortment and recombination of factors established by this second group of experiments, together with Mendel's theory of particulate inheritance, form the heart of his contribution to genetics. Further work, in three-factor crosses in peas, and crosses in French beans, is reported in Mendel's paper. It was fortunate that Mendel's experiments on peas were more or less complete before 1864 for in the following summers infestations of pea beetle (*Bruchus pisi*) made

Fig. 4.3. A Mendelian two-factor cross: pure breeding yellow round peas (*YYRR*) × green wrinkled (*yyrr*). The appearance of the F₂ progeny was:

	Yellow round	Yellow wrinkled	Green round	Green wrinkled
Theoretical phenotype	9	3	3	1
Mendel's result	315	101	108	32
Experimental ratio	9.8	3.2	3.4	1

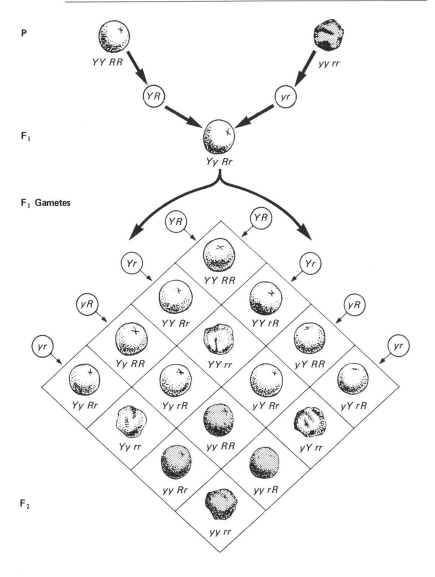

cultivation of peas difficult and finally impossible in Brünn (see letter from Mendel to von Nägeli dated 18 April 1867 in Stern & Sherwood, 1966).

Pangenesis

Perhaps we should now compare the ideas current at the end of the nineteenth century with those of Mendel which superseded them. Darwin, in his astonishingly productive later years, gave a great deal of thought to the problems of heredity and, in 1868, in *The variation of plants and animals under domestication*, he put forward his theory of 'pangenesis'. This theory, in many ways derived from Hippocratean ideas about the direct inheritance of characters, suggested that cells of plants and animals threw off minute granules or atoms (Darwin called them gemmules), which circulated freely within the organism. It was these gemmules which were transmitted from parent to offspring. Blending of gemmules occurred in the progeny. The phenomena of 'segregation' of a recessive plant in an F_2 or subsequent generation Darwin could account for only by suggesting that sometimes the gemmules were transmitted in a dormant state.

The theory of pangenesis, with its notion that gemmules came to the reproductive cells from all parts of the body, provided a mechanism for the inheritance of acquired characters, an idea favoured by Darwin. This view was challenged by Weismann (1883), who, in the words of Whitehouse (1959), disputed the idea that 'something from the substance of each organ was thought to be conveyed to the reproductive elements'. Weismann pictured just the converse situation: 'that the potentially immortal reproductive lineage carried in some mysterious way an exceedingly complex inheritance, which in each generation gave rise to mortal somatic offshoots – the individuals'. According to Weismann, 'the stream [of life] flowed direct from the reproductive cells of one generation to those of the next, and the individuals themselves (apart from their sexual elements) represented side-streams from which there was no return to the main stream'.

It is not necessary to go farther into Darwin's ideas, as they received no support from experiments. Galton, searching for evidence of gemmules, intertransfused blood of different-coloured rabbits and studied the colour of their offspring. There was no evidence that the presence of 'foreign' blood in a female rabbit made any difference to the colour of her progeny (Darwin, 1871; Galton, 1871). For a detailed account of this fascinating episode in the history of genetics, see Pearson's *The life, letters and labours of Francis Galton* (1924).

Galton himself had many ideas about heredity. Those he developed most forcibly were based upon a belief in blending inheritance. He did not carry out any breeding experiments as did Mendel but, as we saw in Chapter 3, he analysed records of human families, and developed the 'law of ancestral heredity'. This 'law' was a statistical statement of general patterns in samples, rather than a genetic analysis. He showed that a quarter of the heredity of an individual was determined by each of its parents, one-sixteenth by each of its grandparents, and so on.

Mendelian ratios in plants

Limitations of space prevent us from giving the fascinating details of the actual re-finding of Mendel's work; for this the relevant chapters in Roberts (1929), Sturtevant (1965), Olby (1966) and Provine (1971) may be consulted. We must concern ourselves here only with the reception of his work.

Towards the end of his paper Mendel wrote: 'It must be the object of further experiments to ascertain whether the law of development discovered for *Pisum* applies also to the hybrids of other plants'. A point neglected in most elementary books is that Mendel did not derive a generalised scheme of heredity for all organisms from his particular experimental results and the hypothesis to explain them. He thought more of his work as demonstrating the method by which the laws of heredity could be worked out. The possibility that there might be different laws seemed high, especially as Mendel did not get the same results in his later crossing experiments with *Hieracium* species. *Hieracium* was recommended as an interesting subject for crossing experiments by von Nägeli (Stern & Sherwood, 1966). We shall see why experiments with this genus gave different results in Chapter 7. When Mendel's results became available in 1900, it was soon realised that his hypothesis of segregating factors, or 'genes' as they came to be called, could explain the results obtained for many plants and animals. Table 4.2, compiled mostly from Bateson (1909), gives a representative list of plants in which Mendelian inheritance was discovered. The characters involved range from those of the general growth habit, to details of the leaf, flower, fruit and seed. It is very interesting that variants known for many years were investigated. For instance, white flower colour in a variety of *Polemonium caeruleum* (described by John Ray in the seventeenth century) was shown to be recessive to the normal blue colour, in experiments of De Vries. Not only did morphological characters show Mendelian inheritance, but so did physiological traits. An example is disease resistance in Wheat (*Triticum*) infected with the fungus *Puccinia glumarum*, where susceptibility was shown by Biffen to be dominant.

Table 4.2. *Plants in which Mendelian inheritance was demonstrated before 1909; in some cases dominance is incomplete (data mostly from Bateson, 1909; R.E.C. = Royal Society Evolution Committee Reports)*

	Gene dominant	Recessive	Material and author
Growth habit	Tall	Dwarf	Pea (*Pisum*): Mendel; von Tschermak. Sweet Pea (*Lathyrus*): R.E.C. Runner and French Bean (*Phaseolus*): von Tschermak.
	Branched	Unbranched	Sunflower (*Helianthus*): Shull. Cotton (*Gossypium*): Balls.
	Straggling	Bushy	Sweet Pea (*Lathyrus*): R.E.C.
	Biennial	Annual	Henbane (*Hyoscyamus*): Correns.
Leaves	Much serrated	Little serrated	Nettle (*Urtica*): Correns.
	'Palm'	'Fern'	Primula (*Primula sinensis*): Gregory.
	Normal	Laciniate	Greater Celandine (*Chelidonium majus*): De Vries.
	Yellow sap	White sap	Mullein (*Verbascum blattaria*): Shull.
	Rough	Smooth	Wheat (*Triticum*): Biffen.
Stems	Hairy	Glabrous	Campion (*Silene*): De Vries; R.E.C.
Flowers	Beardless	Bearded	Wheat (*Triticum*): Spillman; von Tschermak.
	Long pollen with 3 pores	Round pollen with 2 pores	Sweet Pea (*Lathyrus*): R.E.C.
	Normal pollen	Sterile	Sweet Pea (*Lathyrus*): R.E.C.
	Yellow	Brown	*Coreopsis tinctoria:* De Vries.
	Purple	White	Thorn-apple (*Datura*): De Vries.

Table 4.2. (*cont.*)

	Gene dominant	Recessive	Material and author
	Purple disc	Yellow disc	Sunflower (*Helianthus*): Shull.
	Black palea	Straw palea	Barley (*Hordeum*): von Tschermak; Biffen.
	Purple spot	No spot	Opium Poppy (*Papaver somniferum*): De Vries.
	Red chaff	White chaff	Wheat (*Triticum*)
	Red flower	White flower	Clover (*Trifolium pratense*): De Vries (1905)
	Blue flower	White flower	Jacob's Ladder (*Polemonium caeruleum*): De Vries (1905).
	Blue-purple flower	White flower	Sea Aster (*Aster tripolium*): De Vries (1905).
Fruits	Prickly	Smooth	Buttercup (*Ranunculus arvensis*): R.E.C.
	Blunt pods	Pointed pods	Pea (*Pisum*): von Tschermak; R.E.C.
	Two-celled	Many-celled	Tomato (*Lycopersicum esculentum*): Price and Drinkard.
	Dark	Light	Deadly Nightshade (*Atropa belladonna*): De Vries; Saunders.
Seeds	'Long staple'	'Short staple'	Cotton (*Gossypium*): Balls.
	Round	Wrinkled	Pea (*Pisum*): Mendel; Correns; von Tschermak; Lock; Hurst; R.E.C.
	Starchy	Sugary	Maize (*Zea*): De Vries; Lock; Correns.
	Yellow endosperm	White endosperm	Maize (*Zea*).

Cases of independent segregation in two-factor crosses were also discovered. For example, in crossing a white-flowered 'three-leaved' *Trifolium pratense* with a red-flowered, 'five-leaved' variant, De Vries (1905) obtained an approximate fit to an expected 9:3:3:1 ratio, the characters 'red-flowered' and 'five-leaved' being dominant.

Gradually Mendelian explanations for many single discontinuous variation patterns were accepted by most botanists, and a number of useful terms were introduced. The alternative factors *A* and *a*, as, for example, tall and dwarf in peas, were spoken of as *alleles* (allelomorphs) of a gene by Bateson & Saunders (1902) who also introduced the term *heterozygous* (*Aa*) to describe a zygote or individual with two unlike alleles, and *homozygous* (*AA*, *aa*) for one with two alike.

Mendelism and continuous variation

Notwithstanding the success of Mendelian explanations of familiar patterns of variation, universal acceptance did not follow. The biometricians, led by Pearson, remained loyal to the 'law of ancestral inheritance' of Galton, which we have shown is based upon blending inheritance.

Among the criticisms of Mendelism, one of great weight was that in certain crossing experiments no clear-cut segregation occurred in the F_2 generation. As an example, East's (1913) data for corolla length in F_1 and F_2 hybrids of *Nicotiana forgetiana* (♀) × *N. alata* var. *grandiflora* (♂) are given in Table 4.3. Here, a short-flowered plant was crossed with a long-flowered plant; the F_1 was of intermediate corolla length and the F_2, showing wider variation, did not segregate with Mendelian ratios. Is such a situation an example of blending inheritance? Pearson and his school of biometricians considered blending inheritance to be the general rule, Mendelian inheritance only applying in special circumstances. In the early years of the twentieth century, the problem of explaining continuous variation patterns was very urgent. An initial difficulty in understanding continuous variation was in estimating the environmental and genetic components of the variation pattern. This difficulty was largely removed by the work of Johannsen, to which we have already referred.

Yule (1902) was probably one of the first to suggest that many genes were involved in continuous variation. To show what he had in mind, we may take as an example human height, which follows a typical normal distribution and, even though nutritional factors are highly important in determining the height of a person, the fact that Pearson and Galton showed a positive correlation (*r* about 0.5) between the height of parent and offspring provided a *prima facie* case of genetic control of height. Yule considered that a number of genes might be involved in determining

Table 4.3. *Frequency distribution of corolla length in the cross* Nicotiana forgetiana × N. alata *var.* grandiflora *(data of East, 1913)*

	Length of corolla (mm)														
	20	25	30	35	40	45	50	55	60	65	70	75	80	85	90
N. forgetiana	9	133	28	—	—	—	—	—	—	—	—	—	—	—	—
N. alata var. grandiflora	—	—	—	—	—	—	—	—	—	1	19	50	56	32	9
F_1	—	—	—	3	30	58	20	—	—	—	—	—	—	—	—
F_2	—	5	27	79	136	125	132	102	105	64	30	15	6	2	—

continuous variation patterns, and in this case different genes might determine leg length, trunk length, neck length, etc. In order to make this hypothesis credible, it was necessary to demonstrate that the genetics of a single character could be controlled by at least two genes.

Such a situation was discovered in 1909 by Nilsson-Ehle, who studied hybrids between wheats with brown and white chaff (Fig. 4.4). In the F_1 of the cross, brown chaff was dominant. Inter-crossing of the F_1 gave an F_2 generation, not in the expected 3:1 ratio of brown:white, but in the ratio of 15 brown:1 white. This result was confirmed in a second

Fig. 4.4. Chaff colour in Wheat (*Triticum*). (From Nilsson-Ehle, 1909)

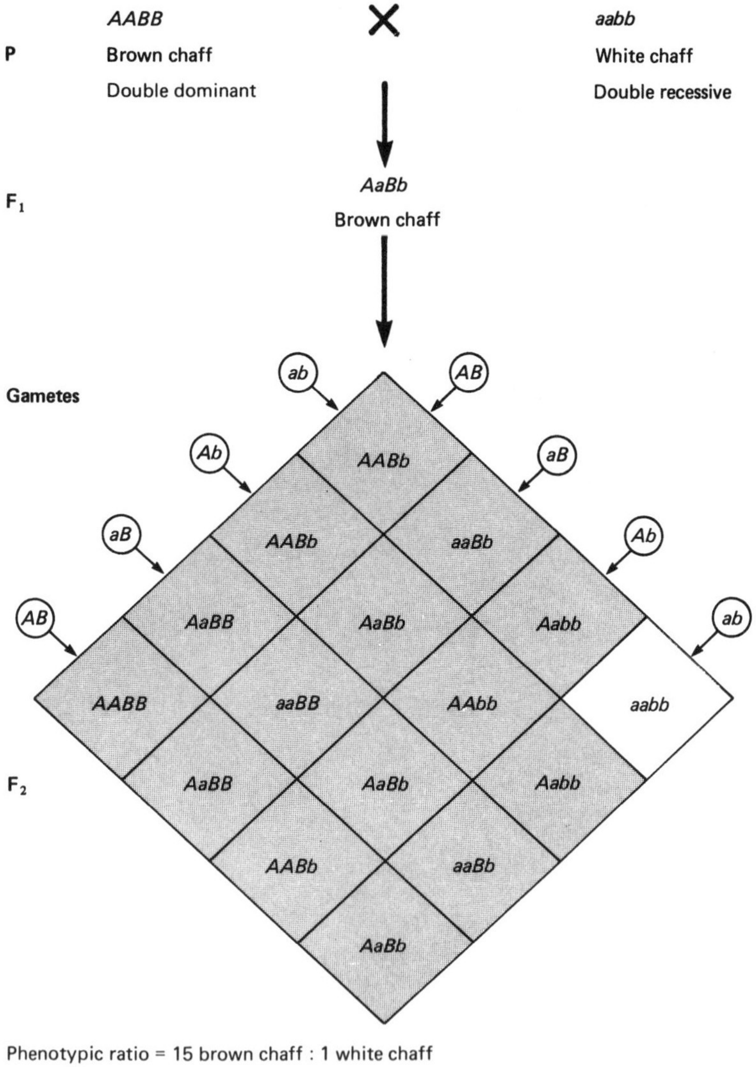

Phenotypic ratio = 15 brown chaff : 1 white chaff

experimental cross. Nilsson-Ehle considered that in this case two different genes were involved in chaff colour and that the 15:1 ratio was in reality a modified 9:3:3:1 ratio. The presence of a single dominant in an individual was sufficient to give brown chaff; only one-sixteenth of the progeny (of genotype *aabb*) had white chaff. Here is a clear case of two genes affecting the same character.

These experiments of Nilsson-Ehle, which were paralleled by the

Fig. 4.5. Flower colour: a hypothetical case.

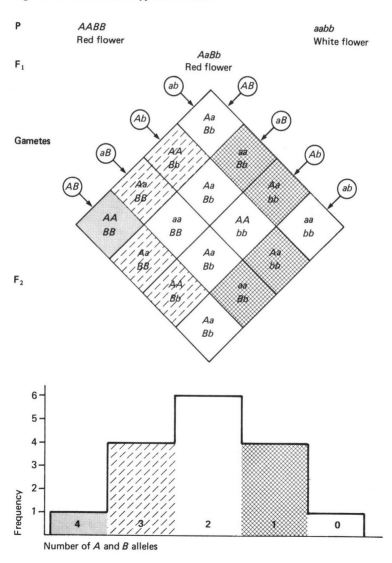

independent work of East, provide the necessary basis for an understanding of the genetics of continuous variation. To demonstrate the principles we will examine a hypothetical case of flower colour (Fig. 4.5). In this model we postulate that two different genes are involved: *A* and *B* being the dominant alleles determining red flower colour, alleles *a* and *b* determining white flower colour. In the example, we assume, however, that the effects of *A* and *B* are additive, the degree of red colour in the flower depending upon the number of *A* and *B* alleles present in an individual. Examination of the F_2 'chequer-board' shows that one-sixteenth of the progeny has four red alleles, four-sixteenths have three, six-sixteenths have two, four-sixteenths have one, and one-sixteenth has none. It should be noted that our example still shows Mendelian segregation of 15 red: 1 white on a broad classification, in detail with four different categories of red. Expressing the frequencies as a histogram, we obtain a distribution which bears a striking resemblance to a normal curve.

Consider now what might happen if a larger number of genes was involved. With six dominant genes, all additive in their effect for red colour, the F_2 would show very many categories of individuals, and a closer fit to a normal curve. Of great importance, too, the parental genotypes *AABBCCDDEEFF* (red) and *aabbccddeeff* (white) would be very infrequently segregated in the F_2. In fact only 1/4096 of the progeny would be *AABBCCDDEEFF* and, even more important, there would be a similar proportion of *aabbccddeeff* which would be the only white phenotype. In actual practice, if the cross were made, even though Mendelian segregation had taken place at gamete formation in the F_1 plants, it is quite likely (especially if the F_2 is represented by a small number of plants) that no *aabbccddeeff* plants would be recovered at all. The F_2 progeny would then all be red-flowered, in different degrees, giving a normal distribution curve.

Turning now to an actual experiment, the *Nicotiana* crosses of East which we referred to earlier (Table 4.3), far from demonstrating blending inheritance, may more satisfactorily be interpreted on the basis of multiple factors affecting corolla length. The two variants of *Nicotiana* used differ in corolla length, and the F_1 from the cross is intermediate in length, indicating the absence or incompleteness of dominance. In the F_2 a wide array of corolla sizes is found, the frequency distribution approximating to a normal curve. Note that the extreme 'parental' corolla sizes are not represented in the data. East considered that there were probably four genes involved in the determination of corolla length (Fig. 4.6).

Many investigations of continuous variation patterns in nature have

Fig. 4.6. East's cross in *Nicotiana*. (From East, 1913)

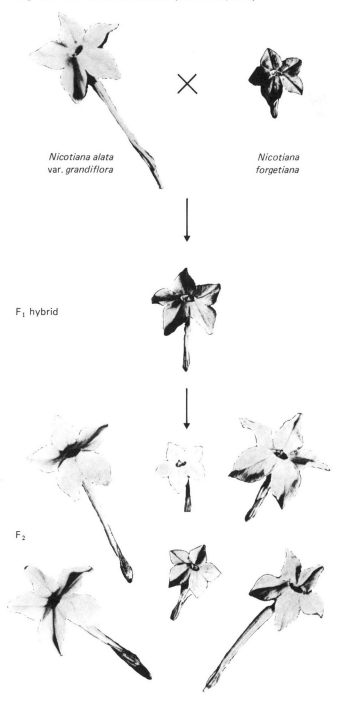

Nicotiana alata
var. *grandiflora*

Nicotiana
forgetiana

F₁ hybrid

F₂

given similar results to those of East, and elaborate genetic and statistical experiments since that time have demonstrated the general validity of the multiple-factor hypothesis. Such systems, in which the character is determined by several genes, are usually called *polygenic*.

Physical basis of Mendelian inheritance

So far we have not discussed the physical nature of Mendel's factors. In Mendel's day little or nothing was known about the physico-chemical basis of heredity but there was plenty of theoretical speculation. Von Nägeli, for instance, postulated a genetically active 'idioplasm'. By the time Mendel's work was rediscovered in 1900 the situation, however, was very different. The latter half of the nineteenth century had seen an enormous increase in interest in the microscopic study of plant and animal cells. Certain technical innovations such as the use of stained material (carmine was introduced in the 1850s, and haematoxylin and anilin dyes in the 1860s) and the perfecting of apochromatic lenses (by Abbé in 1886) enabled biologists to make a close study of all aspects of cell division and development (see Hughes, 1959). It is impossible in the space available to review the results of these studies in any detail, but the main conclusion was that the chromosomes discovered in cell division were clearly very important in heredity.

It was found that each species has a characteristic number – the diploid number – of chromosomes, visible in stained preparations of meristematic cells. The account that follows applies to diploid organisms, i.e. those whose nuclei contain two like sets of chromosomes, one set from pollen and one from egg. There are, however, haploid organisms, e.g. certain fungi, whose nuclei contain only one set of chromosomes. In higher plants there is generally consistency of number, size and form of chromosomes of the meristematic cells of root-tip and shoot apex where chromosomes divide by *mitosis*. Essentially each chromosome divides into two daughter chromosomes and at the end of the process the two groups of daughter chromosomes are separated from each other by a new cell wall (Fig. 4.7).

Studies of the division of chromosomes in young anthers and in ovules revealed a different kind of nuclear division, the so-called *reduction division* or *meiosis* (Fig. 4.8). In this process the chromosome number is halved, the four derivatives having the haploid chromosome number. This halving compensates for the doubling in chromosome number following fertilisation of egg by sperm. Thus in a diploid plant a haploid comple-ment of chromosomes has come from each parent. Microscopic examina-tion of favourable material establishes a most interesting fact; if maternal and paternal haploid complements are examined they are normally found to be exactly alike in appearance (except in the case of certain sex chromosomes). In the early stages of meiosis, homologous chromosomes, from maternal and paternal sources, pair together. Studies as early as those of Rückert (1892) suggested that in this paired state exchanges of chromosome material occurred.

This very brief outline of mitosis and meiosis gives some idea of the sort of knowledge about chromosomes which was available at the beginning of the century. It was only a short time after the discovery of Mendel's work that various biologists, Boveri, Strasburger and Correns among them, saw a possible connection between Mendelian segregation and chromosome disjunction. It was probably Sutton (1902, 1903), however, who first set out with clarity a cytological explanation of Mendel's findings. In his view the separation of maternal and paternal chromo-

Fig. 4.7. The stages in mitosis in an organism with two pairs of chromosomes. (From McLeish & Snoad, 1962)

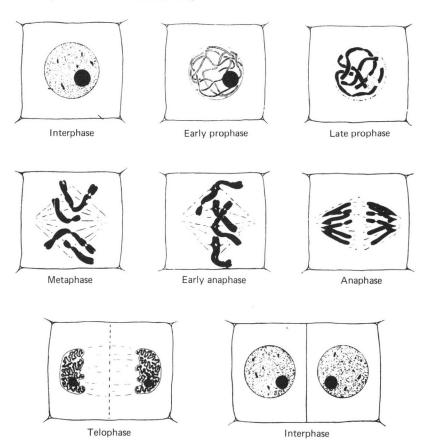

| Interphase | Early prophase | Late prophase |

| Metaphase | Early anaphase | Anaphase |

| Telophase | Interphase |

Fig. 4.8. The stages in meiosis in an organism with two pairs of chromosomes. The formation of one bivalent and its subsequent behaviour are shown diagrammatically above appropriate stages. (From Whitehouse, 1965)

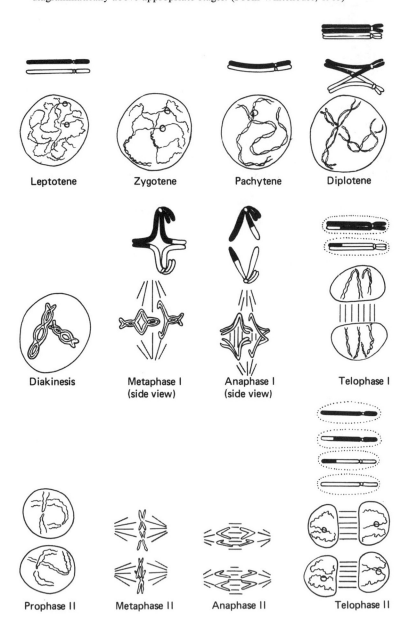

somes of a homologous pair at the end of the first stage of meiosis resembled the postulated separation of factors which Mendel suggested occurred at gamete formation. Further, if the orientation of pairs on the spindle was at random, a number of combinations of maternal and paternal chromosomes would be obtained in the gametes. If the chromosome number was very small the number of combinations would also be relatively small; on the other hand a diploid chromosome number as low as 16 would give 65 536 possible zygotic combinations (Table 4.5, Fig. 4.9) (see Sutton, 1903). As many plants have chromosome numbers higher than this, a huge number of combinations is possible. We have here the beginnings of the chromosome theory of heredity, which is the basis of all modern genetics.

Mendel postulated in his experiments the independent segregation of factors, and this view received support from the early geneticists. There were, however, increasing signs in the first decade of this century that not all genes segregate independently. Bateson, Saunders & Punnett in 1905, working with two-factor crosses in Sweet Peas (*Lathyrus odoratus*), did not get 9:3:3:1 ratios in F_2 families. Similar aberrant results were obtained from many organisms, amongst them the Fruit Fly (*Drosophila*) and the Garden Pea (*Pisum*). Many biologists followed Mendel in experimenting with peas, and up to 1917 an additional 25 character-pairs were examined (White, 1917). A very interesting series of crosses was made by de Vilmorin (1910, 1911) and subsequently by de Vilmorin & Bateson (1911) and Pellew (1913) working with 'Acacia' peas, a variant characterised by the absence of the normal leaf tendrils. The absence of tendrils was associated with wrinkled seed. The cross 'Acacia' × round seed and tendrilled leaf gave an F_1 with round seed and tendrilled leaves. The F_2, instead of segregating to give 9:3:3:1, gave the results in Table 4.4.

Table 4.4. 'Acacia' peas, *results of various experiments as reported in White (1917)*

Source	Wrinkled seed, no tendril	Wrinkled seed, tendril	Round seed, no tendril	Round seed, tendril
de Vilmorin	70	5	2	113
de Vilmorin	99	4	1	170
Bateson	64	1	4	210
Pellew	564	15	20	1466

Table 4.5. *Possible zygotic combinations (after Sutton, 1903)*

Chromosome number		Combinations in gametes	Combinations in eventual zygotes
Diploid	Haploid		
2	1	2	4
4	2	4	16
6	3	8	64
8	4	16	256
10	5	32	1024
12	6	64	4096
14	7	128	16 384
16	8	256	65 536
18	9	512	262 144
20	10	1024	1 048 576
22	11	2048	4 194 304
24	12	4096	16 777 216
26	13	8192	67 108 864
28	14	16 384	268 435 456
30	15	32 768	1 073 741 824
32	16	65 536	4 294 967 296
34	17	131 072	17 179 869 184
36	18	262 144	68 719 476 736

Fig. 4.9. The random orientation of bivalents at meiosis. At fertilisation a male and female gamete fuse, and each has a haploid set of chromosomes. Meiosis takes place in the diploid phase of the life-cycle. Homologous chromosomes form bivalents (maternal chromosomes dark; paternal ones white), which orientate themselves at random about the equatorial region of the cell. The diagram represents one of the possible meiotic metaphase arrangements of the chromosome complement in a diploid cell with 16 chromosomes (8 pairs). As Table 4.5 shows, in a plant with $2n = 16$ there are 256 possible patterns of arrangement of paternal and maternal chromosomes. As homologous chromosomes may carry different alleles, independent orientation means that many different combinations of maternal and paternal genes may be obtained in the gametes.

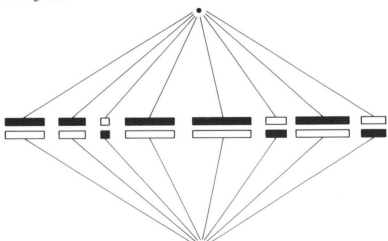

It is quite clear that the two factors are not segregating independently: the grandparental combinations of wrinkled/no tendril and round/tendril are being recovered with too high frequencies. Various explanations were offered for this phenomenon, which came to be called 'partial linkage', later abbreviated to 'linkage'. Bateson & Punnett (1911) favoured an obscure 'reduplication' hypothesis; as time went by, however, the views of Morgan prevailed. He suggested that partially linked groups of factors were together on the same chromosome. There would then be a haploid number of 'linkage groups' in the pea ($n = 7$). In the Fruit Fly, *Drosophila melanogaster*, where $n = 4$, the extensive researches of Morgan and his colleagues established beyond doubt the existence of four such linkage groups. In the formation of a diploid organism the two gametes each carry one set of linkage groups – the haploid chromosome number. The appearance of occasional recombinants in small numbers in a cross such as that in Table 4.5 was accounted for by postulating an exchange of parts by homologous pairs in the first stage of maturation division. Evidence of such an exchange was seen in the chiasmata of prophase (Fig. 4.10).

There is insufficient space to examine further the progress and implications of the chromosome theory. A number of excellent books exist, which discuss chromosome mapping and give full and detailed evidence for all the postulates in the chromosome theory of heredity (for example, Whitehouse, 1973).

Fig. 4.10. A bivalent at diplotene with a single chiasma. The position of some genes (represented by letters) is indicated. Crossing-over has occurred between one chromatid of the maternal chromosome (white) and one chromatid of the paternal chromosome (dark). (From McLeish & Snoad, 1962.)

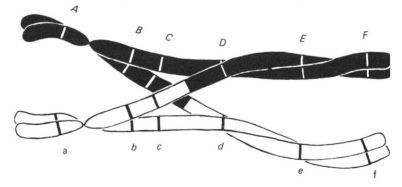

5

Post-Darwinian ideas about evolution

For 40 years after the publication of the *Origin*, Darwin's ideas were a source of tremendous public controversy. For this reason he never received any awards from the state, although he was awarded honorary degrees and decorations in plenty. Despite thousands of sermons attacking the idea of evolution by natural selection, more and more biologists became convinced of the truth of Darwin's view. The biological literature of the period is full of papers speculating about the adaptive significance of various structures, the probable course of evolution in the plant and animal kingdoms, and so-called 'evolutionary trees' showing phylogenetic relationships (Fig. 5.1). Some of this work is of lasting interest; but there was a depressing tendency in the later years of the period for armchair biologists to produce highly speculative theories, and there was a lack of critical experiment with living material. Towards the end of the century, however, there were signs of an increasing interest in the possibility of using experiments for the investigation of evolutionary problems.

Experimental investigation of evolution

A good example of this change in climate is provided by the controversy which enlivened the pages of *Nature* and the editorials of *The Gardeners' Chronicle* in 1895. At a discussion meeting of the Royal Society, the Director of Kew, W.T. Thiselton-Dyer, had shown specimens of the 'feral' type and cultivated variants of what he called *Cineraria cruenta* – the gardener's Cineraria – and an extended account of his ideas was printed in *Nature* (1895). He suggested, as befitted an ardent and orthodox disciple of Darwin, that as far as was known the garden Cineraria was derived from *Senecio* (*Cineraria*) *cruentus* from the Canaries 'by the accumulation of small differences'. Bateson (1895a, b) responded to this view in a lengthy letter to the Editor, questioning the assertions of Thiselton-Dyer. Bateson concluded, after a study of the literature, that modern Cinerarias arose from hybridisation between several distinct species, that selection was practised on variable hybrid progeny, and that 'sports' may have been important, as well as subsequent improvements as a consequence of the selection of small-scale variation. The arguments in *Nature* continued back and forth with four letters from Thiselton-Dyer, three from Bateson and three from the biometrician Weldon. It became clear that argument could not settle the issue of the origin of the garden Cineraria. The possibility that experimental hybridisation might shed light on the variation patterns occurred to Bateson, who enlisted the help of Lynch, Curator of the University Botanic Garden in Cambridge. Lynch raised stocks and made a number

Fig. 5.1. Tentative sketch of a phylogenetic tree from Darwin's notebook (1837) (see De Beer, 1960–61) contrasts with the baroque splendour of Haeckel's highly speculative 'Monophyletic pedigree of the vegetable kingdom' of 1876.

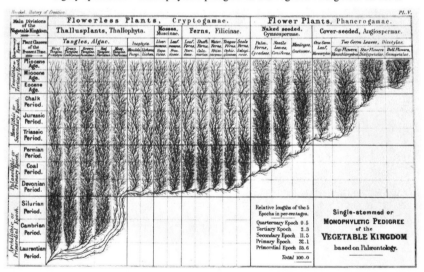

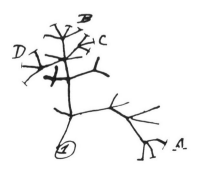

of artificial crosses, some of which were exhibited at a meeting of the Cambridge Philosophical Society in 1897 (Fig. 5.2.). The report of the meeting (Bateson, 1897) says that the experiments 'were entirely consistent with the view that Cineraria was a hybrid between several species'. Lynch's experiments were published in detail in 1900. Here we have a clear case of speculation about evolution leading directly to experiment. It is interesting that a recent review of the Cineraria problem (Barkley, 1966) reveals that it has received little attention since these early experiments.

In the period 1892–1910 some of the first experiments investigating

Fig. 5.2. Hybrid Cinerarias: an early biosystematic study. Three of the variants produced by R.I. Lynch, Curator of the University Botanic Garden, Cambridge, by crossing wild species of *Senecio* and the cultivated Cineraria. Painted by H.G. Moon and published in *The Garden* in July 1897. (Illustrated in colour on the cover.)

natural selection were carried out. Darwin had written in the *Origin* (Chapter IV): 'Can we doubt (remembering that many more individuals are born than can possibly survive) that individuals having any advantage, however slight, over others, would have the best chance of surviving and of procreating their kind?'. In 1895 Weldon wrote: 'The questions raised by the Darwinian hypothesis are purely statistical, and the statistical method is the only one at present obvious by which that hypothesis can be experimentally checked. In order to estimate the effect of small variations upon the chance of survival, in a given species, it is necessary to measure first, the percentage of young animals exhibiting this variation; secondly, the percentage of adults in which it is present' (Weldon, 1895*b*). If the percentage of adults exhibiting the variation proved to be less than that in young animals, then some of the young animals must have been lost before reaching adulthood, and a measure of the advantage or disadvantage of the variation could be obtained. In putting these novel ideas to the test, Weldon (1898) investigated the variation in the crab *Carcinus maenas* at a site on Plymouth Sound. While it is inappropriate to give details of his results, we may note that his findings offered some support for the initial hypothesis. After several years' investigation (1892–8) he also concluded that the population was unstable. He deduced that changes were caused by the increasing amounts of china clay and sewage in the waters of Plymouth Sound, and carried out experiments investigating the death rate of captive crabs subjected to foul water. Crabs survived captivity in clean water but only a portion of the variable population – those with small frontal breadth relative to their carapace size – survived in foul water. While it is true that these experiments may be criticised on grounds of sampling technique (and, indeed, in other ways), they do represent a major step forward in the design of investigations into natural selection and, as we shall see, are a model for some more recent botanical studies.

Weldon was also instrumental in encouraging some of the first field studies of selection. Di Cesnola (1904), a student of Weldon's at Oxford, noticed that in Italy there were green and brown variants of the Praying Mantis, *Mantis religiosa*. The green variant was found in grasses and the brown on vegetation burnt by the sun. In an experiment which lasted several days, individuals were tethered by silk threads as follows:

1. in a green grassy area – 20 green and 45 brown individuals
2. in a brown area – 25 green and 20 brown individuals.

The 25 greens in the brown area and 35 of the browns in the green area were taken by birds or ants. It is significant that the individuals which matched the background vegetation were untouched. These studies were forerunners of many experiments by zoologists to test the supposed adaptive significance of protective coloration (see Cott, 1940). As in the case of the investigations on crabs, it is obvious that the experiment with *Mantis* is not beyond criticism, but it is based, nevertheless, on a novel approach to the study of the force of natural selection.

In plants too, studies of variation led to insights into natural selection. For example, Brand & Waldron (1910) and Waldron (1912), working at Dickinson, North Dakota, USA, cultivated 68 samples of the important legume forage crop *Medicago sativa* (Lucerne or Alfalfa) collected from different parts of the world. Plants from Mongolia proved to be cold-resistant, whilst those from Arabia and Peru, on the other hand, were frost-sensitive. In the severe winter of 1908–9 the pattern of losses due to frost damage provided interesting information. For instance, in a strain from Utah, many plants died of frost damage, but out of the progeny of three specially resistant plants originating from this stock only 3.5% died. Brand & Waldron deduced that a frost-hardy strain could originate from extreme individuals of an otherwise frost-sensitive stock.

There was also interesting research on 'races' in hemiparasitic plants. Thus, three variants of *Viscum album* (Mistletoe) were described by von Tubeuf (1923) from broadleaved trees, Fir and Pine. Each was morphologically distinct in such characters as size and shape of leaves, and colour of berries. Some attempt was made (only partially successful) to test by transfer of seed whether there were three different physiological races of Mistletoe, each adapted to a different host. In 1895, von Wettstein described the phenomenon of seasonal dimorphism in a number of hemiparasitic genera including *Euphrasia, Odontites* and *Rhinanthus*. As a result of careful investigation, many species appeared to have two subspecies, *viz.* an early summer flowering variant (aestival) and a later variant (autumnal). He considered that the practice of haymaking in central Europe was important in the origin of the two types of subspecies. The maintenance of the annual habit was only possible if plants either fruited before midsummer grass-cutting, or elongated and matured after the hay crop had been taken. Plants flowering or in immature fruit at the time of haymaking would fail to reproduce, and selection would therefore favour both early and late flowering. This work provoked a good deal of controversy, especially about which subspecies was ancestral and whether patterns were as simple as was suggested by von Wettstein. For full details of the historical studies in this area, together with a recent elegant study of *Rhinanthus angustifolius* (*R. serotinus*) in different grasslands in the Netherlands, see Ter Borg (1972).

The mutation theory of evolution

The experimental approaches employed by biologists at the turn of the century not only provided insights into biometry and natural selection, but also provoked Bateson, De Vries and others to propose a rival theory to that of Darwin – the mutation theory of evolution. The theory stressed the importance of 'sports' and various other abruptly occurring new variants. While the theory was claimed to be new, it is clear, as we saw in Chapter 2, that it grew out of a longstanding interest in the subject of 'sports'.

Darwin argued that species were ever-changing entities, the products of natural selection; his thesis was descent with modification, involving continual and gradual change. De Vries and Bateson did not deny the existence of natural selection; in fact it still played a key role in their ideas of evolution. What was different was their view that new species arose abruptly by 'mutation'. They confined the significant changeability of species to distinct and probably short periods. They accepted the theory of descent with modification, but thought that the changes occurred abruptly, interspersed with periods of stability.

What evidence could the 'mutationalists' find in support of their theory? First, they examined cases of the apparent abrupt evolution of new persistent variants. Most famous of these were plants of the genus *Oenothera* studied by De Vries (1905). Secondly, they discussed problems of heredity. For the 'mutation theory' it was *discontinuous* patterns which were significant and, with the re-finding of Mendel's work in 1900, Bateson did not fail to point out that this provided a mechanism explaining discontinuous variation. In his view, continuous variation was the product of environmental factors. A third piece of evidence was also forthcoming. In his careful experiments with beans, Johannsen had discovered that selection was ineffective. Try as he might, from a particular line he could not select a strain with larger or smaller beans. Some variations did occur, but Johannsen ascribed this to the effect of the environment. Bateson and others went further and argued that all 'fluctuating variations' found in nature were environmentally based. Darwin had considered this type of variation to be the raw material upon which selection acted, but for the 'mutationalists' natural selection occurred only when the products of mutation were being sorted out.

By the beginning of the century a curious situation had developed. In opposing camps were the 'Mendelian-mutationalists' and the 'Darwinian-biometricians' (Waddington, 1966; Crew, 1966). The Darwinian-biometricians, for the most part, remained loyal to Darwin's theory of gradual change. It is remarkable that this group of mathematically minded scientists opposed Mendel's views, preferring instead Galton's law of ancestral inheritance. As pointed out by Fisher (1958), the lack of mathematical understanding in biologists possibly contributed to the neglect of Mendel's work at the time of its publication, yet, paradoxically, on its re-finding it was the mathematical biologists who opposed it.

It was not until the 1930s that Darwin's ideas triumphed over 'mutation' theory. Gradually, more and more evidence against the 'mutationalist' view was discovered. First, intensive genetic and cytological studies of many species, including species of *Oenothera*, were carried out, and it became obvious that the new persistent variants found in *Oenothera* were of several different sorts. Some were simple mutants; others were polyploid derivatives; and a further group was the result of complex interchanges of chromosome segments (see Chapter 6). Other species did not give the same results, and the *Oenothera* situation was seen to be unique in its complexity.

A re-interpretation of Johannsen's results was also put forward. The ineffectiveness of selection in his case was seen to be due to the genetic invariability of the progeny from a single selfed individual. Continued self-fertilisation is characteristic of French Beans and, as we shall see later, this leads to genetic homozygosity. It is only in the absence of genetic variability that selection is ineffective, a fact attested by many successful selection experiments with other organisms.

The idea of blending inheritance was finally demolished by work carried out at this time. Darwin's idea of the persistence of favoured variants under a regime of blending inheritance necessitated a high mutation rate. Examination of the natural mutation rate showed that its incidence was much lower than that required to support the idea of blending. Further, it became clear that cases of inheritance which were at first explained by blending were explicable in terms of Mendelian genetics – some as instances of systems with no dominance (Fig. 5.3), and others, as we saw in Chapter 4, as examples of inheritance controlled by many factors.

We also find that the problem of the extraordinary amount of variation showed by domesticated animals and cultivated plants which, as Darwin had seen, was highly relevant to an understanding of evolution, had continued to exercise biologists. Darwin had been forced to explain the phenomenon in terms of some direct effect of the changed environment in calling forth a great range of hereditary variations in exceptionally large numbers, on which artificial selection worked quickly and efficiently. He drew the inference that environmental changes suffered by wild species would, from time to time, similarly liberate great bursts of new hereditary

variation on which selection could act, the unselected variants being
rapidly eliminated by blending inheritance. Experiments designed to show
any effects of this type, such as those of the botanist Hoffmann (1881),
who grew a range of plants through several generations in rich garden
soils and in soils treated with salt, lime or zinc, revealed no burst of

Fig. 5.3. *Mirabilis jalapa*. At first sight the production of pink-flowered
offspring from red- and white-flowered parents looks like a case of blending
inheritance, but, as the diagram shows, the situation can be explained in simple
Mendelian terms if we assume the absence of dominance. Reprinted with
permission of Macmillan Publishing Co. Inc. from Strickberger (1976) after
Correns. Copyright, © 1976 Monroe Strickberger.

induced variations. Moreover, it seems likely that the effect, for example, of transfer to rich soil or an abundant food supply would be a purely temporary one, yet domesticated organisms continue to reveal new variations indefinitely. For all these reasons, Darwin's ideas of blending inheritance had to be abandoned.

Thus, with the disappearance of the grounds for believing in the 'mutation' theories of De Vries and Bateson, and the demise of the theory of blending inheritance, the way was clear for a demonstration of how Mendelian genetics could be integrated with Darwin's evolutionary theory.

Neo-Darwinism

Inheritance is particulate and new alleles of genes arise by mutation. Mutant alleles are usually, but not invariably, recessive. This new genetic variation, if recessive, does not immediately find expression in the phenotype. For instance, in a population of plants of genotype AA, mutation in ovules or pollen may give rise to new individuals of constitution Aa; such plants on crossing with the more numerous AA plants will produce a $1AA:1Aa$ ratio. It is only on selfing or inter-crossing $Aa \times Aa$ that plants of constitution aa are produced, the ratio for the cross being $1AA:2Aa:1aa$ (a 3:1 phenotypic ratio, as A is dominant). Another important point to realise from this example is the sheltering effect on allele a in heterozygous individuals. Even if at the time of its origin a is deleterious, or even lethal in double dose, allele a still survives in the heterozygous plants of the population. This survival is very important because some future change in the environment may alter the balance of selective advantage in favour of a against A. In this way Mendelian genetics can account for the origin of new variants and also for their persistence.

This integration of Mendelian genetics and Darwin's evolutionary theory in what is now often termed 'neo-Darwinism' did more than remove the difficulties raised by Fleeming Jenkin and others. It allowed very considerable progress to be made in understanding new facets of evolution. The studies since the 1920s may be divided into two main groups. First, evolution could be examined at its most basic level – that of frequency changes in alleles. This new approach was made possible because particulate inheritance, unlike blending, is amenable to rigorous mathematical treatment. Theoretical models, worked out in detail mathematically, could be tested experimentally. Studies of this sort usually involved short-term experiments, often with stocks of laboratory animals, but increasingly allelic changes are being studied in natural populations of plants. This fascinating new science of population genetics owes its origin to the important studies of Chetverikov (1926), Fisher, Haldane, Wright, and others, and in the following chapters we shall examine some of their findings concerning microevolution.

A second area of investigation, that of the nature and origin of species, was given great impetus in the 1930s by important advances in genetics and cytology. The basic idea behind many of these studies has been outlined by Clausen (1951). It is, quite simply, that studies of the patterns of variation of living plants reveal, on examination, stages in the evolution of species. This hypothesis is open to experiment and, as we shall see in later chapters, an impressive amount of detailed information is now available on the modes of origin of the distinct 'kinds' of organisms which we normally recognise as species. In particular we shall be forced to consider the vexed question of what constitutes a species.

6

Modern views on the basis of variation

In previous chapters we have discussed the variation patterns found in samples taken from plant populations, and suggested that for descriptive purposes it may be useful to distinguish three broad types of individual variation – *viz.* genetically determined, environmentally induced and developmental variation. This notion of three types of variation, however, gives an oversimplified picture of the nature of individual variation. There is strong evidence for the proposition that the phenotype and behaviour of the plant are determined by *interactions* between genotype and environment. As we shall see, different genotypes react differently to a given set of environmental conditions, and plants of identical genotype produce different phenotypes under contrasting environmental conditions. Moreover, the interactions between genotype and environment are further complicated by the complex sequence of changes which occurs as a plant develops from an embryo to the mature fruiting state.

Chemical nature of the gene

In Chapter 4 we discussed the whereabouts of Mendel's factors in the cell, outlining the growth of the idea of genes located at particular points on the chromosomes. The early geneticists visualised genes as 'beads threaded on a string'. Gene action had been thought of from the earliest days of cytogenetics, at least in particular cases, as being connected with the action of enzymes. Detailed studies on biochemical mutants in the lower organisms in the 1940s provided a basis for what came to be called the 'one gene – one enzyme' hypothesis.

In 1944 we find the physicist Schrödinger, in his fascinating little book *What is life?*, suggesting that genes could be complex organic molecules in which endless possible variations in detailed atomic structure could be responsible for codes specifying the stages of ontogenetic development. More recently, as a result of the application of sophisticated biochemical and physical techniques to biological material, spectacular progress has been made in our understanding of the nature of the hereditary materials. The achievements of molecular biology have been so great and the progress so rapid that an enormous body of information now exists. A revolution in biology is taking place – no less of a revolution than that caused by Darwin's theory of evolution by natural selection. Obviously in this short book we cannot survey even the main findings of these new approaches; instead we must be content to stress some of the ways in which this new knowledge of molecular biology illuminates various aspects of plant variation and microevolution (Watson, 1970; Strickberger, 1976; Clark, 1977; and Mays, 1981 may be consulted for fuller details).

The individuality of the gene is now known to reside in the nucleic acid component of the complex nucleoproteins of which the nuclei of higher organisms are partly composed. The name of the particular substance, deoxyribonucleic acid, or DNA, is a household word, and the detailed structure of its molecule is described in many up-to-date textbooks of genetics. It is difficult to remember that only 30 years ago most biologists speculating on gene action were inclined to look to the protein for the key to gene specificity and not to the relatively simple nucleic acid.

It is interesting to note that DNA was discovered in white blood cells by Friedrich Miescher as long ago as 1869 (Mirsky, 1968). It is clear, however, that the role of DNA in heredity has only relatively recently been recognised. Stent (1968), Waddington (1969), Olby (1974) and Portugal & Cohen (1977) have contributed to our understanding of the historical development of researches on DNA which culminated in the establishment of the double helical structure of DNA by Watson & Crick (Watson & Crick, 1953; Watson, 1968).

The molecule of DNA, which carries the genetic information in the chromosomes of every cell of the body of plant or animal, is composed of two polynucleotide chains. Fig. 6.1 shows diagrammatically the sugar–phosphate 'backbone' of the two chains which are held together by hydrogen bonding of nitrogenous bases. As indicated in the diagram, the pairing of these bases is highly specific. Four different bases are found in DNA: adenine always pairs with thymine and cytosine with guanine. The sequence of base-pairs, of which there are many thousands in a DNA molecule, determines the action of the gene, and the DNA replicates by a process in which the bonds between the base-pairs are broken and a new partner is synthesised by each half-molecule acting as a template for the new half.

As a great deal of molecular biological research is carried out on bacteria, fungi and tissue cultures, much remains uncertain of the way DNA operates in higher plants. For instance there has been controversy (see Whitehouse, 1973, (Chapter 17) and Rees & Jones, 1977) concerning

Fig. 6.1. The structure and mode of replication of DNA. (*a*) Structure of DNA, showing the complementary paired strands linked together by hydrogen bonds. Adenine (A) and thymine (T) always pair together, as do guanine (G) and cytosine (C). The backbone of the DNA molecule consists of phosphate (P) and deoxyribose sugar (S) groups. (From Hayes, 1964, and Berry, 1977) (*b*) Replication of DNA as suggested by Watson & Crick. The complementary strands are separated and each forms the template for the synthesis of a complementary daughter strand. Reprinted by permission of Watson, J.D., *Molecular biology of the gene*, Menlo Park, California. The Benjamin Cummings Publishing Company, Inc. 1965.

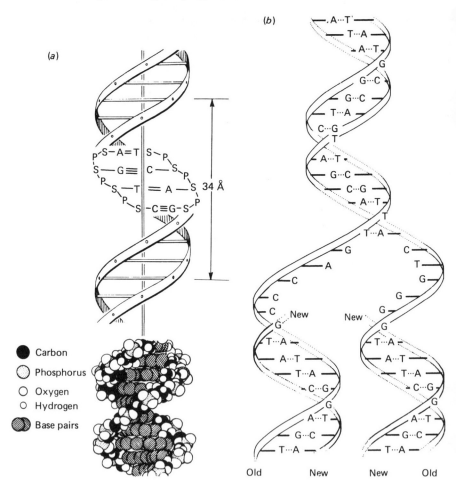

chromosome structure. Evidence suggests that each chromosome in the germ line is made up of a single continuous DNA double helix (in association with protein), rather than two or more DNA molecules lying parallel to each other.

Fig. 6.2. Partially diagrammatic representation of a young cell from the root-tip of a germinating Pea seed, as seen with the electron microscope. D, Golgi body (dictyosome); ER, endoplasmic reticulum; M, mitochondrion; N, nucleolus; P, proplastid with starch; Pi, site of pinocytosis; Pl, plasmodesma; Sph, spherosome or lipid droplet; V, vacuole; ZK, nucleus. The ground plasm contains numerous minute ribosomes. Much of the endoplasmic reticulum is also thickly beset with ribosomes, but some remains quite free of them. (From von Denffer *et al.*, 1971, after an original drawing by E. Perner.)

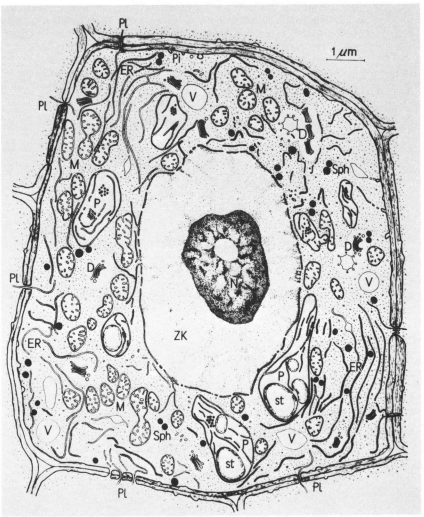

Looked at from the biochemical point of view, the classical Mendelian gene is a length of DNA which carries in its order of base-pairs a specific 'code' of information by which the initiation of enzyme-controlled processes of differentiation is governed. We are beginning to have some idea, in chemical terms, of the first steps in the synthesis of specific proteins governed by the nucleus and we now visualise these as being through the intermediary of another nucleic acid, ribonucleic acid, or

Fig. 6.3. Protein synthesis. One of the DNA strands serves as a template for the production of RNA (which differs from DNA in being single-stranded, and in having ribose sugar in its nucleotides and uracil in place of thymine). RNA passes out of the nucleus into the cytoplasm. Twenty different amino acids are found in proteins and there are 20 different transfer RNA (tRNA) molecules which bring the amino acids to the ribosomes. The primary protein structure is determined by the sequence of nucleotides in DNA, groups of three successive nucleotides specifying particular amino acids in the protein. The coded information in the DNA is accurately transcribed in the messenger RNA (mRNA) molecule, which in the ribosomes acts as a template on which the appropriate amino acids specified by the DNA code are joined together. (From Berry, 1977)

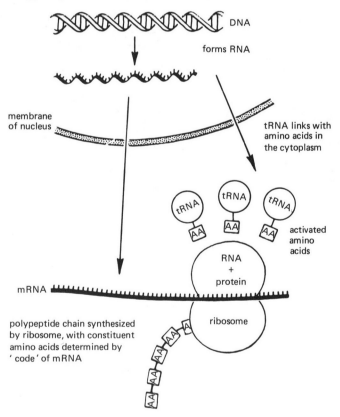

RNA, which differs slightly but significantly in chemical structure from the nuclear DNA. The seat of activity of RNA is at the ribosomes, minute particles which are found in the endoplasmic reticulum of the cell cytoplasm (Figs 6.2 and 6.3).

Mutation and abnormalities

So far we have discussed the gene and its alleles in terms of behaviour and chemical structure. Now we must look at the origin of alleles by mutation. The term mutation has had an odd history (Mayr, 1963). In the seventeenth century it was used to describe changes in the life-cycles of insects, and in the nineteenth it was employed by palaeontologists for markedly new variants in a line of fossils. De Vries (1905) used the term for new phenotypes which arose abruptly in a stock of plants. In particular, as we saw in Chapter 5, he studied mutants in the Evening Primrose (*Oenothera*). The term mutation was used by the early geneticists to describe the spontaneous origin of new variation, much of it Mendelian and allelic in nature. For instance, in *Drosophila* it was established that the alleles of particular genes (for example, for eye colour) arose as rare, spontaneous changes from the normal or 'wild type', and it was eventually demonstrated that any gene has a low but measurable 'mutation rate' by which the particular 'wild-type' allele can change to a mutant allele (Fig. 6.4 and Table 6.1). Moreover, such mutations are reversible.

The phenomenon of spontaneous mutation in plants is little understood. Studies of *Crepis capillaris* have shown a remarkable increase in spontaneous mutation rate with increasing altitude, with the highest values at 3000 m above sea-level (Bruhin, 1950). As Bruhin (1951) was able to increase the mutation rate in *C. capillaris* achenes treated at 45°C for 16–20 days, temperature – especially in the strongly insolated soils of alpine areas – was thought to be an important factor in nature. There is evidence that, as seeds age, the spontaneous mutation rate increases. Such increases may be due to metabolic changes in the seeds, and D'Amato & Hoffmann-Ostenhof (1956) have discussed the role of naturally occurring mutagenic substances (or precursors which under certain circumstances may be converted to mutagens) in producing mutation in seeds and other plant structures.

A large body of experimental work has demonstrated a whole range of physical and chemical 'mutagens' which can bring about the accelerated production of gene-mutants from the wild type. Spontaneous and induced mutants are often 'deleterious' in the sense that the phenotype is defective, and will not survive well in competition with the normal, wild-type genotype.

Although practically all the experiments in this field have been carried out on micro-organisms, it seems reasonable to assume that similar chemical changes must underlie the general phenomena of mutation in higher plants and, indeed, in all plants and animals. The recognition of mutation depends, of course, on the appearance of a significant change in the phenotype and, so far as higher plants (and higher animals) are concerned, the discussion of mutation and its effects cannot be separated from the question of abnormalities or 'sports' – the so-called 'teratological phenomena'. (Heslop-Harrison, 1952, has provided an important

Table 6.1. *Spontaneous mutation rates in Maize (from Stadler, 1942)*

Gene	Number of gametes tested	Number of mutations	Frequency per million gametes
R	554 786	273	492
I	265 391	28	106
Pr	647 102	7	11
Su	1 678 736	4	2
C	426 923	1	2
Y	1 745 280	4	2
Sh	2 469 285	3	1
Wx	1 503 744	0	0

Fig. 6.4. The spontaneous mutation rate $(A \to a)$ may be obtained from the proportion of a gametes produced by an AA individual, and detected by backcrosses as shown in the diagram (from Edwards, 1977). Other methods of estimating mutation rates are given in Strickberger (1976).

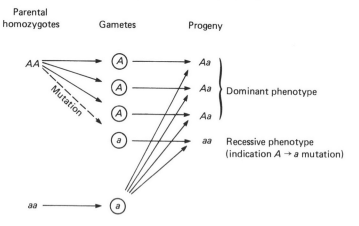

review of the subject.) Some of these abnormalities, like the 'double' roses of our gardens, are so familiar to us that it is perhaps only the botanist who recognises them as abnormal products of the horticultural craft, originally spotted as rare mutants, and preserved by conscious selection. Darwin saw how significant such 'domesticated' plants and animals were to the understanding both of heritable variation and of selection, and he devoted a whole book to these phenomena (1868). There is certainly no lack of examples of mutation affecting all parts of the plant, vegetative and reproductive (Meyer, 1966, for example, gives a very useful review and systematic list of the recorded abnormalities affecting floral structure), but there is as yet relatively little detailed experimental work involving both developmental and genetic studies, on which to base any very clear understanding of the complex phenomenon represented by 'abnormalities'. Indeed, the very idea of an 'abnormality' is difficult: for example, we do not readily think of white-flowered variants of typically coloured species such as Violets (*Viola*) or Bluebells (*Hyacinthoides*) as abnormal, although we would more willingly apply this term to, say, an apetalous variant of a normally petaloid *Ranunculus*. The assumption underlying the term 'abnormality' is usually two-fold: that the 'abnormal' variant is sporadic and nowhere in the majority; and that it is disadvantageous and therefore subject to negative selection. The first criterion is a matter of record and (at least theoretically) simply determinable; but the second presents much more formidable difficulties, as we shall find in discussion later in this book.

What, then, can we say about the origin of 'abnormalities' in plants? It seems to be the case that the same abnormal phenotype (e.g. an apetalous or a unisexual flower in a species normally petaloid or hermaphrodite) may be produced either by environmental stress directly affecting the individual developmental process, or by genetic change. It is the latter case to which we would wish to restrict the term 'mutant'. A study of developmental irregularity is, however, obviously highly relevant to an understanding of mutation, since our picture of gene action must involve the initiation of a sequence of enzyme-controlled developmental processes. The work of Astié (1962) is of great interest in this field. She studied both spontaneous and induced abnormalities in several common flowering plants, including the Soapwort, *Saponaria officinalis*, a perennial herb which occurs commonly as a cultivated 'double-flowered' variant. By applying the hormone 2,4-D to plants in a controlled experiment, she induced a very wide range of abnormalities amongst which, as she points out, occurred variants of inflorescence and flower which illustrated most of the taxonomic differences by which we recognise the different genera and subfamilies of the Caryophyllaceae. Indeed, some of the abnormalities of the inflorescence (for example, a capitulum instead of a lax panicle) are unknown in the family but characteristic of other families. Many, though not all, of these abnormalities have been recorded spontaneously in one or other flowering plant species, as can be verified in Meyer's review mentioned above. It seems, therefore, that gross changes in floral structure can be caused either by developmental manipulation or by mutation, and that characters traditionally used to classify major groups of angiosperms (such as the possession of a corolla-tube or an inferior ovary) are 'liable' to mutate in just the same way as any other apparently trivial character.

Molecular basis of mutation

In recent years molecular biologists have provided a clearer understanding of the chemical basis of mutation. The topic is one of great complexity; however, experiments with micro-organisms have shown that the process of mutation involves a change in DNA base-sequences. Given a starting sequence of DNA bases, mutants have been found to have an altered sequence of bases in some part of the DNA molecule. Chemical and physical mutagenic agents, which greatly increase the mutation rate above its 'natural' level, all interfere with the normal replication of DNA. There is evidence that the base-sequence of DNA may be altered by chemical activity following the application of certain chemical or physical agents, and in some cases molecules of various kinds may become interpolated into the DNA causing mistakes in replication. In the normal course of replication and recombination, errors in DNA base-sequences may arise and substances may owe their mutagenic activity to the fact that they interfere with the enzymic control of repair processes in replication or recombination (Auerbach, 1976). It is worth noting at this point that mutagenic agents such as X-rays can cause a whole spectrum of genetic changes, from small alterations of DNA base-sequence to visible chromosomal changes.

Estimates of gene size (Hartman & Suskind, 1965; Mays, 1981) indicate that the gene consists of 1000 to several thousand nucleotide pairs. In a molecular structure of this size and complexity there are obviously many different structural changes possible at the level of the DNA. This fact illuminates an early finding of genetics, that sometimes in genetic studies many different alleles of the same gene may be found. In the cases that Mendel examined there were only two alternative states of the gene but later it was found that 'multiple alleles' were common. These multiple alleles can now be visualised as different variants in base-sequence at the level of the DNA.

Genetic and chromosomal differences between plants

As we shall see when intraspecific variation is examined, many patterns are understandable in terms of allelic differences between plants and these can be analysed by the classical Mendelian methods. Some situations are relatively simple, others certainly complex. For instance, *Drosophila* geneticists have demonstrated conclusively that the manifestation of a gene in the phenotype can be modified by altering the *arrangement* of the genes in the chromosomes – the so-called 'position effect'. (Position effects have also been demonstrated in *Oenothera* by Catcheside, 1939, 1947.) Two individuals with genotypes with the same alleles may thus have different phenotypes.

Another complication displaces the early view that one gene determines only one characteristic. It is now known that one gene may affect many quite different phenotypic characters – a phenomenon known as *pleiotropy*, as in the case of the recessive mutant 'compacta' of *Aquilegia vulgaris* (Anderson & Abbe, 1933) (Fig. 6.5).

Fig. 6.5. Pleiotropy in *Aquilegia vulgaris*. In comparison with normal plants, mutant 'compacta' individuals are shorter, bushier and more branched, with erect, not drooping, buds, shorter petals and less well-developed sepals. Cross-sections of peduncles at equivalent points (2 cm from bud) of mutant (*b*) and normal (*a*) plants reveal a lack of secondary thickening in the latter. In contrast, 'compacta' plants have marked sclerenchyma development. The manifold effects of the 'compacta' gene would seem to flow directly or indirectly from precocious secondary thickening of cell walls in many parts of the plant. (From Anderson & Abbe, 1933, *American Naturalist*, **67**. Reproduced by permission of the publishers, © University of Chicago.)

(a)   (b)

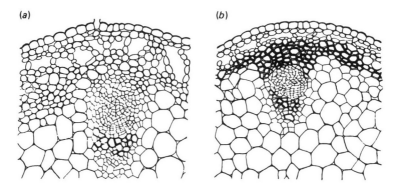

Fig. 6.6. Chromosomes of *Lilium*. (*a*) Mitosis in very young anthers of *Lilium martagon* ($2n = 24$) ($\times 1500$) (From Guignard, 1891) (*b*) Idiograms of species of *Lilium* ($\times 660$). (From Stewart, 1947)

(*a*)

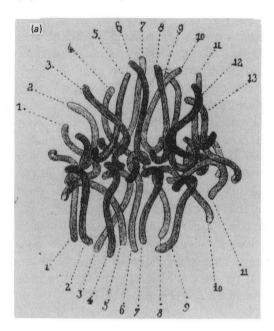

(*b*)

L. japonicum   L. martagon

L. leichtlinii   L. longiflorum

Other studies of genetic variation between plants reveal such relatively large differences that, under certain circumstances, they are detectable at the chromosome level. Early cytologists demonstrated that a particular diploid number of chromosomes was normally characteristic for a species. For example, the French cytologist Guignard (1891) made a clear drawing of the stages of mitosis in very young anthers of Turk's-cap Lily (*Lilium martagon*), establishing the diploid number for that species as 24. Descriptive cytology has made great strides in the present century and it is instructive to compare Guignard's drawings with a diagrammatic representation of the haploid set of chromosomes of *Lilium martagon* – a so-called 'idiogram' – prepared by Stewart (1947) (Fig. 6.6). A close study of this shows that individual chromosomes of the set are distinct in such features as length, length of arms and the presence of secondary constrictions. Fig. 6.6(*b*) also shows that different *Lilium* species have different idiograms.

An examination of *Lilium martagon* root-tip mitosis demonstrates that the diploid chromosome number is 24 and that this is the characteristic number for the species. Studies on very many Linnaean species reveal, however, that different individuals may have different chromosome numbers. In diploid organisms, homologous pairs of chromosomes are found. Occasionally, however, misdivision of the chromosomes at meiosis may give gametes with more or fewer chromosomes than the haploid number. Gametes may contain the unreduced number of chromosomes, or chromosomes may be missing or represented more than once. Gametes with an incomplete haploid complement are usually defective, but those with one or more additional chromosomes may be fertilised and may develop into adult plants. Thus, in a large sample of plants there may occur individuals, called aneuploids, which have the different chromosomes of the set present in different numbers. For example, in a sample of 4000 plants of *Crepis tectorum* ($2n = 8$) Navashin (1926) found:

> 10 plants $2n = 2x + 1 = 9$
> 4 plants $2n = 2x + 2 = 10$
> 4 plants $2n = 2x + 3 = 11$

Fusions between unreduced and 'normal' gametes (and between unreduced gametes) will produce in many cases viable plants with elevated chromosome numbers. For example, *Campanula rotundifolia* plants are divisible into three main groups, $2n = 34$, $2n = 68$ and $2n = 102$. In many genera, individual species form a polyploid series, in which high numbers are simple multiples of the lowest haploid number. The widespread occurrence of polyploidy in higher plants emphasises the extreme importance of the phenomenon, which will be examined in more detail in Chapters 9 and 11.

There remain two other categories of chromosome variation to consider. If individuals of a Linnaean species are examined, in some cases chromosomes additional to the normal complement may be found (Fig. 6.7). Such accessory chromosomes, or 'B' chromosomes as they are often called, are usually small, are often but not always heterochromatic (see below), do not usually pair with the larger members of the set – the 'A' chromosomes, and are frequently present in different numbers in different individuals (Jones, 1975). Further, in some species of plants, which have separate male and female individuals, the sex-determining mechanism involves distinct sex chromosomes (Grant, 1975).

In addition to gross features of chromosome morphology, different parts of the chromosomes may show differential staining. A combination of cytological and genetic studies indicates regions of the chromosomes, called heterochromatic segments, different from 'normal' euchromatic

Fig. 6.7. Metaphase plates from root-tips of different plants of *Festuca pratensis*. In some plants B chromosomes are absent, as in (*a*) (2*n* = 14). In others B chromosomes are present, as in (*b*) (2*n* = 14 + 7B) and (*c*) (2*n* = 14 + 16B). B chromosomes may be the same size, as in (*c*), or different sizes as in (*d*) (2*n* = 14 + 5B). All × 2700. (From Bosemark, 1954)

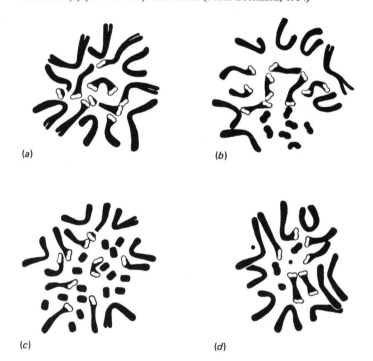

(a)   (b)

(c)   (d)

regions. Heterochromatin has several distinctive cytological properties. In its coiling behaviour it is often out of phase with euchromatin and in some taxa its presence may be accentuated by cold treatments before staining (Darlington, 1956); particular cytological techniques such as staining with Giemsa dye (Vosa, 1975) may be helpful in revealing heterochromatic patterns. In some species there is a consistent pattern of distribution of heterochromatin (so-called constitutive heterochromatin), but in many taxa there is variation in the pattern and a class of 'facultative' hetero-chromatin has been distinguished (Rees & Jones, 1977). The significance of heterochromatin is still largely unknown but recent work suggests that the relationship between euchromatin and heterochromatin may hold the key to an understanding, in biochemical terms, of the nature of the control of one gene by its neighbours. We know enough at present to suspect that the apparent inertness of heterochromatic regions of the chromosome is really a measure of our ignorance.

Studies of meiosis also reveal cytological differences between plants. Fig. 6.8 shows diagrammatically the main kinds of chromosomal change which have predictable genetic effects and visible cytological peculiarities at meiosis in the heterozygote. Such chromosomal rearrangements, incidentally, provide the most convincing proof that the genes are

Fig. 6.8. Diagrams to show how chromosome breakage and reunion can give rise to the four principal changes which chromosomes undergo. (After Stebbins, 1966)

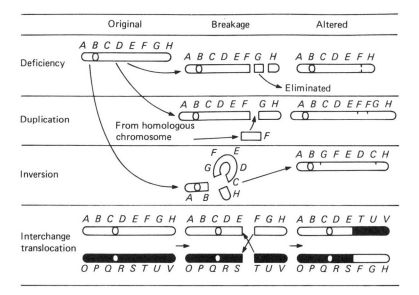

arranged in a linear order on the chromosome, for the order of genes in the linkage group determined by purely genetic means provides a basis for predicting the genetic effects of a particular cytological change. Such inversions, translocations, deletions and duplications of segments of chromosomes are to be found in naturally occurring individuals as well as experimental material; they clearly provide a further important basis of variation with great evolutionary significance. The 'mutations' of De Vries were of different kinds: many arose from a peculiar situation in *Oenothera*, where the species is a complex heterozygote in which all the chromosomes exhibit interchanges, but occasionally more or less homozygous 'mutant' individuals are produced.

Non-Mendelian inheritance

Correns, one of the rediscoverers of Mendel's work, showed as early as 1909 that in the familiar American garden flower *Mirabilis jalapa* ('Four O'Clock' or 'Marvel of Peru') some variants with yellowish-green or variegated leaves showed normal Mendelian segregation, while a particular variant, *albomaculata*, with yellowish-white variegation, did not. Plants of the variant *albomaculata* produced occasional shoots which were wholly green and others which were white; flowers on green shoots gave only green progeny whether pollinated from flowers on green, variegated or white shoots, and conversely flowers on white shoots, whatever the pollen parent, gave only white progeny (which died in the seedling stage). Variegated shoots gave all three kinds of progeny. In two respects this inheritance was clearly non-Mendelian: first, because the offspring resembled closely the female parent and the reciprocal crosses gave entirely different results; and secondly, because there was no regularity in the proportions of the phenotypes in the segregating families.

A similar case involving variegation in Maize illuminates the difference between Mendelian and non-Mendelian inheritance. The variant 'iojap', in which the phenotype has striped leaves, was found by Jenkins in 1924 to be caused by a single recessive gene. Using 'iojap' plants as male parents, he found that the F_2 segregated in the expected 3:1 ratio of normal:'iojap'. Rhoades, however, showed in 1943 that female 'iojap' plants gave quite different results: widely varying proportions of green, white and 'iojap' phenotypes were found in the offspring of these plants, the particular result being apparently unrelated to the constitution of the male parent. Rhoades explained his results by postulating that the gene for 'iojap', when homozygous, causes striping of the leaves by initiating a process which is then inherited through the cytoplasm. Since the cytoplasm is for all practical purposes entirely derived from the female parent, this condition shows maternal inheritance.

Other cases of non-Mendelian inheritance of leaf variegation have been investigated, notably in the Garden Geranium (*Pelargonium*) and in the Evening Primrose (*Oenothera*). Although the details differ, they are generally open to the explanation that there are hereditary particles in the cytoplasm which can replicate, sometimes indefinitely. In the case of variegation effects, the plastids themselves, which contain the green colouring matter chlorophyll, are self-replicating structures in the leaf cells and might therefore contain hereditary particles. Not all cytoplasmic inheritance concerns chlorophyll-containing plastids, however; it has been shown by repeated back-crossing with species of Willow-herb (*Epilobium*) that 'alien' cytoplasm of one species can persist and give a variety of genetic effects with the nucleus of another species (Michaelis, 1954).

In order to explain these phenomena it had been postulated that the cytoplasm contained hereditary particles which replicated themselves through several generations. In the case of variegation it was supposed that there were self-replicating particles in the plastids (see Jinks, 1964 for historical review). It is of considerable interest that DNA has now been discovered in the chloroplasts and mitochondria of plant cells (Beale & Knowles, 1978; Cummings *et al.*, 1979). The role of nuclear DNA in controlling cytoplasmic DNA is being studied by many research workers. That there may be different degrees of control by nuclear DNA is suggested by the existence, not only of long-lasting cytoplasmic effects, but also of short-term influences, as perhaps in the cases of delayed Mendelian inheritance (see Fig. 6.9).

Phenotypic variation

We have discussed so far the genetic basis of individual variation and must turn now to an examination of the development of the phenotype. In the growth of an organism from fertilised zygote it seems clear that a minute quantity of DNA in the cell plays a vital role in determining the characteristics of the mature phenotype (perhaps a tree 30 m high), which is organised from raw materials drawn from outside the plant. There are complex interactions between genotype and environment at the level of the cell and of the whole plant.

Concerning the genetic control of cellular processes, it seems likely that they are of such interlocking complexity that it is no longer possible to claim that one gene determines a particular character. A gene provides information which, in an appropriate environment, will contribute to a particular phenotype. There is no certainty that a particular gene will even manifest itself – 'become penetrant' – in all environments. For example, in *Lotus corniculatus* certain plants known to possess alleles appropriate to the production of HCN when their foliage is crushed, do not in fact produce the cyanogenic reaction in every circumstance (Dawson, 1941). Also it has been found that a particular barley variant produces albino phenotypes out of doors, yet the same genotype grown at higher temperatures in a glasshouse has normal foliage (Collins, 1927).

In general, plants show greater phenotypic variability than is found in species of higher animals. In animals individual variation is apparently held within very tight bounds by the early precision of mechanisms determining irreversibly the form and relationship of the main organs. To some extent this is true of plant structures, but there is a very important difference between the plant and the animal, which resides in the fact that there is a persistent meristem or growing-point tissue in even the most short-lived of ephemeral plants, on which a succession of organs of limited growth is initiated. The result of this difference is that the individual plant is open to much more environmentally induced variation over a much greater part of its life than is the animal.

Here it may be helpful to consider the causes of phenotypic variation in plants. If material of a given genotype is divided into separate pieces (ramets) and grown in two or more different environments, different genotype–environment interactions may be produced. While the plant responds to the environment as a whole, it is possible, by appropriate experiments involving changes in particular factors, to deduce that

Fig. 6.9. Pollen shape in *Lathyrus odoratus* in experiments of Bateson, Saunders & Punnett (1905). The shape of pollen was determined by a single allelic difference, with the factor 'long pollen' (*L*) dominant to the recessive factor 'round pollen' (*l*). In F_1 plants it could be inferred that the pollen was of two genotypes, *L* and *l*, with equal frequency, but on examination all pollen proved to be of long phenotype. There is therefore a delay in the manifestation of the nuclear change, which may be due to cytoplasmic effects of the mother plants persisting after genetic segregation.

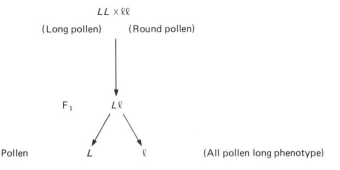

certain elements of the environment are of particular importance. Test environments may differ in soil properties (e.g. water table, base status, level of toxic ions), and aspects of climate or microclimate (e.g. temperature extremes, wind exposure, rainfall). In their competition with one another in experiment or natural communities, plants show many diverse interactions, for example, to the effects of shading. Also of importance may be the influence (at present often hypothetical) of chemical substances produced by one taxon in suppressing the growth of others (so-called allelopathic effects: see Rice, 1974; Harper, 1977). The availability of organisms to form the natural symbioses characteristic of many plants is a critical factor in natural and experimental situations, as are the presence and severity of grazing, pest attack and disease. Given the variety of different environments a great diversity of genotype–environment interactions is possible. Sometimes, but perhaps rarely, plants may be growing in ideal conditions. Often, however, the stresses of the environment may kill the plant or restrict its growth and perhaps prevent reproduction.

Developmental variation

Phenotypic variation should be viewed within the context of developmental variation. In most flowering plants the cotyledon stage of the seedling is very different from the adult – a fact which has, of course, excited the interests of botanists from early times, and which provided the basis for John Ray's inspired division of all flowering plants into the monocotyledons and dicotyledons, a division which we still use as a primary grouping. The developmental transition between the simple cotyledon and the often lobed or dissected mature leaf is generally rather abrupt, and provides the most familiar example of a phenomenon which Goebel (1897) called 'heteroblastic development', or the change from a juvenile to an adult phase accompanied by more or less abrupt changes in morphology. Strictly speaking, the cases which Goebel and others have mainly called 'heteroblastic' are those in which a juvenile leaf other than the cotyledons contrasts more or less clearly with an adult one, as, for example, in the case of Gorse (*Ulex europaeus*), in which the seedlings produce trifoliate leaves (of a type which is normal in related genera) before the simple ones (Fig. 6.10). There seems, however, much to be said for extending the term to cover all ontogenetic phase changes (in leaf-shape, etc.), including the cotyledons, and including also changes between early and late phases which are more gradual.

If we investigate closely the detailed development of a plant with an adult leaf clearly different from the juvenile, the commonest situation is likely to be as shown in Fig. 6.11 for *Ipomoea caerulea*, the Morning Glory, a common tropical climbing herb. In this illustration, taken from the work of Njoku (1956), the top line shows the shape of the first 10 leaves of a plant grown in the shade, and the second line shows the same series from a plant in full daylight. Here two things are evident: first, that the development of the adult three-lobed leaf-shape is gradual; and secondly, that the onset of the three-lobed leaf-shape in the developmental

Fig. 6.10. Juvenile (top two rows) and adult (bottom row) foliage in *Ulex europaeus*. The leaves are numbered in sequence. (From Millener, 1961)

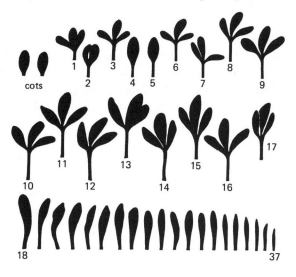

Fig. 6.11. Njoku's experiment with *Ipomoea caerulea*. The first 10 leaves are shown of (*a*) a plant grown in the shade; (*b*) a plant grown in the light; (*c*) a plant transferred from shade to light as the second leaf unfolded; and (*d*) a plant transferred from light to shade as the second leaf unfolded. (× 0.125)

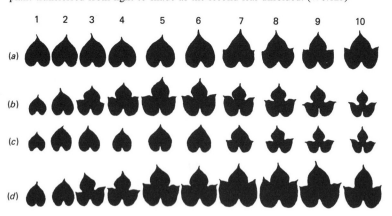

series is greatly modified by the environment, in this case by light. Fig. 6.11 also illustrates the effect of transferring the plant from shade to light at the stage of the unfolding of the second leaf, and also of transferring it from light to shade. In both cases there is a 'time-lag' in reaction which lasts until the sixth or seventh leaf, suggesting that the developmental processes which determine the form of the mature leaf are operating at an early stage in the differentiation of leaf primordia on the growing point, and that once these have reached a certain stage the effect of environmental factors is no longer operative. The important point to note is the irreversibility of change at a certain stage in the development of mature structures and its relation to developmental variation.

Another example of a situation where changes are generally irreversible is found in Ivy (*Hedera helix*). The wild plant, common in Britain and Atlantic Europe, shows a very marked heterophylly, the familiar lobed 'ivy-leaf' being produced exclusively on non-flowering shoots which are normally flattened or adapted for growth attached to trees or on the woodland floor. The flowering shoots, in contrast, are erect, branch more or less radially, and bear simple leaves. Intermediate shoots and leaves are rare (Fig. 6.12). Seedlings, as would be expected, produce lobed leaves and quickly assume the vegetative phase. If, however, portions of *either* vegetative or reproductive shoots are detached and rooted separately, the plants so produced normally continue to grow in the manner characteristic of the particular phase. This is apparently true even of intermediate shoots. In this way, whole plants of *Hedera* with a more or less erect habit and simple leaves can be propagated, apparently indefinitely, though 'reversion' to the juvenile vegetative phase can be induced, for example, by repeated cutting, by grafting on to the juvenile stock, or by spraying with the growth-substance gibberellic acid. In such a case, we appear to have a condition which is explicable neither in terms of genetics nor in terms of direct effect of the environment on the phenotype. Brink (1962) and others have postulated self-replicating factors in the cytoplasm to explain such phenomena, and have pointed out that, although *Hedera* is a specially familiar case, there is a whole group of phenomena in woody plants which are not essentially different and which are not yet understood. A useful review is provided by Doorenbos (1965).

We must now look briefly at the phenomenon of flowering. Here there is a great deal of information, derived in part from studies of cultivated plants in which flower and fruit production are economically important (see the reviews by Lang, 1965 and Wareing & Phillips, 1981). The reproductive phase in the development of the individual plant is usually marked by an abrupt change in pattern of growth at the apex. The floral parts originate, like the leaves, as lateral outgrowths on the growing-point, but the internodes, which were very obvious between the successive leaves, are suddenly greatly reduced or suppressed, so that whorls or spirals of tightly packed floral parts are produced. There is great variation between different plants as to the influence of the environment upon the initiation of flowering, but there is usually some detectable effect, and in the cases of photoperiodic response the effect can be very great indeed.

Fig. 6.12. Heterophylly in *Hedera helix*. Note the simple leaves on flowering shoots. (From Ross-Craig, 1959) (× 0.75)

Certain species apparently require exposure, often for very brief periods only, to a particular daylength before they will pass into the reproductive phase which, once initiated, can continue whether the particular daylength conditions are still present or not. This kind of adaptation to daylength has, of course, obvious importance in terms of wild populations; it may mean (as, for example, with certain plants of American origin in Europe) that the plant may be unable to flower and fruit when introduced into a country where the particular daylength conditions are not present, and in this way the spread of a species may be restricted. Other species may require cold treatment before flowering is initiated, a subject discussed by Wareing & Phillips (1981) who provide a useful survey of the present state of knowledge on flowering and its physiological control.

The control of phenotypic plasticity

We have discussed various general aspects of the development of the phenotype and suggested the possibility of different genotype–environment interactions. In conclusion a number of important points must be raised. Whilst the subject of phenotypic plasticity has been strangely neglected by botanists, there is limited evidence from observations and experiments (see Bradshaw, 1965) for the following hypotheses:

1. The extent of phenotypic plasticity apparently differs in different taxa.
2. Different characters of the phenotype show different degrees of plasticity.
3. Phenotypic plasticity is under genetic control.

The variable extent of plasticity in different related taxa is beautifully illustrated by the group of species of *Ranunculus* subgenus *Batrachium* which exhibit heterophylly. Species inhabiting shallow water produce both types of leaf, those from mud or very shallow water produce only floating leaves, while the taxa from deep water or swiftly flowing water produce finely divided leaves (Table 6.2; see also Cook, 1974).

Experiments with *Polygonum amphibium* have also yielded valuable information on phenotypic plasticity. Plants growing in water and on land have very different phenotypes (Fig. 6.13). If ramets of cloned material are separately grown in conditions simulating land, waterlogged and submerged conditions, the degree to which individuals may produce both the 'land' and 'water' phenotype may be investigated. Studies by Turesson (1961) and Mitchell (1968) suggest that the different individuals (presumably different genotypes) show different degrees of modifiability. The extent of phenotypic plasticity would appear to be under genetic control.

Table 6.2. *The occurrence of heterophylly in British species of*
Ranunculus *subgenus* Batrachium *(from Bradshaw, 1965)*

		Leaves	
Species	Habitat	Floating	Submerged
R. hederaceus	Mud or shallow water	Many	—
R. omiophyllus	Streams and muddy places	Many	—
R. tripartitus	Muddy ditches and shallow ponds	Some	Many
R. fluitans	Rapidly flowing rivers and streams	—	Many
R. circinatus	Ditches, streams, ponds and lakes	—	Many
R. trichophyllus	Ponds, ditches, slow streams	—	Many
R. aquatilis	Ponds, streams, ditches, rivers	Some	Many
R. peltatus	Lakes, ponds, slow streams	Many	Many
R. baudotii	Brackish streams, ditches, ponds	Some	Many

Fig. 6.13. *Polygonum amphibium.* (*a*) Terrestrial form, growing on land near water; (*b*) aquatic form, with submerged stems and floating leaves; (*c*) form growing in damp sand-dunes. (From Massart, 1902) (× 0.5)

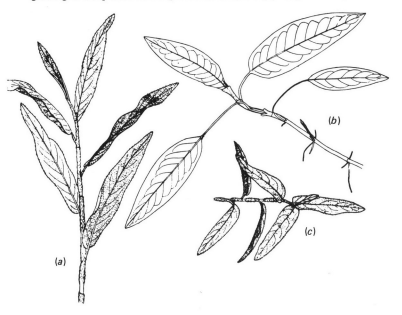

Cultivation tests with *Trifolium repens* (Table 6.3), in which the number of vascular bundles in petioles was investigated in clonally propagated material grown under contrasting conditions, lead to similar conclusions. Different plants are capable of different responses to the same environment and show different degrees of 'stability' in different environments. For instance, Clone A varies rather widely; Clone E on the other hand is almost invariable. While full analysis of the situation has not been attempted, it seems likely that such responses are under genetic control.

Table 6.3. *Average number of vascular bundles per petiole in 10 petioles of each of 24 clonal lines of white clover growing in three different environments (Gibson & Schultz, 1953)*

	Environment		
Clone	Greenhouse	Cool, moist field	Hot, dry field
Ladino			
A	8.7	10.1	5.7
B	8.9	10.8	8.6
C	6.4	7.5	5.0
D	5.6	6.5	5.0
E	5.1	5.1	5.0
F	6.8	6.2	5.7
G	6.2	6.0	5.0
H	6.5	8.6	5.8
I	7.6	8.2	5.9
J	6.0	6.5	5.0
K	6.2	7.9	5.2
L	6.9	7.6	5.5
Mean	6.7	7.6	5.6
Intermediate white			
M	5.0	6.6	4.7
N	5.0	5.0	3.7
O	5.0	5.0	4.3
P	5.0	5.1	3.4
Q	5.1	6.4	5.0
R	5.1	5.6	4.3
S	5.0	5.0	4.0
T	5.0	6.1	3.1
U	5.0	5.1	3.3
V	5.0	5.0	4.8
W	5.3	6.8	4.1
X	5.0	5.1	5.0
Mean	5.0	5.6	4.1

Our analysis of phenotypic responses has stressed the morphological aspects of variation, but differences in the flexibility of underlying physiological processes must not be overlooked. Finally it could be argued that the consequences of genotype–environment interaction are adaptively significant. We will examine this hypothesis in a later chapter.

7

Breeding systems

The study of a range of plants in the wild or on display in a Botanic Garden reveals a bewildering array of floral types. Many books describe in some detail a selection of pollination mechanisms, often discussing in highly technical botanical language the variety of floral structure. Special terms have quite properly been devised by botanists to enable them to write concise, accurate plant descriptions but, in our opinion, although the botanical literature reports extensively on the structures involved in reproduction, it does not pay sufficient attention to the variety of breeding systems in plants, systems of which complex structures and pollination mechanisms are only a part.

We have noted in earlier chapters the role of the internal sources of variation, namely mutation and recombination, and have seen that a vast number of gametic types is theoretically possible as a consequence of these factors. Which gametes are actually brought together to form the zygotes, however, depends upon the breeding system of the plant concerned. In this chapter, as a prelude to our discussions of variation within and between populations, we will show how studies of breeding behaviour have developed, and make it clear how a knowledge of breeding systems provides an indispensable framework for understanding the complexities of patterns found in nature.

There are three basic breeding mechanisms, which we may examine in turn.

Outbreeding

In many animal groups outbreeding – crossing between different individuals – is rendered likely by sexual differentiation. In higher plants separation of the sexes is the exception. Only 2% of the indigenous plants of Britain are dioecious. South of the Equator the frequency of dioecy may be higher (e.g. 12% of the 1800 native plants of New Zealand are dioecious) and there is increasing evidence that dioecy may be frequent in the tropics (Richards, 1979). However, most higher plants are hermaphrodite. Thus the typical angiosperm flower has a zone of pollen-bearing stamens (androecium) surrounding a gynoecium containing one or more ovules and, even in the case of simple unisexual flowers of the catkin-bearing woody plants (Amentiflorae), male and female flowers are usually found on the same individual. Such juxtaposition of stamens and ovules suggests that self-fertilisation would be the most likely mode of reproduction and it is most interesting to see the historical development of ideas which force us to conclude that many plants are adapted to facilitate cross-fertilisation (i.e. crossing between different individuals) and to minimise or prevent self-fertilisation (Whitehouse, 1959).

The sexual function of flowers was established in the seventeenth and eighteenth centuries (Proctor & Yeo, 1973). In 1793 Sprengel published his classic book in which he produced excellent descriptions of the wind and insect pollination of plants, but he was apparently unaware that flowers are primarily adapted for cross-fertilisation. It was Darwin who concluded, after a detailed study of many orchid species, that the orchid flower was 'constructed so as to permit of, or to favour, or to necessitate cross-fertilisation'. Reflecting further on the orchid studies, Darwin wrote (1876):

> It often occurred to me that it would be advisable to try whether seedlings from cross-fertilised flowers were in any way superior to those from self-fertilised flowers. But as no instance was known with animals of any evil appearing in a single generation from the closest possible interbreeding, that is between brothers and sisters, I thought that the same rule would hold good with plants ...

However, in 1866 Darwin was studying inheritance in *Linaria vulgaris*, and raised two large beds of seedlings of self-fertilised and cross-fertilised individuals. To his surprise the 'crossed' plants, when fully grown, were taller and more vigorous than the 'selfed' progeny. Darwin's interest was thoroughly aroused and he investigated over many years the effect of cross- and self-fertilisation in a number of species, e.g. *Ipomoea purpurea, Mimulus luteus, Digitalis purpurea, Zea mays*. He gave great attention to experimental design; for example, his basic comparative test was devised as follows. Seed, from cross- and self-fertilised plants, was germinated and a 'crossed' seedling was matched against a 'selfed' seedling, several such comparisons being made in each of a number of pots. A partition was placed between the two sets of seedlings but in such a way as to make sure that both sets of plants were equally illuminated. Other types of pot experiment and garden trial were attempted and the effect of crossing and selfing in some cases was studied for a number of generations. With some exceptions, his results revealed that the progeny of cross-fertilised plants were superior in performance when compared with the progeny from self-fertilisations. (He examined one or a number of measures of performance such as height, weight or fertility.) An example of his results with Maize is set out in Table 7.1.

As a result of his experiments, which were a landmark in the study of breeding behaviour, Darwin was able to see how the enormous range of floral types and physiological differences in behaviour, such as different times of maturity of stamens and stigma on the same flower, could be viewed as adaptations to ensure cross-fertilisation. Why it might be beneficial for progeny to be crossbred rather than the product of self-fertilisation Darwin was unable to decide. This most interesting subject is discussed below.

In his experiments on 'cross' and 'self' progenies, Darwin discovered that in some cases the attempt to produce progeny by self-fertilisation failed – certain plants were self-sterile. Examples of self-sterility had also been noted by other botanists. Again Darwin was unable to account satisfactorily for this phenomenon.

Great progress has been made this century in our understanding of self-sterility, which may arise from a number of causes. In most cases self-incompatibility is involved. A fertile hermaphrodite plant is incapable of producing zygotes following self-pollination. There is a mechanism operating at the pre-zygotic stage preventing self-fertilisation.

The first, and inconclusive, studies of self-incompatibility were made by Correns (1913) in experiments with *Cardamine pratensis*. He suggested that a genetic mechanism was involved but it was not until the experiments of East & Mangelsdorf (1925) working with *Nicotiana* that the genetics of self-incompatibility became clearer. The elements of the scheme they proposed – a gametophytic incompatibility system – have now been found to apply to very many, but not all, incompatible species. The stages are as follows:

1. Each plant is heterozygous for a gene S; e.g. $S_1 S_2$. (Many, perhaps scores, of alleles occur: say S_3, S_4, S_5, etc.) The style and stigma in the flowers of a plant are maternal tissue containing nuclei with the diploid chromosome number – in consequence style and stigma contain $S_1 S_2$ in a plant of genotype $S_1 S_2$.

2. At meiosis prior to pollen formation, segregation occurs and pollen, which contains nuclei with a haploid chromosome number, receives one of the two S alleles. The pollen hereafter behaves in pollination in accordance with its S allele genotype.

Table 7.1. *Darwin's experiments with Maize*
(Zea mays): *height (in inches) of young plants
raised from seeds obtained by cross- and self-
fertilisation*

	Cross-fertilisation	Self-fertilisation
Pot I	$23^4/_8$	$17^3/_8$
	12	$20^3/_8$
	21	20
Pot II	22	20
	$19^1/_8$	$18^3/_8$
	$21^4/_8$	$18^5/_8$
Pot III	$22^1/_8$	$18^5/_8$
	$20^3/_8$	$15^2/_8$
	$18^2/_8$	$16^4/_8$
	$21^5/_8$	18
	$23^2/_8$	$16^2/_8$
Pot IV	21	18
	$22^1/_8$	$12^6/_8$
	23	$15^4/_8$
	12	18

While we might accept Darwin's (1876) general
conclusions from his experiments that progeny from
cross-fertilised plants were generally taller, etc. than
progeny from self-fertilised plants, it is interesting
to discover, as Darwin himself realised, whether the
results of a particular experiment were statistically
significantly different. Darwin consulted his cousin
Galton who, employing the crude statistical
techniques available at the end of the nineteenth
century, had to conclude that the experiment was
based on too few plants. It was not until 1935 that
Fisher made a close analysis of Darwin's
experiments. He concluded that Darwin's
experimental design was fundamentally sound –
comparing the growth of an outcrossed with that of
a selfed seedling. (However, planting several plants
in each pot must have led to competition and
perhaps a better planting arrangement could have
been devised.) Fisher used statistical tests first
devised in the early years of the twentieth century
for studying small samples, and discovered that the
Maize result above was just statistically significant.

3. Pollen arrives at the stigma, which may be specially adapted for pollination by wind, insect or other means.

4. The pollen grain germinates to give a pollen tube which grows through a special tract of tissue or along a canal in the style.

5. If the *S* allele present in the pollen is also found in the style, the growth of the pollen tube is arrested in the style; the tip of the incompatible tube grows abnormally and becomes occluded with callose. If pollen and stylar tissues have dissimilar *S* alleles, growth of the pollen tube proceeds normally and fertilisation may occur. (In some plants, however, for example the grasses, the incompatibility reaction appears to occur at or near the stigmatic surface rather than in the style.) The identity of the substances involved in incompatible reactions is the subject of intense research (see De Nettancourt, 1977, for details).

Given such a genetic mechanism it is now clear why certain self-fertilisations fail to produce offspring. The *S* alleles segregating in the pollen (say $S_1 S_2$) are both represented in the stylar tissue of the same plant and all pollen is arrested in its growth down the style. The consequences of self-pollination and pollination by individuals of various '*S*' genotypes is displayed in Fig. 7.1. It is important to note that a given

Fig. 7.1. Gametophytic self-incompatibility. A pollen parent of genotype $S_1 S_2$ will be infertile, semi-fertile or fully fertile according to the genotype of the female plant. In most species with this system, incompatible pollen tubes are inhibited in the style. ES = embryo-sac. (From Heslop-Harrison, 1978.) Note: In some families gametophytic self-incompatibility is more complex genetically, being controlled by two or more loci (De Nettancourt, 1977).

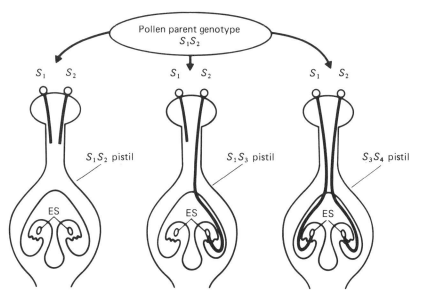

plant will not necessarily be fully fertile with all other *individuals* of the same species even if they differ substantially genetically. The *S* alleles represented in pollen and style are decisive. The stylar tissue may be seen to be acting as a selective contraceptive device preventing free access of pollen tubes to the ovule, permitting pollen tube growth in certain cross-fertilisations and preventing pollen tube growth on self-pollination.

In 1950 American botanists studying *Parthenium argentatum* (Gerstel, 1950) and *Crepis rhoeadifolia* (Hughes & Babcock, 1950) discovered what has come to be known as the sporophytic incompatibility system. The following are the important differences from the gametophytic system just outlined:

1. Individuals are heterozygous for the *S* gene, say $S_1 S_2$, and, as before, segregation occurs at meiosis preceding pollen formation. The pollen, notwithstanding the 1:1 ratio of segregation of *S* alleles, behaves as if it had the genotype of the plant which produced it. Evidence suggests that this effect is produced by the loading of the outer layers of the pollen grain with materials produced by the tapetal cells of the anther (Heslop-Harrison, 1975).

2. In general, pollen is rejected if its pollen parent shares the same *S* allele as the stigma and style of the female parent. (However, in some species there are more complex reactions including dominance of certain alleles.) The results of certain cross-pollinations are set out in Fig. 7.2.

3. The site of pollen tube failure is often different in sporophytic systems, occurring on or close to the stigmatic surface. As in gametophytic systems callose production may indicate incompatible pollinations.

Darwin's list of self-sterile plants was quite short, but geneticists are now suggesting that perhaps half the species of angiosperms, both monocotyledons and dicotyledons, may have self-incompatibility mechanisms (De Nettancourt, 1977).

Another finding of considerable interest is that plants having gametophytic incompatibility systems are all homomorphic, that is, there are no structural differences associated with different *S* alleles. In contrast, some plants with sporophytic incompatibility systems are heteromorphic. The existence of different forms of flowers in the same species was the subject of some of the most famous studies made by Darwin. He carried out crossing experiments with heterostylous plants of both the distylic and tristylic type (Fig. 7.3). In the case of the Cowslip, *Primula veris*, he showed that there are two forms of the flower, 'pin' and 'thrum', each with a characteristic syndrome of characters (Fig. 7.4). The pollinations 'pin' × 'thrum' and 'thrum' × 'pin' yielded good seed set; selfing 'pins' or 'thrums' or crossing 'pin' × 'pin' and 'thrum' × 'thrum' yielded very much less seed (Table 7.2). In view of the comparative self-sterility of 'pin' and 'thrum' plants, Darwin concluded that distyly was a device favouring outcrossing. Pollinators visiting *P. veris* were likely to pick up pollen in a pattern related to anther position. Subsequent visits would transfer pollen to stigmas at 'an equivalent height'.

Darwin also studied Purple Loosestrife, *Lythrum salicaria*, in which three kinds of flowers are found corresponding to different levels of anthers and stigmas in the flowers; 18 different interpollinations were needed to investigate the effects of six kinds of anther and three kinds of style, and these Darwin carried out. The number of seeds in six pollinations between anthers and stigmas at the same height was much greater than those produced in other pollinations (Fig. 7.5).

Since Darwin's time, further investigation has revealed that *Primula veris* and *Lythrum salicaria* have sporophytic incompatibility systems, and

Fig. 7.2. Sporophytic incompatibility. In this diagram the S alleles are presumed to act independently. Other relations are known, including dominance and mutual weakening. ES = embryo-sac. (From Heslop-Harrison, 1978)

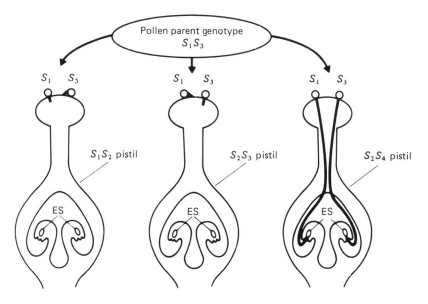

it seems possible that the ancestral stocks from which heterostylous plants arose were homomorphic. The heterostylous condition may be of second-ary origin, increasing the chance of a stigma receiving compatible pollen (Yeo, 1975). It is, however, difficult to see selective advantage in all the hypothetical steps necessary to convert the homostylous plant into the fully fledged distylous condition. Speculation about the evolution of the tristylous condition from a monomorphic sporophytic incompatible ancestral stock is likely to be the hobby of the man who plays three-dimensional chess!

Inbreeding

The second main breeding method involves self-fertilisation. In his account of 'cross'- and 'self'-fertilisation of plants, Darwin (1876) reported a list of species which yielded seed when covered by a net to exclude insect visitors; in 1877 he discussed the presence in some species of flowers which never open, pollination occurring automatically in the closed bud, which gives rise directly to a fruiting structure. Such flowers – which are of course self-fertilising – are referred to as cleistogamous.

Fig. 7.3. Symbolic representation of distyly and tristyly. In each system, the compatible pollinations only involve anthers and styles at the same level, and therefore the following are incompatible combinations: pin × pin, thrum × thrum, long × long, mid × mid, and short × short. (From De Nettancourt, 1977)

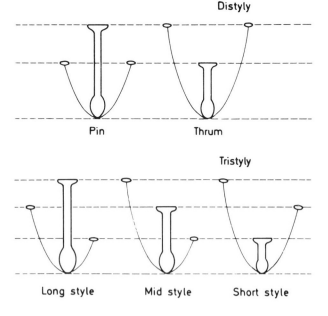

Table 7.2. *The results of Darwin's experiments in crossing 'pin' and 'thrum' plants of* Primula veris

Nature of the union	Number of flowers fertilised	Total number of capsules produced	Number of good capsules	Weight of seed in grains	Calculated weight of seed from 100 good capsules
Long-styled by pollen of short-styled. Legitimate union	22	15	14	8.8	62
Long-styled by own-form pollen. Illegitimate union	20	8	5	2.1	42
Short-styled by pollen of long-styled. Legitimate union	13	12	11	4.9	44
Short-styled by own-form pollen. Illegitimate union	15	8	6	1.8	30
Summary The two legitimate unions	35	27	25	13.7	54
The two illegitimate unions	35	16	11	3.9	35

Fig. 7.4. Heterostyly in *Primula veris*. (× 2). Darwin (1877*a*) discovered that the long-styled (pin) form always had a much larger pistil with a globular rough stigma standing high above the anthers. Pin pollen was oblong in shape. In contrast the short-styled (thrum) plants always had a short pistil about half the length of the corolla, with a smooth stigma (depressed on the summit). Pollen in the anthers, which stood above the stigma, was spherical and larger than in the pin.

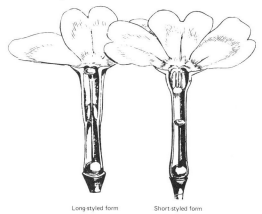

Long-styled form   Short-styled form

Fig. 7.5. Heterostyly in *Lythrum salicaria*. Diagrams of the three forms of flowers, in their natural position, with petals and calyx removed on the near side. The dotted lines with the arrows show the directions in which pollen must be carried to each stigma to ensure full fertility. (From Darwin, 1877*a*)

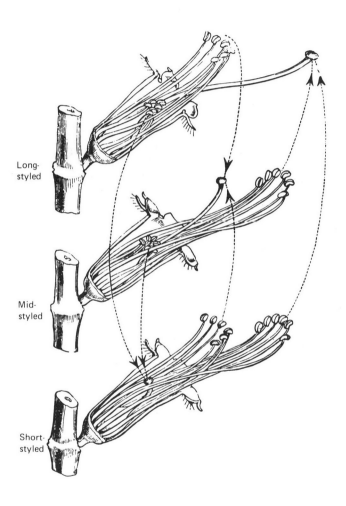

Long-
styled

Mid-
styled

Short-
styled

Following a literature search and as a result of his own researches, Darwin listed 55 genera in which cleistogamous flowers had been found, and investigations by botanists over the years have added to the list (Uphof, 1938). In some cases cleistogamy appears to be more or less obligatory, as in several grass species (McLean & Ivimey-Cook, 1956), but, as we shall see below, cleistogamous flowers occur seasonally in some species.

Observations and experiments by botanists have shown that some species are predominantly self-fertilising, but in many cases this is a presumption based on the following types of evidence. It may be readily demonstrated that some species are self-compatible and it can be observed that some plants, visited by insects in summer, can flower and fruit in autumn, winter or spring when insects are absent. Further, progenies of some species are remarkably uniform in appearance and it is assumed that such uniformity results from persistent inbreeding.

If some plants are predominantly or obligately self-fertilising, how is this to be equated with the demonstration by Darwin of general superiority of outbred plants? This paradoxical situation will be examined below.

Apomixis

In this third reproductive mode, reproduction is achieved without fertilisation, the sexual process being wholly or partly lost (the term and its definition are according to Winkler, 1908). Two types of system are found: vegetative apomixis and agamospermy.

Vegetative apomixis

While for a time plants are usually rooted to the precise spot where germination and establishment has occurred, radial growth by means of rhizomes, stolons, runners, etc., is characteristic of many perennial species.

Bulbils (small, readily detachable propagules, often borne on aerial structures, Fig. 7.6) are found in some species. Plants arising from such propagules will, of course, have the same genotype as the parent plant. Certain crop-plant varieties, e.g. potatoes, and such familiar garden plants as *Pelargonium* (the gardener's 'Geranium') are regularly propagated by vegetative means.

Agamospermy

In certain plants normal seed is set but no sexual fusion has occurred in its production. Offspring have the genetic constitution of the plant which produces them. A plant reproducing by seed apomixis or agamospermy has all the advantages of the seed habit (dispersal of propagules and a potential means of survival through unfavourable seasons) without the risks which may be associated with pollination. As there is no essential genetic difference between simple agamospermy and asexual reproduction, Winkler grouped these two types of reproduction under the common term of apomixis.

Agamospermy was first described in 1841 by J. Smith in plants of the Australian species *Alchornea ilicifolia* growing at Kew Gardens. The deduction that seed development had occurred without fertilisation could safely be made, as the Kew collections consisted entirely of female plants. Embryological studies which revealed some of the underlying mechanisms of seed apomixis were carried out by Murbeck and Strasburger on *Alchemilla* and *Antennaria* at the turn of the century. Since those classical

Fig. 7.6. Bulbils. (*a*) *Saxifraga cernua:* 1. (× 1); 2. a cluster of bulbils; 3. bulbils in various stages of development. (From Kerner, 1895) (*b*) *Poa alpina:* 1. *P. alpina* with bulbils replacing its flowers (× 1); 2. a portion of the inflorescence; 3. a miniature grass-plant developed between the glumes of a spikelet of *P. alpina*.

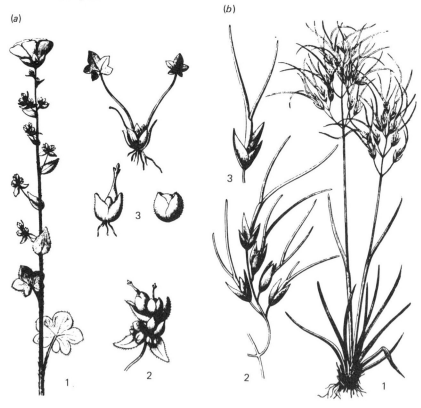

studies, a wealth of detailed examples has accumulated and several very important generalisations can now be made. First, apomictic species occur very widely in higher plants, both in the ferns and the angiosperms; but no apomictic gymnosperms are known. Secondly, there are certain flowering plant families which show a great deal of apomixis affecting several genera; the outstanding familiar examples in the Northern Temperate flora are in the Rosaceae and the Compositae. Thirdly, there is in these families a rather obvious correlation between taxonomic difficulty and the occurrence of apomixis; many of the so-called 'critical' genera or species-groups of the nineteenth century, in which the taxonomists found extreme difficulty in reconciling the points of view of the 'splitters' and the 'lumpers' as to specific delimitation, turn out to be agamospermic. Familiar examples are *Rubus* and *Sorbus* in the Rosaceae, and *Hieracium* and *Taraxacum* in the Compositae.

Complete agamospermy is not difficult to detect. For example, in most plants of the Common Dandelion (*Taraxacum officinale*), emasculation of all the florets, if performed carefully, with the exclusion of all foreign pollen by covering the capitulum, will still result in a perfect head of fruit. Partial apomixis, on the other hand, may be very difficult to detect merely from emasculation experiments, for a low proportion of seed set could so easily be due to chance contamination with pollen. Even more difficult are the cases of *pseudogamy*, in which pollination is necessary for seed formation but nevertheless the embryo is not formed by sexual fusion. Indeed, the detection of pseudogamous situations is so difficult that we may well suspect them to be more common than we know at present. Maternal (matroclinous) inheritance is usually an indication of pseudogamy; if an apparent cross between two plants differing obviously in easily scored characters produces a rather uniform F_1 resembling very closely the female parent, one should look for pseudogamy in the details of embryo formation. This is how the phenomenon was first suspected in the case of *Ranunculus auricomus* (Fig. 7.7); and it can readily be demonstrated in 'crosses' in *Potentilla* (Fig. 7.8).

Embryology of apomixis
It is not possible to give within the restricted space available here any detailed account of the embryological and cytological complexity of apomictic groups. For this, reference must be made to the standard works on apomixis by the Swedish botanist Gustafsson (1946, 1947a, b) and to later papers, particularly an excellent review of the subject by Battaglia (1963) which forms a chapter in *Recent advances in the embryology of*

128

Fig. 7.7. *Ranunculus auricomus*. Pseudogamy in this species produces many distinct variants to which specific and intraspecific rank can be given. Two such variants are illustrated. (×0.2)

Fig. 7.8. Leaves of *Potentilla* species to show maternal (matroclinous) inheritance – evidence that the attempted 'cross' between two species has produced only pseudogamous offspring. Top two rows: *P. tabernaemontani* (*a*). Middle two rows: offspring of attempted cross between (*a*) and (*b*), showing leaf shape of (*a*). Bottom two rows: *P. arenaria* (*b*), a variant with only three leaflets, used as pollen parent (× 0.2)

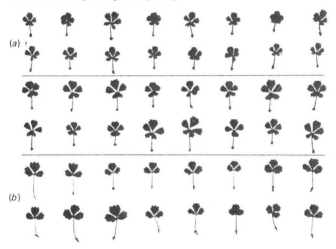

Angiosperms edited by Maheshwari. Nevertheless, the subject is interesting and, as apomictic plants are found in all parts of the world, a general account is necessary.

When we come to look at the causes of apomixis at the embryological level, we find that we are not dealing with a single, standard pattern of development, but with a whole range of possible situations having in common only one feature, namely that they involve abandonment of the fusion of gametes in the normal sexual process as a necessary preliminary to embryo and seed development. Like many biological topics where our knowledge has accumulated rapidly, we find terminological difficulties and must do the best we can in this situation. The terms here used generally follow Gustafsson; where Battaglia differs substantially, we have indicated the alternative term in brackets.

Before apomictic situations can be understood, we must know the normal pattern for a sexually reproducing flowering plant. (In ferns, apomixis is necessarily somewhat different because the gametophyte is a free-living plant separate from the sporophyte; most fern apomixis is technically *apogamy*, in which vegetative cells of the gametophyte give rise to a new embryo directly, thus omitting the stage of gamete production.) Fig. 7.9 illustrates in diagrammatic form the essential features of the

Fig. 7.9. Diagram of development of embryo-sac in angiosperms. (*a*) Typical case. (*b*) *Lilium type*. (From Robbins, Weier & Stocking, 1962)

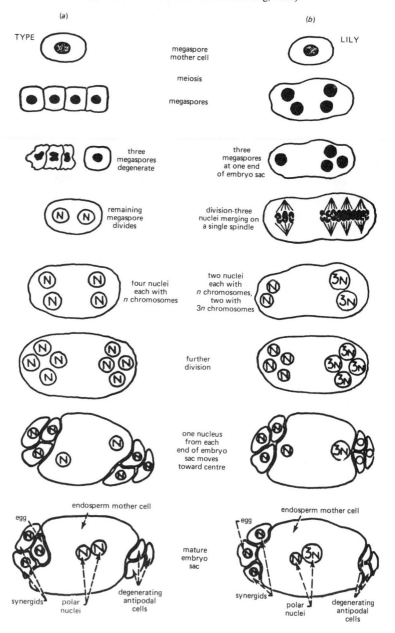

development and fertilisation of the ovule of a typical angiosperm.* Note the following points:

1. A mature ovule ready for fertilisation contains a single embryo-sac, which corresponds to the free-living gametophyte generation in the ferns (where it is called a 'prothallus').

2. This embryo-sac contains eight nuclei, one of which is an egg-cell or female gamete; the embryo-sac has developed by means of three ordinary mitotic nuclear divisions from a single cell, the megaspore.

3. The megaspore originated by meiotic division as one of four products of an initial megaspore mother-cell; the other three nuclei degenerated early.

The great majority of apomictic deviations from this pattern involve the production of an apparently normal embryo-sac, from which, however, an egg-cell develops directly, without fertilisation by a male nucleus from a pollen tube.

In the normal sexual process, the meiotic division which occurs in the formation of the megaspore (and also a similar meiotic division in the formation of the microspore or pollen grain) results in the production of a cell with a single set of chromosomes – the so-called haploid state. Subsequent mitotic divisions replicate this haploid set, so that the gametophyte generation, and the male and female gametes produced, are all haploid. The sexual fusion of egg-cell (female gamete) with pollen-tube generative nucleus (male gamete) restores the diploid state in the zygote, which then divides mitotically to give the embryo sporophyte; this grows eventually, after the germination of the seed, into a mature diploid plant. This cycle of haploid gametophyte generation succeeded by diploid sporophyte generation is of fundamental significance in the plant kingdom, and can be traced from the more complex algae right through to the flowering plants. The cytological differences between the generations are accompanied in all land plants by very obvious morphological differences. The apparent simplicity of the life-cycle of the flowering plant, involving pollination, seed setting and dispersal, disguises a complex evolutionary history of suppression of the free-living gametophyte generation and the free-swimming gametes, which are still present in the more primitive members of the land flora such as the ferns. Returning to apomictic flowering plants, we find it is in the production of an unreduced embryo-sac with the normal diploid number of chromosomes that their deviation

---

* Not all Angiosperms follow the same pattern of development. To emphasise this important point we also illustrate the embryology of *Lilium* (Fig. 7.9(*b*)).

normally shows. (There is some objection to the use of 'diploid' in such contexts, for most apomicts are polyploid; for that reason 'unreduced' is perhaps a preferable descriptive term.) Such an embryo-sac has naturally a diploid egg-cell, which requires no complement of chromosomes from a male gamete to restore the normal sporophyte number. In this way, the commonest kinds of apomixis cut out the meiotic stages from the life-cycle, so that the possibilities of variation generated by sexual reproduction are lost. It is for this reason that apomictic reproduction is genetically equivalent to vegetative propagation.

If the origin of the unreduced gametophyte is investigated, it is generally possible to distinguish between two situations. In the first, which is called *diplospory* (*gonial apospory*), the gametophyte arises from an unreduced megaspore; whereas in *apospory* (*somatic apospory*) it arises from an ordinary somatic cell of the sporophyte, which is, of course, also unreduced. Obviously the diplosporous condition involves a less radical departure from the normal sexual pattern than does the aposporous. In plants where the megaspore mother-cell is clearly differentiated in an early stage in the ovule (that is, the archaesporium is unicellular), there is no difficulty in deciding between these two possibilities if apomixis is present; this is the case, for example, in the Compositae (Fig. 7.10). However, there are some groups (for example, Rosaceae) in which there is a multicellular archaesporium, and in such cases a decision as to whether a particular cell which has undergone apomictic development is or is not part of the archesporial tissue may be difficult or even almost arbitrary. Thus the fact that both diplospory and apospory occur in *Potentilla tabernaemontani* is not indicative of any fundamental difference in this case (Smith, 1963*a*, *b*).

Diplospory and apospory are the commonest apomictic situations, but we must briefly mention the range of possibilities which Battaglia calls aneuspory. In such cases the megaspore mother-cell undergoes a more or less irregular meiosis to form the megaspore. In apomictic *Taraxacum* the first division of meiosis, instead of producing two nuclei, results in a single 'restitution nucleus'; the second division then produces a dyad of unreduced cells (instead of the normal tetrad) and the lower one of these functions as a gametophyte initial, giving the normal eight-nucleate embryo-sac. The importance of such behaviour is that it may allow some crossing-over and reassortment of genetic material which is not possible in the simple cases of diplospory and apospory. 'Sub-sexual' complexities of this kind may be more widespread – and more important in their effect on variation patterns – than has yet been established.

Fig. 7.10. Diplospory in *Taraxacum*. (*a*) Restitution nucleus; (*b*) dyad of unreduced cells (megaspores) formed after division of the restitution nucleus; (*c*) degeneration of the upper cell, and development of the lower one to a functional megaspore: (*d*) functional megaspore: (*e*) binucleate embryo-sac formed from megaspore; (*f*) mature embryo-sac, with egg-cell (ovum) and synergidae. In the formation of restitution nuclei, irregular chromosome behaviour may give rise to nuclei, and ultimately to seeds, with chromosome numbers different from that of the parent plant. For example, Sørensen and Gudjónsson (1946) produced from triploid *Taraxacum* ($2n = 3x = 24$) some plants with $2n = 3x - 2 = 22$, $2n = 3x - 1 = 23$ and $2n = 3x + 2 = 26$. Aberrants were also found with unaltered chromosome numbers. In these cases it seems likely that some pairing of chromosomes occurs in the embryo-sac mother-cells and crossing-over takes place, giving rise to plants of different genotype from that of the parent. (Drawings from Osawa, 1913) ((*a*) ×1100; (*b*), (*c*), (*d*) ×860; (*e*) ×400; (*f*) ×270)

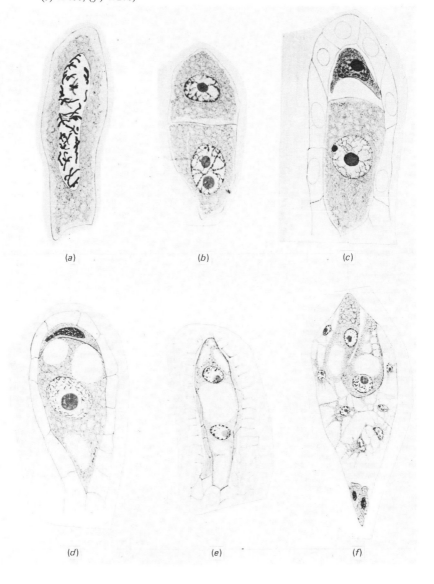

(*a*)   (*b*)   (*c*)

(*d*)   (*e*)   (*f*)

One final question concerns the function of pollination in pseudogamous species. In most cases, it seems likely that the characteristic fusion of one of the male nuclei with the polar nuclei of the embryo-sac to form the endosperm does take place in pseudogamous apomicts, although it is understandably difficult to demonstrate the actual fusion process. This would explain why pollination remains necessary for proper seed formation in spite of the apomictic origin of the embryo, for we could assume that, in the absence of nutritive endosperm tissue, the normal embryo development could not take place. In the case of *Orchis*, however, where there is no endosperm in the ripe seed, Hagerup (1947) showed that pollination was necessary for embryo development to start, even though it was clear in some cases that there was no penetration of the embryo-sac by the pollen tube. Such cases remain obscure and emphasise how complex are the apomictic phenomena in flowering plants and how much we still have to discover.

Consequences of different reproductive modes

Having discussed the three main modes of reproduction, we may now examine the consequences of reproduction in each mode.

What happens as a result of repeated self-fertilisation is highly important in our understanding of breeding systems. (For an interesting review of the history of this subject, see Wright, 1977.) Studies of inbreeding in Maize by East and Shull at the beginning of the century provided important insights, which have been confirmed in many other studies of crop plants. In the heterozygous diploid the dominant allele often 'shelters' recessive alleles which are deleterious in the homozygous state. Self-fertilisation quickly results in the segregation of lethal or sublethal types as homozygous recessives are produced. (Unless specially looked for in the seedling stage these types, which may die at a very early stage of growth, may be undetected even in garden or glasshouse culture.) Further selfings produce rapid separation of the material into uniform lines – often called pure lines – differing from each other in various vegetative and reproductive characteristics (Fig. 7.11). The continued selfing of uniform lines may be rendered impossible as some plants may become weak or sterile. Surviving lines may be characterised by plants of reduced vigour and fertility. If plants of pure lines originating from different parental stocks are crossed together, hybrid vigour – so-called heterosis – may be demonstrated. Such hybrid plants are characteristically of great vegetative vigour and high fertility (Table 7.3). It is important to note that crossing genetically closely related plants from lines derived by repeated self-fertilisation from the same original parental stock will not give heterotic plants.

The phenomenon of heterosis, so pronounced in experimental crosses with inbred lines, was not a new discovery, being often reported in the studies of early plant hybridists (Roberts, 1929). The underlying causes of loss of vigour or fertility on repeated selfing and the heterotic effects in products of crossing inbred lines have been the subject of intense study. For accounts of this controversial subject the reader is referred to Strickberger (1976) and Wright (1977).

In considering generalisations which might be made on the effects of inbreeding, it is important to note that our ideas are based on results with a few crop-plant species. The simultaneous study of selfing rates and inbreeding effects has yet to be made on wild species (Charlesworth & Charlesworth, 1979). However, unless the heterozygous state is favoured in selection, the deduction that repeated self-fertilisation could yield complete homozygosity in a few generations is clearly sound. Thus in a weedy diploid species, with several generations per year, complete homozygosity could be established very quickly indeed (Fig. 7.12). However, many plants are polyploids, having more than two representatives of a gene, and a greater number of generations will be required to produce complete homozygosity in such plants.

One possible advantage of repeated self-fertilisation might be that well-adapted genotypes could be replicated without change. A further advantage, especially in extreme or marginal habitats where crossing between plants might be hazardous or fail altogether, is that self-fertilisation is a safe method of producing offspring.

Fig. 7.11. Diagram showing the effect of selfing on a heterozygote. The proportion of heterozygous individuals rapidly declines in successive generations, and 'pure lines' are established. (From Wilmott, 1949)

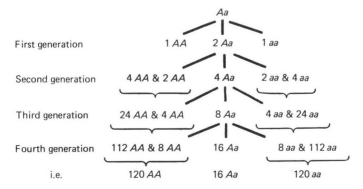

Table 7.3. *Hybrid vigour in Maize (Zea). Crossing inbred lines of Maize P_1 and P_2 yields F_1 plants showing hybrid vigour. Repeated self-fertilisation through several generations results in diminution in height and loss of yield (From Jones, 1924)*

| | Parents | | Successive generations | | | | | | | |
	P_1	P_2	F_1	F_2	F_3	F_4	F_5	F_6	F_7	F_8
Number of generations selfed	17	16	0	1	2	3	4	5	6	7
Mean height (inches)	67.9	58.3	94.6	82.0	77.6	76.8	67.4	63.1	59.6	58.8
Mean ear length (cm)	8.4	10.7	16.2	14.1	14.7	12.1	9.4	9.9	11.0	10.7
Mean yield (bushels per acre)	19.5	19.6	101.2	69.1	42.7	44.1	22.5	27.3	24.5	27.2

An appreciation of the long-term disadvantages of inbreeding enable us to recognise the advantages of the outbreeding mode of reproduction. As we have seen, structural features or physiological mechanisms prevent or discourage self-fertilisation and lead to crossing between different individuals. Such breeding will tend to generate a good deal of genetic variation.

The role of incompatibility mechanisms is very important in considering breeding within populations. Lewis (1979) has recently drawn attention to the important point that while some fruits (or seeds) from a given parent may be dispersed some considerable distance, many fall close to the parent, developing and flowering as a family group. This group, which may include parents and other relatives in plants with a long life-cycle, is made up of genetically related individuals. Crossing between close relatives leads to inbreeding depression, although complete homozygosity is not achieved so swiftly by such matings (Fig. 7.12). Lewis has shown that incompatibility systems will restrict crossing between close relatives, the degree of restriction depending on the genetic mechanism of the incompatibility system.

In general terms, obligate outbreeding would appear to have advantages, but there may be costs to be borne. Reproduction may be rendered uncertain or unlikely by environmental factors influencing, for example,

Fig. 7.12. Graph showing the relation between the number of generations of inbreeding and the percentage of homozygous genes. (From Lewis, 1979)

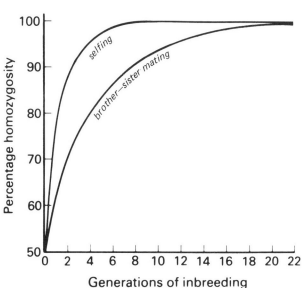

cross-pollination. Moreover, given that plants which survive to reproduce successfully in a habitat are well adapted, outbreeding might seem to offer only the possibility of loss of such variants as each generation produces new variability.

The third mode of reproduction – apomixis (either by vegetative (asexual) means or by agamospermy) – facilitates the reproduction of well-adapted genotypes. It may also offer the possibility of reproduction by seed in plants with 'odd' or unbalanced chromosome numbers, such plants being unable to produce viable products at meiosis and likely to be totally or partially seed-sterile in sexual reproduction. Apomixis would also appear to be important at the edge of the range of many species, allowing populations to persist in territory in which various factors exclude the possibility of sexual reproduction (e.g. at high altitude or latitude).

While we might postulate various advantages of the apomictic mode of reproduction, it is clear that there are some disadvantages also. The generation of variability would seem to be restricted or prevented in obligately apomictic plants. Whilst higher animals usually have a clearly defined lifespan, it is not clear whether there are ageing processes in perennial plants which would restrict or prevent natural asexual reproduction in the long term. Observations on cultivated *Citrus* plants suggest that repeated vegetative propagation leads to senescence (Frost, 1938). Moreover, reproduction by vegetative means carries with it the possibility that virus or disease might build up in the plant, and there is evidence that the sexual cycle provides a means of 'purging' the plant system of certain viruses and other disease organisms (see Smith, 1977).

An analysis of the three modes of reproduction reveals, therefore, that each has its advantages and disadvantages. Consideration of the conditions experienced by plants over many thousands of years would suggest that, while a capacity to reproduce well-adapted genotypes unchanged might be important in the short term, a lineage lacking the capacity to produce variation might be at a serious selective disadvantage in competition with lineages capable of change. A lack of variation might prevent a lineage from withstanding the selection pressures associated with, say, an ice age. Important too, and perhaps less well known, are systematic changes for example in climate in the short term. Thus there is evidence, from documents, plant remains, works of art, etc., for fluctuations in climate in recent historical times (Lamb, 1970). Three examples can be given:

1. *The freezing of rivers*. The River Thames regularly froze in the

period 1540–1814, indicating a period of severe winters.

2. *Agriculture on marginal land.* Aerial photographs and other evidence suggests that arable farming has been practised on marginal land and at altitudes above the present cultivation zone, in many parts of the world.

3. *The tree line.* Trees have at various times grown successfully above the present natural tree line.

Given fluctuations in climate, as well as the marked differences in weather in successive years, and seasonal changes, it is not surprising that detailed studies of the breeding behaviour of particular species reveal that, instead of reproducing entirely in one of the three modes outlined above, many plants reproduce by several methods. Thus a plant may produce lineages of progeny showing relatively little variation and others exhibiting considerable variation. A different balance of variance and invariance may be seen in the products of different species or lineages within species (Mather, 1966).

Breeding systems in wild populations

In order to understand and discover the actual breeding system in the field, detailed studies at the population level are necessary, investigating natural lineages of plants and their production. Inferring the breeding behaviour from flower structure or pollinator activities would seem to be fraught with difficulty. For instance, the flowers of certain taxa of *Calyptridium* (Portulacaceae), which are regularly visited by insects, and therefore on logical grounds likely to be cross-pollinated, are in fact regularly self-pollinated by the insects which visit them (Hinton, 1976).

We may now examine, in outline, a number of different situations, showing how many plants have a capacity to produce both variance and invariance, and discuss a number of experiments which have shown how environmental factors provide a 'trigger', switching the plant from one mode of reproduction to another. (See Heslop-Harrison, 1964.)

Outbreeding combined with vegetative reproduction

Individuals of many self-incompatible species, producing variable progeny by outbreeding, are capable of considerable lateral spread. Decay of plant connections yields clonal patches of well-adapted genotypes often of considerable size and age, e.g. *Trifolium repens* (Harberd, 1963) and *Lysimachia nummularia* (Dahlgren, 1922).

As we indicated in Chapter 6, the switch from vegetative to reproductive mode has been studied in many plants, such environmental factors as daylength and winter temperatures being very important.

Outbreeding in association with vivipary

Some species of genera such as *Agrostis, Allium, Deschampsia, Festuca, Poa* and *Saxifraga* have the capacity to reproduce not only by the sexual processes but also by vivipary, a condition in which tiny plantlets are produced in the inflorescence instead of (or mixed with) ordinary florets. In normal sexual reproduction, the generation of variation is possible, whilst the viviparous propagules reproduce the genotype of the plant which produces them (Fig. 7.6).

In experiments with *Poa bulbosa* (Youngner, 1960) it was shown that conditions of long daylength and a short cold period followed by high temperatures yielded sexual inflorescences. In contrast, short daylength and low temperatures yielded viviparous inflorescences, and mixed panicles of sexual and viviparous products resulted from long day/low temperature and short day/high temperature combinations.

Rigorous outbreeding combined with occasional self-fertilisation

Often it is not clear whether self-incompatibility mechanisms totally prevent selfing in nature. However, there are many cases known where largely self-incompatible species are capable of producing seed on selfing, for example the *Primula veris* stocks studied by Darwin. There is some experimental evidence which suggests that self-fertilisation may occur in 'self-incompatible' species under certain conditions, for example, in material subjected to high temperatures, in situations where pollination of ripe stigmas is long delayed, or at the end of the flowering season (De Nettancourt, 1977). The rigidity, or otherwise, of incompatibility systems under field conditions requires further study.

Outbreeding combined with regular self-fertilisation

In some species (e.g. *Viola*) the spring-formed insect-pollinated flowers allow the possibility of outbreeding (Fig. 7.14). In the summer, cleistogamous (closed) flowers are produced in which self-fertilisation is automatic. Daylength is critical in the regulation of flowers. Borgström (1939) has shown that plants grown under 13–15 hours light per day produce normal flowers, whilst longer days (> 17 hours) typical of early summer induce the formation of cleistogamous flowers. Cleistogamy has been regarded as a rather rare phenomenon, but the capacity to produce cleistogamous flowers may be more widespread than hitherto realised (Richards, 1979). In many plant species, male-sterile individuals occur as rare mutants but in some species, perhaps in a greater number than previously acknowledged, populations have a high proportion of female plants together with hermaphrodite individuals. Such species, which are particularly frequent in the Labiatae, are referred to as gynodioeceous, a term coined by Darwin (1877*a*). Darwin's early studies of *Thymus* (Fig. 7.13) are particularly interesting. For a review of present knowledge of the biology and genetics of gynodioeceous plants, Lewis & Crowe, 1956, Lloyd, 1975 and Dommée, Assouad & Valdeyron, 1978, may be consulted, together with references cited therein. For our purposes it is sufficient to note that gynodioecy permits the generation of variation in the crossing of hermaphrodite and female plants, while allowing the possibility of selfing in hermaphrodites.

Predominantly self-fertilisation with occasional outbreeding
Many crop plants, including the familiar cereals such as Wheat (*Triticum*), are predominantly self-fertilising but the occasional crossing between varieties occurs, its extent depending on distance between plots, as well as genotypic and environmental factors. It is supposed that many wild plants which are predominantly self-fertilising are also occasionally outcrossed, e.g. *Senecio vulgaris* with 1% outcrossing (Hull, 1974).

Fig. 7.13. Gynodioecy in *Thymus vulgaris*: larger hermaphrodite flower (left), and two smaller female flowers with reduced stamens. (From Darwin, 1877*a*)

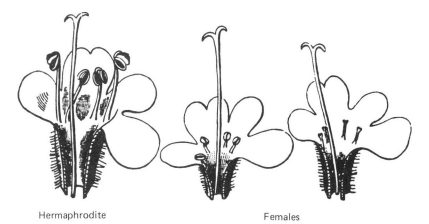

Hermaphrodite   Females

Thymus vulgaris (magnified)

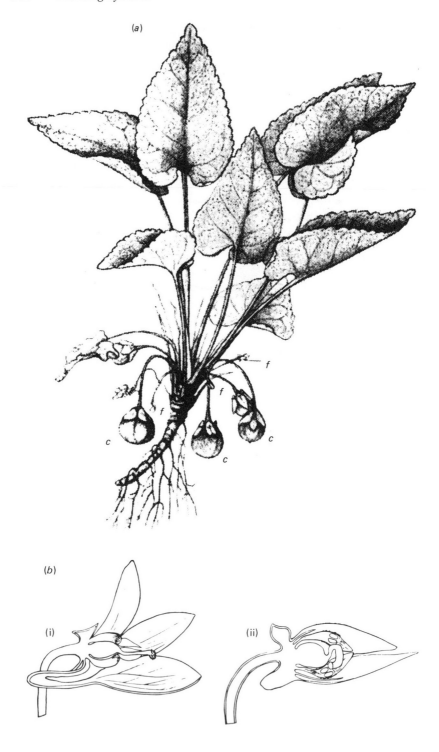

Facultative and obligatory apomixis

In some genera, apomixis seems to have replaced completely the sexual processes in the great majority of species. In the Lady's Mantles (*Alchemilla*), for example, plants of the common northern European species-group to which Linnaeus gave the general name *A. vulgaris* show defective pollen, often degenerating in the tetrad stage, and precociously ripening fruit – sure indications that pollination is not necessary for seed formation. Indeed, so far as is known, all *Alchemilla* species in Europe are apomictic, with the exception of a very dwarf alpine species *A. pentaphyllea* and a very few alpine taxa belonging to another subsection of the genus. It is interesting that Robert Buser, the Swiss expert on *Alchemilla* who achieved an unrivalled knowledge of the plants in field, in herbarium and in cultivation, had rightly suspected from field evidence that certain puzzling intermediate populations in the Alps were hybrids of *A. pentaphyllea* and other species before anything was known of their genetical complexity. Obligatory apomixis, as is shown by all the 'vulgaris' Alchemillas, is accompanied by a relatively straightforward pattern of variation; the collective Linnaean species *Alchemilla vulgaris* and *A. alpina* consist (in Europe) of some 300 taxonomically distinguishable microspecies, many of which are wide ranging and no more difficult to identify than many sexual species in other genera. It is the number of microspecies involved, and the relative complexity of the detailed morphological differences between them, which make such critical groups the hobby of the few (Fig. 7.15).

Total or obligatory apomixis is, however, much less common than partial or facultative apomixis; indeed, since we can never be certain that sexual reproduction is quite ruled out even in cases such as *Alchemilla*, it may be that strict obligate agamospermy does not occur. In facultatively apomictic plants, amongst them the taxonomically difficult genera of *Rubus* and *Potentilla* in the Rosaceae and *Pilosella* in the Compositae, apomictic embryos develop automatically without sexual fusion. In addition some egg-cells are produced with the reduced number of chromosomes. Such eggs may be fertilised giving rise to sexually produced embryos. Obligate apomixis, as we have seen, may yield little variation. In

Fig. 7.14. Cleistogamy in *Viola*. (*a*) *Viola hirta:* plant with cleistogamous flowers, *f*, and developing capsules, *c*. (× 1.0) (*b*) *Viola riviniana:* open (i) and cleistogamous (ii) flowers in longitudinal section, showing in the latter the crumpled style in contact with the developing anthers. (× 2.0) (From McLean & Ivimey-Cook, 1956)

contrast, facultative apomixis combines a capacity to reproduce successful genotypes unchanged with a mechanism allowing the generation of
variation on an occasional or regular basis.

A good example of how occasional sexual crossing might yield an even
more complex pattern is provided by the British representatives of the
Series Aureae of *Potentilla*, which are, so far as is known, all pseudogamous apomicts. Most British plants can be classified as either *P.
tabernaemontani*, a rhizomatous, mat-forming perennial of chalk and
limestone mainly in the lowlands, or *P. crantzii*, a non-rhizomatous
perennial with an unbranched woody stock, typically found on calcareous
cliffs in upland or mountain areas in Britain. In some parts of upland
Britain, especially in northern England, however, puzzling intermediate
plants occur which obscure the otherwise fairly clear distinction between
the two species. These plants mostly have higher chromosome numbers
than normal specimens of either species. During a detailed study of the
group in the experimental garden in Cambridge, Smith (1963*a*, *b*, 1971)
showed that if large numbers of progeny were raised from 'crosses',
occasional aberrant individuals could be detected because they differed
from the normal offspring which resembled closely the female parent. One
such individual, raised from the 'cross' between a *Potentilla tabernaemontani* with $2n = 49$ from Fleam Dyke, Cambridgeshire, as a female parent,
and *Potentilla crantzii* with $2n = 42$ from Ben Lawers, Scotland, as pollen
parent, was found to have 70 chromosomes. There seemed little doubt
that, in this case, an unreduced egg-cell with 49 chromosomes had been
fertilised by a normal reduced pollen grain with 21 chromosomes, to give
a zygote, and eventually a mature sporophyte plant, with $2n = 70$. A
hybrid apomict had been synthesised. Such a plant can produce both
vegetatively (to a limited extent at any rate) and agamospermously, and
might have established a more or less uniform population in nature.
Moreover, since this sexual reproduction does not seem to be very rare,
the event could take place many times, giving populations differing subtly
from each other according to the exact genetic constitution of their
parents. Such events are particularly likely in pseudogamous species,
where the pollen is largely normal. Further, occasional sexual crosses may
be brought about by pollen from a plant which is itself an obligate
apomict pollinating a sexual or facultatively apomictic plant. Even the
very limited sexuality of *Alchemilla* may for this reason be of far greater
significance than we think.

A further interesting example of facultative apomixis is provided by the
genus *Hieracium*. After studying the genetics of peas, Mendel attempted

experiments with *Hieracium*. Notwithstanding the technical difficulties of emasculating and crossing plants with such small flower parts, Mendel was successful in crossing some *Hieracium* species but, to his surprise, hybrids bred true. We can now see that the combination of occasional sexual reproduction with the regular production of apomictic offspring makes *Hieracium* an impossible subject for the study of segregation ratios. It is of interest that Mendel supplied von Nägeli with material of his crosses. In making a special study of the *Hieracium* group, von Nägeli & Peter (1885, 1886–9) produced herbarium specimens which were sent out to the major herbaria of Europe: one of these specimens of Mendel's hybrids from the Cambridge University collections is shown in Fig. 7.16.

Two other points of interest may be made concerning facultative apomicts. First, it has been found in experimental studies of *Dichanthium aristatum*, for example, that daylength may be important in determining

Fig. 7.15. Leaves of apomictic *Alchemilla* species, showing a range of form from the simple, lobed 'vulgaris' shape, (*a*), to the compound 'alpina' shape, (*f*). (*a*) *A. fulgens* (Pyrenees); (*b*) *A. faeroensis* (Faeroes, East Iceland); (*c*) *A. conjuncta* (Jura, West Alps); (*d*) *A. plicatula* (Pyrenees, Alps, Balkan Mountains); (*e*) *A. subsericea* (Alps); (*f*) *A. alpina* (Arctic, North European Mountains, Alps). *A faeroensis*, with about 220 chromosomes, is almost certainly of ancient allopolyploid origin from a 'vulgaris' and an 'alpina' species. Nearly all traces of sexual reproduction are now lost in the present-day representatives of these groups. (× ⅓)

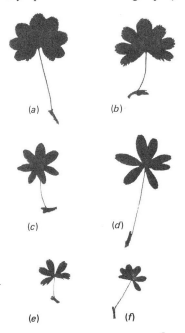

Fig. 7.16. Herbarium specimen of a *Hieracium* cross made by Mendel in the famous monastery garden in Brünn (now Brno), cultivated by von Nägeli and Peter in the Munich Botanic Garden in the 1880s and called by them *Hieracium monasteriale*. Photographed in the Cambridge University Herbarium.

the balance between apomictic and sexual reproduction. Under continuous short days, up to 79% of embryos produced were apomictic; under long days, after floral induction in short days, only about 47% of embryos were aposporously produced (Knox & Heslop-Harrison, 1963). Secondly, there are cases where genetic differences within a species are of great importance. For example, in studies of *Agropyron scabrum* very considerable differences were discovered between different populations (Hair, 1956), as follows:

1. Plants completely and normally sexual.
2. Plants facultatively apomictic.
3. Plants predominantly apomictic, meiosis suppressed in ♀, not in ♂.
4. Plants obligatorily apomictic, suppression of meiosis in ♀ and ♂.

The present chapter has surveyed the generation of variation associated with different breeding systems. In stressing the complexities so far discovered and indicating our lack of knowledge about breeding systems in nature, we hope our account will serve as an antidote to the more complacent accounts of breeding systems available to the student of evolution. In particular we would stress the difficulty of generalising about breeding systems and the variation they generate. We do not wish to restrict the flow of hypothesis and speculation about plants, but point to the difficulties in generalising about 'inbreeders', 'apomicts', etc.

Our discussions of variation have brought us to the point where we have explored the *potential* variation generated by different breeding systems. Plants grow in populations and we may now turn our attention to a review of our knowledge of variation between and within populations. A discussion of the taxonomic treatment of different groups exhibiting diverse variation patterns is deferred until some of the evidence about patterns of variation and processes of change has been examined.

8

Infraspecific variation and the ecotype concept

We saw in Chapter 7 how different breeding systems can be expected to produce different patterns of variation. If we are to understand the variation patterns actually found in nature and the processes which give rise to these patterns, we must discover how the potential variation in seeds is revealed in the variation of actual reproducing plants. Historically, the first advances in this field were made by means of comparisons between plants belonging to the same species but from different populations. Taxonomists, biometricians and, later, geneticists became interested in genetic variation in the wild and many of their studies converged at one point, namely, the controversy over the 'reality' of the infraspecific groups which could be distinguished in nature, whether they were the subspecies or varieties of the taxonomist or the 'local races' of the biometrician. A new look at this old question was provided by the famous researches of Turesson published in the early 1920s.

Turesson's pioneer studies and other experiments

At the time of Turesson's experiments, the question of the reality of 'local races' was combined with another controversial issue, namely, how much of the observed variation in natural populations was the result of the direct modification of plants subjected to severe environmental stresses? By the end of the nineteenth century many botanists reasoned that distinctive infraspecific variants were merely 'habitat modifications'. Turesson, however, pointed out that in all previous cases known to him, only a partial test of the 'habitat modification' hypothesis had been carried out. For example, he considered the studies of *Lathyrus japonicus* (*L. maritimus*) undertaken by Schmidt (1899). Baltic populations of this plant have dorsiventral leaves, whilst on the North Sea coast of Denmark the plant has isolateral leaves. Schmidt showed by experiment that watering the Baltic variant with sodium chloride solutions induced a leaf structure typical of Danish plants. Given that the North Sea has a higher percentage content of salt than Baltic waters, Schmidt deduced that the leaf structure of the plants on the North Sea coasts of Denmark was merely a habitat modification.

The logic of this type of deduction did not satisfy Turesson. His approach to the problem was to grow infraspecific variants of many species in a standard garden, to see if the 'distinctiveness' was retained or lost. He collected living plants (and in certain cases seeds) of many common plants from a variety of natural habitats in southern Sweden and grew them in experimental gardens first at Malmö (1916–18) and subsequently at the Institute of Genetics at Åkarp. In this way he studied, for example, shade variants, dwarf lowland plants from coastal habitats, and succulent variants, in most cases growing these plants alongside collections of the same species collected from ordinary inland habitats (Turesson, 1922*a*, *b*).

In some cases the distinctness of the variants was lost in cultivation in an inland garden, but usually the distinctive plants originating from extreme habitats retained their characteristics in cultivation even in the absence of shading, salting, etc. These observations were clearly at odds with the notion that extreme variants were nothing more than habitat modifications, and the persistence of distinct variants under standard conditions suggested to Turesson that the variation had a genetic basis.

Many of Turesson's early experiments were carried out on the Composite *Hieracium umbellatum*. This plant is common in southern Sweden where its principal habitats – woodland, sandy fields, dunes and cliff tops – may all be found. In each of these habitats a distinctive plant was discovered in the field. By careful sampling and cultivation, Turesson found that, with few exceptions (for example certain prostrate plants from sandy fields) distinctive variants retained their characteristics in cultivation. The results of these experiments were consistent with those obtained in studies of other species, and again Turesson considered that patterns of residual difference had a genetic basis.

Hieracium umbellatum is a common plant in southern Sweden and Turesson was able to collect many samples from each habitat type. A close study of his extensive collections after a number of years of cultivation suggested to him the exciting possibility that habitat-correlated patterns of genetic variation were present, that is to say, in a particular habitat of *H. umbellatum* a certain race of characteristic morphology was invariably present. In the appropriate habitat there was to be found a dune race, a woodland race, etc. Turesson called these local races 'ecotypes' and described five, as follows (note that in these descriptions he considered anatomical and physiological traits (e.g. flowering times) as well as morphological features):

1. *An ecotype from shifting dunes*
   Narrow leaves and slender, less erect, sometimes more or less prostrate stems. Marked power of shoot regeneration in autumn. Leaves tough and thick with three to four layers of palisade cells. Fruiting in early September.
2. *An ecotype from sandy fields and stationary dunes*
   As 1, but power of shoot regeneration in autumn weak or lacking. Extremely prostrate in growth habit.
3. *An ecotype from western sea cliffs*
   Broad leaves, and more or less prostrate stems. Growth form contracted and bushy. Cells of leaves more or less distended. Fruiting late September to early October.
4. *An ecotype from eastern sea cliffs*
   As 3, but plants tall and almost as erect as in 5.
5. *An ecotype from open woodland*
   Stout, erect plants with lanceolate leaves of intermediate width. Leaves thinner with two or, at most, three palisade layers. Fruiting in September.

Turesson notes that additional ecotypes might be discovered in future studies.

Hieracium umbellatum is a member of a genus famed for its apomictic reproduction. In considering Turesson's results it seems essential, therefore, to take into account the breeding behaviour of the plant. In a partial examination of the breeding system of his material, Turesson performed castration experiments, removing the upper half of unopened flowerheads with a razor. No fruits developed. This evidence supports the view that reproduction is sexual and not obligately apomictic. Plants of *H. umbellatum* proved in fact to be self-incompatible, and artificial crosses with plants within a population of the dune ecotype and within a population of the cliff ecotype produced plants in which the ecotypic characteristics of each were perpetuated, confirming the genetic basis of the discovered differences. Lövkvist (1962) has resampled at many of Turesson's *H. umbellatum* sites, and found broadly similar patterns of variation in cultivation trials. He also re-examined the breeding system of southern Swedish material of *H. umbellatum*, and found no evidence of apomixis. However, apomixis *has* been reported in this species (e.g. Bergman, 1935, 1941, and references cited therein) and may influence patterns of variation elsewhere.

Considering the origin of ecotypes, Turesson made two important deductions. He concluded, first, that the finding of widespread habitat-correlated genetic variation does not support the view that the variation patterns are largely governed by chance; rather the evidence suggests that natural selection operates in natural populations, well-adapted genotypes being selected in each habitat. This idea is expressed many times in Turesson's writings; for example, he says (1925) 'Ecotypes ... do not originate through sporadic variation preserved by chance isolation; they are, on the contrary, to be considered as products arising through the sorting and controlling effect of habitat factors upon the heterogeneous species-population'. Turesson further concluded that a close study of the variation within and between ecotypes of *H. umbellatum* revealed patterns of leaf morphology which suggested a 'local' origin for coastal ecotypes from the widespread inland populations. It was possible that an appropriate ecotype could be produced many times, that is to say polytopically, and it was not necessary to postulate the invasion of Sweden by fully formed standard ecotypes after the last glaciation.

In a series of long papers published from 1922 onwards, Turesson eventually described ecotypes in more than 50 common European species. His first papers were about the plants of southern Sweden, but later (1925, 1930) he experimented with material collected from distant localities in all parts of Europe and further showed physiological differences between some of his stocks (1927*a, b*). Analysis of the behaviour of his extensive collections in cultivation enabled him eventually to distinguish two kinds of ecotypes, namely edaphic and climatic ecotypes, where the most important environmental effects were soil type (as in the case of *Hieracium umbellatum* in southern Sweden) and the climatic influences, respectively.

As early as the beginning of the eighteenth century there was a considerable amount of observational evidence that common species did not flower at the same time in different localities. For example, Linnaeus (1737) noted the different flowering times of Marsh Marigold (*Caltha palustris*) (March in the Netherlands, April to May in different parts of Sweden, June in Lappland). Quetelet (1846), having studied the dates of first flowering of Lilac (*Syringa vulgaris*) in different parts of Europe, came to the conclusion that there was a retardation of 34 days for each advance of 10° northwards in latitude. He also compared flowering at different altitudes above sea-level, and discovered a retardation of 5 days for every 100 m increase in elevation. The important environmental factor controlling flowering was thought to be temperature. Turesson, studying the behaviour in cultivation of a large number of spring-flowering species, clearly demonstrated the importance of persisting genetic differences between plants originating from different climatic regions. Southern plants of such species flowered earlier in Turesson's experimental garden than plants of the same species collected from northern latitudes. He suggested that this group of plants is adapted to flower in the period immediately preceding the leafing-out of trees, a phenomenon which occurs earlier in the year in southern latitudes than in northern Europe.

In the botanical literature of the nineteenth century there are scattered reports that alpine plants flower earlier than lowland ones when both are cultivated in lowland gardens. Turesson's extensive experiments with species such as *Campanula rotundifolia* (Table 8.1) and *Geum rivale* enabled him to demonstrate that alpine ecotypes were smaller and retained their early flowering habit in cultivation. He also carried out researches upon summer-flowering plants, showing that northern eco-types were early flowering and of moderate height, while southern plants were late flowering and tall. Western Europe was characterised by late-flowering plants of low growth; from Eastern Europe, on the other hand, came taller early-flowering ecotypes.

Turesson's contribution to our understanding of the patterns of variation within species is of very great importance: he demonstrated clearly the widespread occurrence of infraspecific habitat-correlated genetic variation. Adaptation to the environment was sometimes by plastic responses, but more frequently it had a genetic basis. Such studies were grouped together under the name of 'genecology' and the work was the model for many studies by other botanists. The work of Stapledon (1928) is of special interest. Using the common pasture grass *Dactylis glomerata*, he studied the influence of hay cutting and animal grazing, and described a third class of ecotype, namely the 'biotic ecotype'. His work is summarised in Table 8.2.

Scandinavian botanists have made many notable contributions to genecology, and it is appropriate at this point to give an example of the important experiments of Bøcher. He used the Turessonian technique of cultivation in a standard garden to examine the variation and flowering behaviour of collections of many European plants, and carried the analysis of variation into an important new area, namely the study of the timing of flowering in relation to the life history of the plant. For example, he discovered in cultivation experiments with *Prunella vulgaris* (1949) that there were two main growth types in Europe, namely plants with a short vegetative phase, flowering in their first year, and plants with a longer vegetative phase, flowering in their second year. This latter group was further subdivided into plants which were short lived and perennial

Table 8.1. *Geographic variation in* Campanula rotundifolia
(a) Results of transplant experiments from Turesson (1925) (means of five measurements given)

	Field no.	Transplanted from	Length of stems (mm)	Width of middle-stem leaves (mm)	Number of flowers on stems	Length of corolla (mm)	Width of corolla in the middle (mm)	Width of corolla at mouth (mm)	Length of corolla lobes (mm)	Length of calyx lobes (mm)	Power of regeneration of basal rosette-leaves	Year of collection	No. of plants
Norway and Sweden	99	Vitemölla	547.75	2.18	23.25	18.63	16.50	22.45	7.33	6.33	none–weak	1920	8
	206	Åhus	650.54	2.16	27.49	19.93	16.45	22.82	7.65	7.29	none–weak	1922	13
	270	Ulriksdal	334.30	1.86	20.33	17.13	14.99	20.2	7.02	5.63	none–weak	1921	14
M	298	Åre	308.43	2.97	11.5	22.12	21.0	25.91	9.54	5.88	mostly strong	1921	17
	349	Bergen	378.67	2.73	9.19	20.56	20.53	25.67	8.41	6.36	weak–strong	1922	7
	240	Trondhjem	336	2.03	15.97	21.0	18.34	25.06	8.39	5.80	weak–strong	1922	14
Central Europe M	19–25	Abisko (seeds)	250.10	1.99	13.97	24.47	20.48	27.68	9.32	7.90	strong	1921	seeds
	770	Freiburg	278.56	2.12	19.86	20.44	18.89	24.54	8.5	6.56	none–weak	1923	16
M	796	Feldberg	224.66	4.29	6.88	23.45	21.82	25.32	8.76	7.89	strong	1923	14

(b) Progeny trial, from Turesson (1930)

Field no.	Source	No. of plants	Height (cm) Mean	σ	m±	Earliness of flowering[a] Mean	σ	m±
770	Freiburg	20	68.9	5.89	1.32	±1.60	0.35	0.29
796 M	Feldberg	20	29.5	2.41	0.54	±5.00	0.00	0.00
270	Ulriksdal	20	47.1	5.47	1.22	±2.80	0.44	0.10
298 M	Åre	20	33.4	3.75	0.84	±5.00	0.00	0.00

[a] note that a larger mean corresponds to earlier flowering
M = montane localities

Table 8.2. *Biotic ecotypes in* Dactylis glomerata. *Stapledon (1928) discovered that grassland use determined the type of* Dactylis *present in a particular area*

| | | Per cent growth type | | | | Per cent over 100 cm | Per cent flowering behaviour | | | |
		Hay	'Cup'	Tussock	Pasture		Early 1	2	3	Late 4
Commercial hay stocks	A	59	36	2	3	78	40	50	9	1
	B	66	31	1	2	78	61	32	6	1
Old pastures		15	23	6	56	15	11	35	38	16
Hedgerows and thickets		26	35	25	14	31	17	35	34	14

Hay types with their taller early-flowering plants were distinct from the shorter, later-flowering plants characteristic of grazed pasture. Pasture types had many more tillers than hay types and a smaller percentage of tillers produced inflorescences. Plants from hedgerows had a wide range of variants. Even though this experiment did not reveal a discontinuous pattern of variation, Stapledon was content to interpret his results in terms of 'biotic ecotypes'. (See Warwick & Briggs, 1978a,b, for a partial review of recent work on 'hay' and 'pasture' ecotypes.)

types (Fig. 8.1). The distribution of the two main types – first- and second-year flowerers – proved most interesting; for example, in Mediterranean regions subject to summer drought, only short-lived annual plants were found, whilst in areas with different climatic conditions biennial or perennial types were characteristic. Such patterns are likely to be the result of natural selection: only those plants whose life-history 'fits' the growing season of a particular area will survive in the long term.

Fig. 8.1. Distribution in Europe of first-year flowering and second-year flowering types of *Prunella vulgaris*: the first (with short rosette stage) indicated by open rings, the second (with long-lasting rosette stage) by filled circles. All 75 samples were sown and cultivated simultaneously in 1950–51. On the map on the left are 51 lowland samples. On the map on the right are 24 samples from montane stations. The tendency towards second-year flowering in the northerly direction and from the lowland to the highland areas is evident. (From Bøcher, 1963)

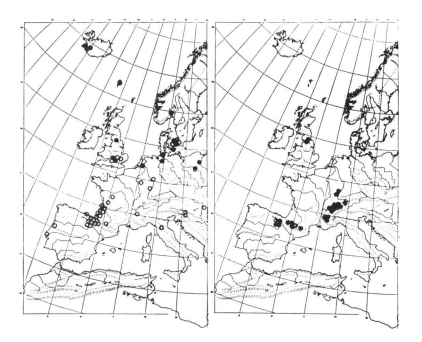

Fig. 8.2. Map and climatic details for Stanford and Timberline, sites in central California, used for the famous transplant experiments of Clausen, Keck & Hiesey. (*a*) Diagrammatic transect showing heights above sea-level. (*b*) Graphs showing annual variation in temperature and precipitation (US Weather Bureau data for 1925–35 inclusive) near Stanford and near Timberline. In the lowland site with 'Mediterranean-type' climate (Stanford) active growth is possible throughout the year, whereas at Timberline (*c*. 3000 m) the active growth period is restricted to July and August. At Stanford, average annual precipitation was 31.7 cm; there was no snowfall except for traces in 1931 and 1932. At Timberline, average annual precipitation was 74.1 cm. Maximum temperature = average of the highest monthly temperatures. Mean temperature = average of the mean monthly temperatures obtained from daily readings. Precipitation = average monthly precipitation. Minimum temperature = average of the lowest monthly temperatures. (From Clausen *et al.*, 1940)

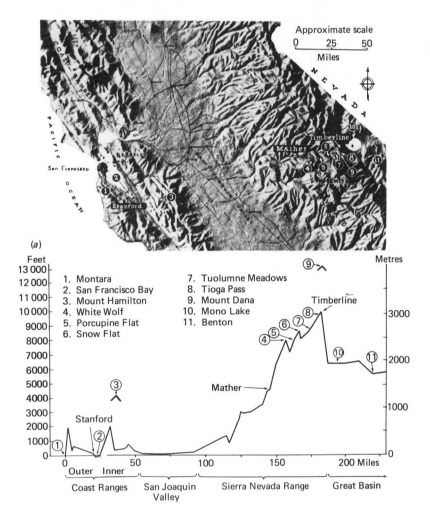

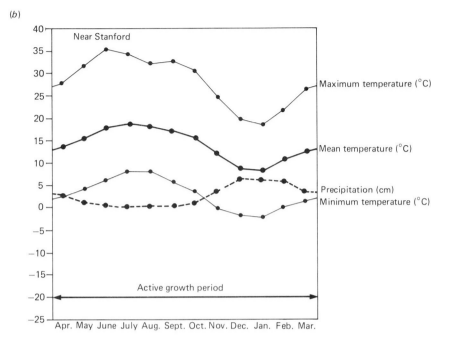

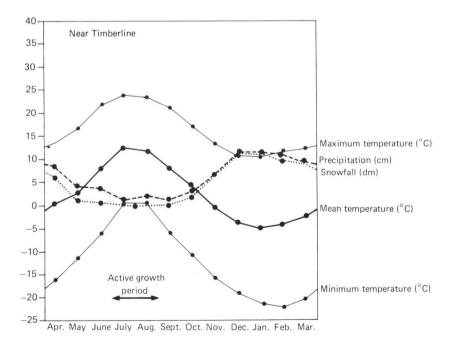

Experiments by American botanists

Some of the most famous experiments on ecotypes were carried out by Clausen, Keck & Hiesey (1940) on different species of plants collected on a 200-mile transect across Central California, from a 'Mediterranean' climate in the west to an 'alpine' climate in the east. Turesson's method of studying ecotypes was to grow all his collections in a lowland garden. Such a method has the limitation that it may not allow certain traits to be revealed (e.g. tolerance or sensitivity to frost, or drought). In an attempt to overcome this difficulty, Clausen and his co-workers carried out experiments with many gardens, and finally used three: at Stanford (30 m above sea-level), Mather (1400 m) and Timberline (3050 m). To illustrate the very different conditions in the gardens, Fig. 8.2 gives climatic details for sites near Stanford and Timberline. Of especial importance are the extremes of temperature and the differences in the length of the growing season. In each garden plants were grown spaced out in weed-free plots protected from grazing. The experimental plantings consisted, in the main, of clone-propagated stocks, each individual being grown and divided, and a ramet of each planted in each garden. Thus the growth and performance of an individual could be studied in a 'Mediterranean', an intermediate and an 'alpine' garden. Climatic ecotypes were studied in many species, particular attention being paid to *Potentilla glandulosa*, a species found from the coastal hills near the west coast of California to high altitudes in the Sierra Nevada. Their experiments made it possible to test the behaviour of diverse stocks in very different standard gardens. For example, they discovered that most lowland stocks died in the harsh climate of the alpine garden, and at the Stanford garden plants originating from high altitude remained winter-dormant under conditions which stimulated growth of lowland samples. Clausen and his associates (1940) decided that there were four distinct climatic ecotypes in *Potentilla glandulosa*, corresponding to the taxa subspecies *typica* (lowland), subspecies *reflexa* and subspecies *hanseni* (intermediate altitudes), and subspecies *nevadensis* (alpine), (Table 8.3). In later writing, e.g. Clausen & Hiesey (1958), it was suggested that each subspecies was in fact made up of two or more ecotypes. The hypothesis that ecotypic variants of *Potentilla* differed genetically received support from a comprehensive series of crossing experiments carried out by Clausen & Hiesey (1958).

Other American botanists made studies of ecotypes using the transplant stations at Stanford, Mather and Timberline. Lawrence (1945), for example, studied ecotypes of *Deschampsia caespitosa*, discovering differences in survival in different stations (Fig. 8.3). Of especial interest were

Table 8.3. *A summary of the characteristics of the ecotypic subspecies of* Potentilla glandulosa *along the Central Californian transect. (Data from Clausen & Hiesey (1958) as summarised by Heslop-Harrison, 1964)*

	typica	*reflexa*	*hanseni*	*nevadensis*
Distribution	Coast ranges and lower Sierra Nevada	Low and middle altitudes of Sierra Nevada	Meadows, midaltitudes of Sierra Nevada	High altitudes of Sierra Nevada
Habitat	Soft chaparral and open woods	Dryish, open timbered slopes	Moist meadows	Moist, sunny slopes
Climatic tolerances as experimentally determined	Coastal to middle altitudes	Coastal to middle altitudes	Middle and high altitudes (poor survival near coast)	Middle and high altitudes (poor survival near coast)
Seasonal periodicity at Stanford (alt. 30 m)	Winter- and summer-active	Winter-active or -dormant; summer-active	Winter-dormant, summer-active	Winter-dormant, summer-active
Internal variation	Wide, probably several 'ecotypes'	Wide, probably several 'ecotypes'	Wide, at least two 'ecotypes'	Moderate, at least two 'ecotypes'
Self-compatibility	Self-fertile	Self-fertile	Undetermined	Self-sterile

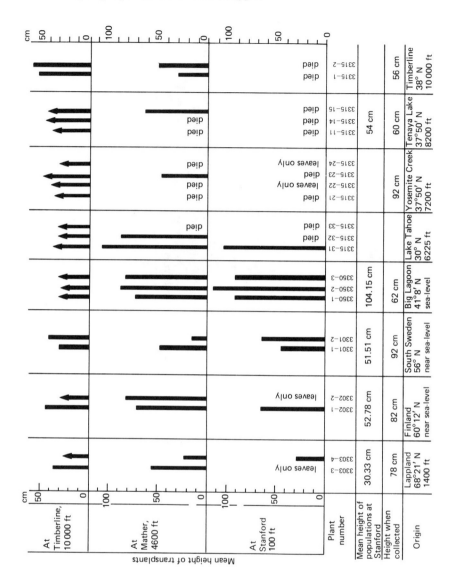

Fig. 8.3. Mean heights of plants of *Deschampsia caespitosa* of diverse origin grown at the three transplant stations Stanford, Mather and Timberline, Central California. (The arrowheads on certain columns at Timberline signify that these individuals did not reach maturity in any year.) (From Lawrence, 1945)

his studies of reproduction in the different transplants; although all individuals survived at Timberline, only the stocks native to that area were able to produce seeds in the short growing season. Such a finding, which is of crucial importance in understanding the genecology of the species, could not have been revealed in a lowland garden. A further point of general interest is revealed by their results with plants of *Deschampsia caespitosa* from Finland (latitude 60°N) and South Sweden (latitude 56°N). When these plants were grown at low altitudes at Stanford (38°N) many of them became viviparous, a character not expressed in their native habitats. Growth in a garden with very different climatic characteristics may provoke an unusual response from the plants.

Experiments with several gardens separated by great distances are expensive to maintain, and botanists have devised ways of investigating ecotypes by varying the conditions in a single garden or laboratory. Turesson's experiments were carried out in a lowland garden on fertile soil and in describing edaphic ecotypes he inferred the importance of soil differences in the wild. A more direct approach to the study of patterns of variation in relation to edaphic factors was made by Kruckeberg (1951, 1954). In one experiment fruits of *Achillea borealis* were collected from serpentine and non-serpentine sites in California. (Serpentine is a rock type which gives rise to soil with high levels of magnesium and low levels of calcium). Two tons each, of a serpentine and a fertile soil, were collected and transported to the University of California Botanical Gardens, and stocks were grown from seed in soil bins, or pots, of the two soil types. Stocks raised from seed of plants native to serpentine soils grew well on the serpentine test soil, but in contrast plants from other soil types (shales, basalt, etc.) generally (though not always) grew badly or died (Fig. 8.4). Kruckeberg's results on *Achillea borealis* and other species are consistent with the idea that a common species found on different soil types may be made up of a number of edaphic ecotypes.

A second example of the way in which diverse stocks may be presented with different environments in one garden or laboratory is provided by the use of glasshouses, growth chambers, etc., in which daylength, temperature and other factors may be varied. Samples may be tested in a variety of artificially controlled environments, in which, for instance, the responses of different stocks may be monitored under different daylengths. In the first experiments studying the effect of different daylengths, plants were grown on movable trucks. After a period of natural daylight plants were moved into light-proof structures where they could be either in total darkness or given supplementary light from artificial sources. A good example of this type of experiment is provided by Larsen (1947), who studied *Andropogon scoparius*, a widespread and important forage grass in North America. Plants were collected from 12 localities from 28°15′N in Texas to 47°10′N in North Dakota. The grasses were given constant daylengths of 13, 14 and 15 hours of light. None of the 12 samples flowered at 13 hours. Plants from the southern USA required a 14-hour photoperiod for floral induction, but a photoperiod of 15 hours was necessary for flowering in many northern plants. Fig. 8.5 illustrates the relation between latitude and daylength at different times of year. *Andropogon* plants growing in the southern USA naturally come into flower after receiving a photoperiod of 14 hours. Northern plants, with longer summer days, need a 15-hour day to come into flower.

As more sophisticated equipment became available, growth chambers

Fig. 8.4. Experiments with *Achillea borealis*, grown on serpentine soil (above) and non-serpentine soil (below). All eight samples grew well on the fertile, non-serpentine soil, whilst three of the four samples from non-serpentine soils (161, 125, 198) grew badly on serpentine soil. The fourth sample from non-serpentine soil (206) grew unexpectedly well on serpentine soil however. (From Kruckeberg, 1951)

were constructed in which many environmental factors, (e.g. temperature, daylength) could be controlled. Adjacent chambers could be used to subject plants to different conditions. A splendid example of such studies is provided by the experiments of Mooney & Billings (1961) who studied *Oxyria digyna* collected from sites between 38°N and 76°N in North America. Other botanists have continued to be fascinated by the different photoperiodic responses of plants from different geographic areas. The work of McMillan (1970, 1971) on *Xanthium strumarium* provides an impressive example. Physiological studies are advancing our understanding of ecotypes, and the reviews of Heslop-Harrison (1964), Hiesey & Milner (1965) and Bannister (1976) may be consulted for further details.

Fig. 8.5. Relation between latitude and daylength at different times of the year. Daylength includes twilight of that intensity receivable when the sun is 6° or less below the horizon, thus adding about 1 h to the daylength between sunrise and sunset. M = Miami, Fla., latitude *c*. 26°N; S = San Francisco, Calif., *c*. 37°N; I = Ithaca, NY, *c*. 42°N; C = southern Canada, 50°N. (From Curtis & Clark, 1950)

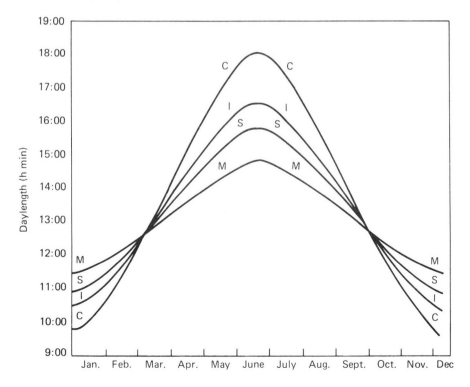

The widespread occurrence of ecotypes

As a result of experiments in which plants have been grown in gardens or under controlled conditions, ecotypes have been described in hundreds of species. There is evidence that ecotypes occur not only in outbreeding species but also in species apparently predominantly inbreeding. There are also numerous studies of facultative apomictic plants in which ecotypic patterns have been described, for example *Poa pratensis* (Smith, Nielsen & Ahlgren, 1946) and *Potentilla gracilis* (Clausen, Keck & Hiesey, 1940).

Of special interest is the finding of genetic heterogeneity in plants which are apparently obligately apomictic. Turesson (1943) discovered, within collections of European *Alchemilla glabra, A. monticola* (*A. pastoralis*) and *A. filicaulis*, that plants from Lappland and montane areas were earlier flowering in cultivation than lowland stocks. The patterns of variation appeared to be ecotypic, but Turesson called the variants 'agamotypes' in recognition of the breeding system of *Alchemilla*. Bradshaw (1963*a, b*, 1964) and Walters (1970) have described dwarf variants of an ecotypic nature in *Alchemilla*, the origin of which is plausibly due to selection in response to grazing by sheep (Fig. 8.6).

Clines

In the experiments outlined above, the researchers were content to describe their material in terms of distinct local races, often using the term 'ecotype'. However, the ecotype concept was not without its critics. Langlet (1934), for example, pointed out that the most important habitat factors, such as temperature and rainfall, commonly varied in a continuous fashion, and thus one would expect graded variation in many widespread species rather than discontinuous variation.

Support for this view was provided by Gregor (1930, 1938) who made an intensive study of *Plantago maritima* in northern Britain. Representative seed collections were made and plants were grown in an experimental garden of the Scottish Society for Research in Plant Breeding. Table 8.4 gives an example of the sort of results obtained by those studies. In this case all three sample zones are from the Forth estuary in eastern Scotland. If collections of *Plantago maritima*, taken from different sites along a gradient from high to low salt concentration, are compared, a progressive increase in scape height is found. In a similar fashion there are increases in scape volume and thickness; in leaf length, breadth and spread; and in seed length. Figure 8.7 illustrates the different growth-habit types found in *Plantago*. As Table 8.4 shows, it is only in the upper marsh that erect plants predominate.

Fig. 8.6. *Alchemilla* plants from closely grazed grassland in the North Pennine Hills, England, after 9 months' cultivation. Top row: *A. minima* from Ingleborough. Bottom row: *A. filicaulis* from Mickle Fell and the Moor House National Nature Reserve. (From Bradshaw, 1964)

Table 8.4. *Results of soil analyses (air-dried samples) and cultivation experiments with* Plantago maritima *(Gregor, 1946)*

Habitat	Mean scape length (cm)	Habit grades (Percentage of sample in each grade)				
		1	2	3	4	5
Waterlogged mud zone (salt concentration 2.5%)	23.0 ± 0.58	74.5	21.6	3.9	—	—
Intermediate habitats with intermediate salt concentrations	38.6 ± 0.57	10.8	20.6	66.7	2.0	—
Fertile coastal meadow above high tide mark (salt concentration 0.25%)	48.9 ± 0.54	—	2.0	61.6	35.4	1.0

Fig. 8.7. Variation in *Plantago maritima*. For purposes of classification, Gregor divided his material into five grades, illustrated diagrammatically here. There was, however, no sharp line of demarcation between one grade and the next. (From Gregor, 1930, 1938)

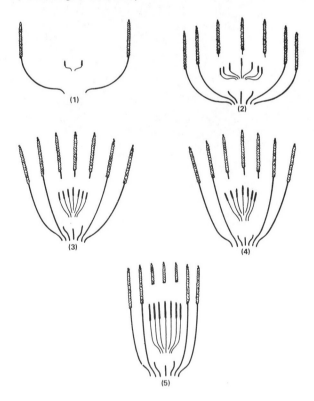

In 1938, Huxley, after surveying the literature, coined the useful term 'cline' for character variations in relation to environmental gradients. Thus a graded pattern associated with ecological gradients is referred to as an ecocline (a good example of this is Gregor's *Plantago maritima* result). If the pattern is correlated with geographical factors, the term topocline can be employed. Clinal variation has been described in a large number of species, and a small selection of examples is given in Table 8.5 and Fig. 8.8.

How far are intraspecific patterns of variation explicable in terms of ecotypes and clines? Experiments, for example, by Bradshaw (1959*a*, *b*, *c*, 1960) on *Agrostis capillaris* (*A. tenuis*), has shown that much more complex patterns may be found in nature. Careful collections of living specimens of this grass were made mostly from localities in Wales. The stocks were grown, and then cloned material was planted into a number of experimental plots in North and Mid-Wales, with an altitudinal range from sea-level to about 800 m. A wide range of different responses was demonstrated by these experiments. Not only were plants different morphologically but there were also physiological differences. For example, certain plants grew well on soils containing lead and other heavy metal residues; others, indistinguishable from them morphologically, died on this type of soil (we shall return to this interesting phenomenon of tolerance of heavy metal ions in Chapter 14). At this point it is important to note that Bradshaw could not delimit ecotypes in *Agrostis capillaris* (*A. tenuis*). This was not because extreme variants were not found in extreme habitats. On the contrary, many very distinctive plants were discovered: for instance, dense cushion plants from the exposed Atlantic cliffs at West Dale, South Wales. The problem was that even though habitat-correlated variation could be demonstrated, the fact that all kinds of intermediate plants were discovered made it utterly impossible to decide where one 'ecotype' ended and another began.

Does the concept of clines help in this situation? Bradshaw studied his material closely with this idea in mind. In many areas, even though clines might be described, he decided that the environmental gradients and the associated variation were too complex.

What then, determines the patterns of intraspecific variation found in the wild? How can one reconcile the distinct ecotypes of Turesson and Clausen with the complex variation found by Bradshaw and other more recent workers (for example Cook, 1962; Warwick & Briggs, 1979)?

Table 8.5. *Some examples of clinal variation*

Species	Variation	Reference
Alnus glutinosa	Increase in leaf and cat-kin size north-west to south-east in Britain	McVean (1953)
Anthoxanthum odoratum	Clines for various characters at mine/pasture boundary	Antonovics & Bradshaw (1970)
Asclepias tuberosa	Clines for flower colour and leaf-shape in North America	Woodson (1964)
Dactylis marina	Increase in papillosity of leaf epidermis from south to north in Portugal	Benson & Borrill (1969)
Eschscholzia californica	Clines in California for various features	Cook (1962)
Eucalyptus spp.	Graded patterns of leaf glaucousness with extreme 'waxy' types in exposed habitats	Thomas & Barber (1974)
Geranium robertianum	Clines for hairiness	Baker (1954)
Geranium sanguineum	Decrease in leaf-lobe breadth west to east in Europe	Bøcher & Lewis (1962)
Holcus lanatus	First-year flowering in South-east Europe. Second-year flowering in northern Europe	Bøcher & Larsen (1958)
Juniperus virginiana	Clines in terpenoid content North-east Texas to Washington D.C.	Flake, von Rudloff & Turner (1969)
Pinus strobus	Decrease in leaf length and number of stomata, increase in number of resin ducts, with increasing latitude in North America	Mergen (1963)
Silene alba and *S. dioica*	Clines for a number of characters across Europe	Prentice (1979)
Ulmus spp.	Increase in leaf breadth west to east in Britain	Melville (1944)
Veronica officinalis	Increase in leaf size northward and eastward in Europe	Bøcher (1944)
Viola riviniana	Clines in plant size	Valentine (1941)

Fig. 8.8. Clinal variation in *Geranium sanguineum*. (Bøcher & Lewis, 1962). At
first sight there seems to be a more or less simple topocline for leaf-lobe width
across Europe, plants from North and West Europe usually having broad-
lobed leaves (leaf index 4 and 5), as in inset A. On the other hand, material
from continental Europe often has narrow leaf-lobes (leaf index 1 and 2), as in
inset B. The distribution map of leaf index values for herbarium material
suggests, however, that the variation is more complex. It seems likely, in view
of the occurrence of broad-lobed plants on the east coast of Sweden and in the
Mediterranean area, that this leaf type is associated with coastal climatic
conditions. Narrow-lobed plants, found in dry limestones of inland Britain and
Sweden, seem to be found wherever continental climatic conditions occur.

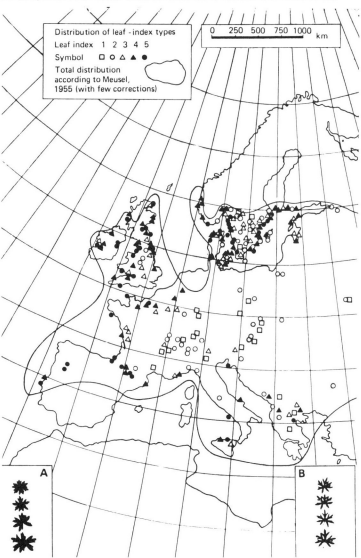

Factors influencing the variation pattern

Of first importance is the type of sampling technique used. Turesson and many other botanists collected widely spaced samples, whereas Gregor & Bradshaw carried out intensive sampling in small areas. Widely spaced samples taken from extreme habitats may exhibit a pattern of distinct 'ecotypes'. Samples taken from along smooth, regular gradients of soil or altitude, in contrast, may well give a pattern of clinal variation in the experimental garden. If, however, sampling is carried out in small areas, the plants being collected at random rather than along particular gradients, then experiment might reveal very complex patterns. Thus, in a very real sense, the mode of sampling largely determines the patterns 'discovered' in cultivation experiments.

Another aspect of sampling is important. An experimenter, in providing himself with material, can choose either to collect a representative seed sample or to dig up mature plants. If both types of sampling are carried out on a single population, different patterns of variation might well be found. This is because mature plants have survived the rigours of selection. Seed collections, on the other hand, give an estimate of potential rather than actual variation. If several adjacent populations in different environments are examined, in a case where pollen can be transported from one population to another, sampling of mature individuals might well reveal a pattern of more or less distinct 'ecotypes'. On the other hand, because of gene flow between populations, seed samples will seem to reveal a more complex pattern in the same case.

Ecological, historical and geographical factors also influence the patterns discovered in experiments. If a species is found as small, non-contiguous populations, or if it has populations inhabiting two or more very different types of habitat, then the pattern of variation in the wild is more likely to be that of distinct 'ecotypes'. In contrast, common species, which throughout their geographical range are more or less continuously distributed over many habitats, will in all probability exhibit complex patterns of continuous variation. Also of prime importance is the breeding system. Small populations of insect-pollinated species often exhibit ecotypic discontinuities, but these are less likely to occur in widespread wind-pollinated species.

Since Turesson's time there has clearly been a change of outlook. Ecotypes are now regarded as nothing more than prominent reference points in an array of less distinct ecotypic populations (Gregor, 1944). In more recent studies, experimenters have been reluctant to designate ecotypes: they have instead carefully recorded the patterns of ecotypic differentiation found in particular experiments (see, for example, Quinn, 1978).

With hindsight one can see in Turesson's own results the possibility that, in common species, variation patterns were more complex than the ecotype concept implied. For instance, where sandy fields and dunes were found as adjacent habitats, a considerable number of intermediate *Hieracium umbellatum* plants were found linking the two ecotypes. Similarly, in *Leontodon autumnalis*, Turesson (1922*b*) found a complex situation where meadows and pastureland ran down to the sea.

9

Species and speciation

The species concept

The investigations of Turesson not only stimulated research into variation between populations but also reactivated an interest in the nature of species and the process of speciation. Studies of ecotypes often included genetic investigations, the most extensive experiments, and justifiably the most famous, being the crosses between ecotypes of *Potentilla glandulosa* which were mentioned in Chapter 8. The genetic basis of ecotypic differentiation was investigated in crosses between alpine and coastal plants, and between plants from subalpine habitats and those from the foothills in California. Further details of plants and the results are given in Table 9.1. In analysing their data, Clausen & Hiesey (1958) used the simplest of models, patterns of F_2 variation being analysed on the basis of pairs of alleles at unlinked loci. This mode of interpretation takes no account of the possibility of linkage or dominance of alleles, however, which could have been studied only with more elaborate crossing experiments. Despite these limitations, Clausen & Hiesey's conclusion that there was evidence of considerable genetic differences between ecotypes seems unassailable. While the genetic differences between, say, coastal and alpine plants did not prevent the formation of fertile hybrids, the differences were of sufficient magnitude to invite consideration of the past and future evolution of the group. First, it seems highly likely that the different ecotypes, which were recognised taxonomically as subspecies, had a common origin in one ancestral stock. Secondly, we might reflect on the possibility that the coastal, foothills, subalpine and alpine ecotypes might become separate species in the course of time. In considering this possibility we must first try to decide what is meant by 'separate species'.

Table 9.1. *Estimate of minimum number of pairs of alleles governing the inheritance of 19 characters in two ecotypic hybrids of* Potentilla glandulosa *(from Clausen & Hiesey, 1958). The majority of the data refers to a cross between a subalpine race of subsp.* nevadensis *from Timberline, 38° 0′N, 119° 15′W, alt. 3050 m, and a foothill race of subsp.* reflexa *from c. 175 km south at Oak Grove, c. 36° 15′N, 118° 50′W, alt. 760 m. From this cross, F_1, F_2 and F_3 plants were grown. The data for characters 6–8, 10, 13, 15 and 17 refer to a cross between an alpine race of subsp.* nevadensis *from Upper Monarch Lake, c. 36° 30′N, 118° 30′W, alt. 3240 m, and a coastal race of subsp.* typica *from c. 250 km southwest near sea-level at Santa Barbara, 34° 25′N, 119° 40′W. This cross was taken only as far as the F_2 generation.*

	Character and action of genes	Estimated number of pairs of alleles at unlinked gene loci
1.	Orientation of petals: 2 erecting, 1 reflexing	3
2.	Petal notch: 1 producing notch, 2 inhibiting	3
3.	Petal colour: 2 whitening, 2 producing yellow, 1 bleaching	5
4.	Petal width: 4 widening, 1 complementary,[a] 1 narrowing	6
5.	Petal length: 4 multiples[b]	c. 4 (plus possible inhibitors)
6.	Sepal length: 3 or 4 multiples for lengthening, 1 for shortening, 1 complementary	c. 5
7.	Achene weight: 5 multiples for increasing, 1 for decreasing	c. 6
8.	Achene colour: 4 multiples of equal effect	4
9.	Branching, angle of	c. 2 (also genes for strict to flexuous branching)
10.	Inflorescence, density of	c. 1 (plus modifiers)
11.	Crown height	c. 3 (also genes for presence or absence of rhizomes and for thickness of rhizomes, to which crown height is related)
12.	Anthocyanin: 4 multiples (1 expressed only at Timberline), 1 complementary	5
13.	Glandular pubescence: 5 multiples, in series of decreasing strength	5
14.	Leaf length: transgressive[c] segregation; many patterns of expression in contrasting environments; possibly different sets of multiples activated	c. 10–20
15.	Leaflet number in bracts	c. 1 (plus modifiers)

Table 9.1. (*cont.*)

Character and action of genes	Estimated number of pairs of alleles at unlinked gene loci
16. Stem length: transgressive segregation, 5–6 multiples plus inhibitory and complementary genes; many patterns of expression in contrasting environments	*c.* 10–20
17. Winter dormancy: 3 multiples of equal effect	3
18. Frost susceptibility: slight transgression toward resistance	*c.* 4
19. Earliness of flowering: strongly transgressive; many patterns of altitudinal expression; possibly different sets of genes activated	many

[a] *Complementary effects:* factor A or B no effect but combination A + B produces phenotypic difference.
[b] *Multiples:* factors with comparable effects.
[c] *Transgressive effects:* in F_2 segregants values for 'extreme' individuals exceed parental values.
Clausen & Hiesey (1958) provides a summary of other studies of the genetics of ecotypic differentiation, for instance: Clausen (1922, 1926, 1951) made crosses between a prostrate perennial coastal variant of *Viola tricolor* and an inland, erect annual variant; and Müntzing (1932) and Bernström (1953) studied the genetics of winter and summer annual races of *Lamium purpureum*.

The word 'species' has different meanings for different botanists. Consider first the species described by taxonomists. Naming, description and classification are based largely upon morphological details of herbarium specimens, supplemented by geographical and sometimes ecological information. The aim of the taxonomist is to provide a convenient general-purpose classification of the material, a classification which will serve the needs of biologists in diverse fields. It is quite obvious that in order to communicate experimental findings to others the experimentalist, like any other botanist, must be able to name his material unambiguously. To this end an International Code of Botanical Nomenclature has been agreed in the present century. The development of this Code has a fascinating history (Smith, 1957). By 1900 four rival codes of practice were employed in different herbaria. Discussions of the problem occupied taxonomic sessions at International Botanical Congresses in Vienna (1905), Cambridge (1930), and Amsterdam (1935), and the successive Congresses, now at approximately 5-yearly intervals, are the occasion for continued revision of the Code. The agreements leading to a unified Code must be recognised as a major achievement.

One meaning of the word 'species' is thus clarified. We may say that species are convenient classificatory units defined by trained biologists using all the information available. Clearly there is a subjective element in their work and we must therefore face the fact that there will sometimes be disagreements between taxonomists about the delimitation of particular species.

During the 1920s and 1930s Turesson, Clausen and other botanists suggested a number of new terms for units at or about the level of species based upon breeding behaviour, an idea with a long history as we saw in Chapter 2. It was, however, the eminent zoologist Mayr who produced the most often quoted definition of what he called 'biological species', a definition which is now found in botanical as well as zoological works. Biological species are (Mayr, 1940): 'groups of actually or potentially interbreeding natural populations which are reproductively isolated from other such groups'. Thus groups of related plants which are distinct at the level of biological species do not interbreed when growing in the same area in nature. They are said to pass the test of sympatry, that is of growing together without losing their identity through hybridisation. The mechanisms which keep biological species separate have been closely studied for many years, and will be examined in detail in Chapters 10–12, but in general, as Table 9.2 shows, isolating factors fall into three groups.

Table 9.2. *A classification of isolating mechanisms in plants (Levin, 1978*b*)*

Premating
Spatial
1. Ecological

Reproductive
2. Temporal divergence
  (a) Seasonal
  (b) Diurnal
3. Floral divergence
  (a) Ethological
  (b) Mechanical

Postmating
4. Reproductive mode
5. Cross-incompatibility
  (a) Pollen–pistil   *Pre-zygotic*
  (b) Seed   *Post-zygotic*
6. Hybrid inviability or weakness
7. Hybrid floral isolation
8. Hybrid sterility
9. Hybrid breakdown

In some cases pollination may be prevented; for instance, biological species found in the same area may grow in slightly different habitats, or flower at different times of day or in different seasons, and/or, for reasons of flower structure or pollinator behaviour, cross-pollination may not be successfully achieved. Even if cross-pollination occurs pollen may fail to grow down the style. It is also possible, if plants are regularly and automatically self-pollinating, that cross-pollination may be prevented or its frequency may be greatly reduced. Experimentalists also recognise a group of so-called 'post-zygotic' mechanisms. The seed from a cross between two biological species may fail to develop properly as a consequence of incompatibility between embryo, endosperm and maternal tissues (Valentine, 1956). The third kind of situation arises when hybrids are produced but show various signs of defective development or reduced fertility in crosses. Hybrids may be viable but sterile, or may be weak as well as sterile. Moreover, any defects in a cross between two putative biological species may be revealed only in F_2 or later generations.

In an important review of isolating mechanisms, Levin (1978*b*) suggests that entities distinct as biological species are not generally separated by only one isolating mechanism and, furthermore, that barriers to crossing are not encountered simultaneously, but may be seen as a series of resistances which have to be overcome if crossing is to be effected. Only if a considerable number of barriers are successfully surmounted will the viability of the hybrid product be put to the test.

The notion of a 'biological species' began to catch the imagination of experimentalists, particularly following the publication of a number of important books on microevolution. The most important were: Dobzhansky (1937), Huxley (1940), Huxley (1942), Mayr (1942), Simpson (1944), Stebbins (1950), and Clausen (1951). From this ferment of discussion the following ideas emerged:

1. As the definition of biological species (and more or less equivalent groupings described in the 1920s and 1930s) involved a test of breeding behaviour, experimentalists considered that entities defined thereby were more objective than the 'species' of the herbarium taxonomist (cf. Gregor, 1931; Müntzing, Tedin & Turesson, 1931; Clausen, Keck & Hiesey, 1939).

2. The species and classifications produced by taxonomists – the so-called 'alpha taxonomy' – should be modified in the light of experiments to give a more perfect system eventually leading to an 'omega taxonomy' in which all the knowledge of biologists reached proper synthesis (Turrill, 1938, 1940).

3. Crossing experiments and other information such as chromo-

some numbers might reveal something of the phylogeny of groups, and the classificatory systems, which are a mixture of convenient arrangement and phylogenetic speculation, could be modified to allow classification to reveal the evolutionary pathways leading to present-day patterns (Darlington, 1956, 1963).

These ideas were attractive to some botanists and provided a stimulus for researches of various kinds. To give the correct historical perspective, however, it is important to state that the ideas were not accepted by all, and a great deal of argument ensued. We will examine the present views of the relation of experimental and taxonomic categories in Chapter 12, after presenting some of the ideas and results produced by the work of experimentalists who came to be known as 'experimental taxonomists' or 'biosystematists' (Camp & Gilly, 1943).

We may now turn our attention to the origins of species. The plural 'origins' is important, as there are a number of modes of speciation which may conveniently be grouped under two heads, namely 'gradual speciation' and 'abrupt speciation'.

Gradual speciation

We have earlier considered what might happen in the future evolution of the *Potentilla glandulosa* group. Could ecotypes evolve to give separate biological species? According to our models of gradual speciation such a possibility exists.

On theoretical grounds, and in the light of experimental evidence from many organisms, population geneticists point to four processes of importance in the evolution of local populations (groups of interbreeding organisms), which we shall call 'gamodemes' (Gilmour & Gregor, 1939).

Mutations. As genetic mutations are the result of random events, it is likely that patterns of genetic change will be different in different gamodemes.

Other effects of chance. Apart from the random effects involved in genetic changes, chance influences population variation in a number of ways. The genetic constitution of the founder members of isolated populations is likely to be very important in their future evolution. As pointed out by Baker (1955, 1967), in a self-compatible species a new population may result from the progeny of a single immigrant. Chance may have a profound effect in restricting the variation in gamodemes, for example, which gametes fuse, which embryos develop fully, which seeds are dispersed to suitable habitats, which adults reach reproductive maturity, the efficiency of pollination. In fact all the stages of the life-cycle are profoundly influenced by chance. As different gametes, different zygotes and different adults may be different genetically, accident can greatly modify the breeding population in the next generation. This is particularly true if the gamodeme is reduced by accident to a very small size. Wright (1931) was the first to point out that alleles might be completely lost by accidents of sampling, or in other cases rare alleles might by chance become more frequent. Such random changes in gene frequency are known as 'genetic drift' or the 'Sewall Wright' effect (Fig. 9.1).

Selection. Different gamodemes grow in different environments and the action of selection is therefore likely to lead to genetic differences between gamodemes.

Migration. The extent and direction of change could be greatly influenced by the degree to which gamodemes are effectively insulated from immigration from other gamodemes. Wholesale breakdown of geographical

Fig. 9.1. Genetic drift: the effect of a severe reduction in numbers of individuals in a population. (*a*) Recovery in numbers: the new population may not differ phenotypically from the original one. (*b*) No recovery in numbers: if from the available seedlings, as an effect of chance only, a small random selection of adults survives at each generation, then the population may become less variable not only in qualitative traits, such as flower colour, but also in the breadth of variation in quantitative characters such as height, etc. While no population biologist disputes the importance of chance effects, it is impossible to isolate such effects from those of natural selection. (*c*) Extinction. Many endangered species illustrate (*b*) or (*c*) (see Chapter 16).

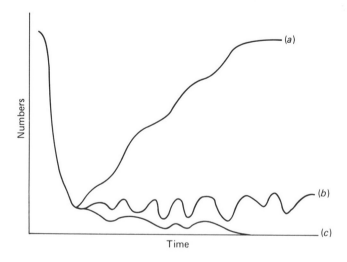

isolation might occur at some point, with important consequences for population variation.

By virtue of the independent changes possible in different gamodemes of a widespread taxon it seems probable that, given a long enough period of geographical and genetic isolation, genetic differentiation in a group of gamodemes of common origin might proceed first through an ecotypic phase and that later, with the gradual evolution of genetic isolating mechanisms, derivatives at the rank of biological species might be formed. As we shall see in Chapter 10, such gradual speciation is thought to be highly important in the evolution of biological species.

Abrupt speciation

In many cases polyploidy is involved in abrupt speciation, but other chromosome changes are possible and likely to be very important. As this chapter is designed to provide an introduction to a more thorough treatment in later chapters, at this point we shall confine our attention to polyploidy.

According to Rieger, Michaelis & Green (1976), the term 'polyploid', for a plant containing more than the normal number (two) of sets of chromosomes, seems to have been first defined by Strasburger (1910). Winkler (1916), in an important paper, describes what is probably the first clear case of experimental production of polyploids in the Tomato (*Lycopersicum esculentum*) and the related Nightshade (*Solanum nigrum*) (Fig. 9.2). Before this, a great deal of interest had centred on the work of De Vries with the Evening Primrose (*Oenothera*) (to which we have already alluded in Chapter 6) and in particular upon a 'mutation' which was called '*gigas*' because it was generally larger than the parent *Oenothera erythrosepala* (*O. lamarckiana*). This '*gigas*' mutant had been shown to possess twice the normal somatic chromosome number of 14, and there was argument between De Vries and Gates as to the significance of this difference, Gates (1909) holding the view that the chromosome doubling was itself a cause of the differences in morphology between '*gigas*' and the normal plant. Winkler's demonstration that his experimentally produced polyploid Tomatoes, with double the normal chromosome complement, differed also from the 'parent' diploid in similar ways to the '*gigas*' mutant strongly supported Gates' interpretation, and subsequent work showed that artificial polyploids generally differed in the larger size of all their parts, from the mean cell size to the size of the whole plant.

Soon after Winkler's paper, Winge (1917) made an important contribution in distinguishing between this kind of polyploidy where, at least in theory, the simple doubling of the chromosome number in a single individual was all that was involved (autopolyploidy), and a more complicated situation where polyploidy succeeded hybridisation (allopolyploidy). (The terms auto- and allopolyploidy were coined by Kihara & Ono in 1926.)

Fig. 9.2. One of the earliest studies of polyploidy was made by Winkler (1916), who investigated this complex phenomenon in experimentally produced chimaeras between different species of *Solanum*. The figure shows one of his drawings of high-polyploid cells side by side with diploid cells in the tissue of the anther wall in one of his experimental chimaeras. (× 2000)

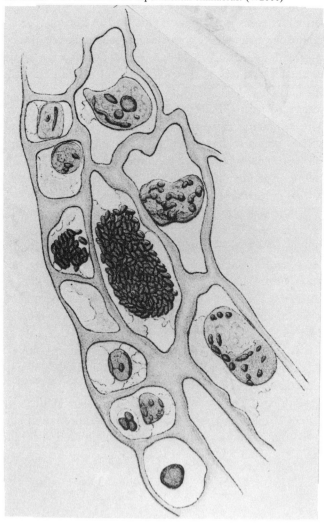

Autopolyploidy can be explained as follows. A diploid plant receives a haploid set of chromosomes (a genome) from each parent. Thus its constitution can be represented as AA. If the plant is subject, for example, to temperature shocks, the regular process of mitosis may be disturbed and instead of two cells each with the diploid number of chromosomes, a single diploid cell with four times the haploid number may be formed (AAAA):

$$\text{AA} \xrightarrow{\text{doubling}} \text{AAAA}$$

In this way polyploid cells arise and may give rise to polyploid branches on diploid plants. Experimentally polyploid cells can be produced with the drug colchicine which acts as a spindle inhibitor preventing regular disjunction of chromosomes. Thus chromosome replication in colchicine-treated material is not combined with the proper division of the products into two daughter nuclei. The formation of a nuclear membrane around all the replicated chromosomes yields a polyploid cell. Seeds from polyploid tissue may give rise to autopolyploid plants.

Let us now consider the origin of allopolyploids. Two related diploid species, which have diverged by gradual speciation from a common ancestor, may be different both chromosomally and genetically, and may be represented as AA and BB. A hybrid between the two species, AB, may very well be highly infertile, as there is insufficient homology between the A and B genomes for proper pairing at meiosis. Often, instead of unbalanced haploid meiotic products, a very small but significant percentage of unreduced AB gametes may be produced, which on fusion can give a plant with the constitution AABB in which the chromosome number has been effectively doubled:

$$\begin{array}{c} \text{AA} \\ \times \longrightarrow \text{AB} \longrightarrow \text{AABB} \\ \text{(hybrid)} \\ \text{BB} \end{array}$$

In this simple case we are dealing with a tetraploid with twice the normal diploid number. Such plants are sometimes referred to as 'amphidiploids' or 'amphiploids'. Other kinds of polyploids with extra genome sets are described with the appropriate term – 'triploid', 'hexaploid', etc. The level of 'ploidy' can be represented as the multiple of the 'basic number' x, which is the haploid number of the presumed original diploid or diploids; thus a triploid can be represented by $3x$, a tetraploid by $4x$, etc. (If this notation is used, it is then possible to retain n and $2n$ to indicate the functional haploid and diploid numbers, as distinct from the presumed polyploid relationships within a whole genus or group of species.)

Meiosis in the new allopolyploid is more normal than in the diploid hybrid, as genomic pairing – A with A and B with B – can occur. If, however, there is still a high degree of homology between A and B genomes (they were derived from a common ancestor by gradual speciation), then more complex pairing of the chromosomes may occur and groups of three and four chromosomes may be found (Fig. 9.3).

Polyploid derivatives are reproductively isolated from their parents, as can be seen by examining what happens when an allopolyploid AABB (with gametes AB) is crossed with one of its 'parental' species, AA, (with gametes A). Triploid individuals of constitution AAB are produced. Even though A genomes may pair at meiosis, there is no pairing partner for the B genome, and highly irregular meiosis occurs which leads to infertility in the hybrid. An isolating mechanism now exists between diploids and their

Fig. 9.3. (*a*) Meiosis (metaphase I) in autotetraploid watercress ($2n = 4x = 64$) prepared by colchicine from *Nasturtium officinale* ($2n = 2x = 32$). ($\times 1600$). (*b*) Meiosis (metaphase I) in *Primula kewensis* ($2n = 4x = 36$). Note the three quadrivalents. ($\times 2500$). (*c*) Meiosis (metaphase I) in wild tetraploid watercress (*N. microphyllum*) ($2n = 4x = 64$). ($\times 1600$). ((*a*) and (*c*) from Manton, 1950; (*b*) from Upcott, 1940.) In tetraploids, a range of different cytological behaviour is found. In autotetraploidy (of type AAAA), quadrivalents are frequently found, as in (*a*). In allotetraploids (of type AABB), where each chromosome has a pairing partner, normal bivalent pairing is found as in (*c*). Sometimes a mixture of quadrivalents and bivalents is discovered as in (*b*). These complex situations are discussed in Chapter 11.

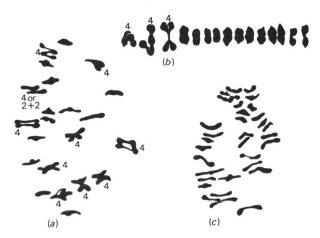

derived polyploid (Fig. 9.4). It is by the abrupt origin of an isolating mechanism in this way that new biological species arise by the process of polyploidy. In contrast to gradual speciation, the new groups may originate within a single gamodeme, that is, sympatrically. Moreover, as they are produced by single, abrupt events, we have here a mechanism whereby, within a man's lifetime, new, fertile species may arise – species satisfying all the criteria of morphological difference and of reproductive isolation.

As we saw in Chapter 2, one of the controversies stimulated by Darwin's work concerned the possibility of 'saltations' or abrupt changes in the course of evolution. Darwin himself saw evolution as a continuous, gradual process, with no place for sudden events, but those who argued for 'saltations' have been vindicated by the cytogenetic research of the present century, at least to the extent that abrupt evolutionary change is now seen

Fig. 9.4. Meiosis in a triploid hybrid. Note the mixture of bivalents (black) and univalents (white) at metaphase I of meiosis in the triploid. (From Manton, 1950) (× 1600)

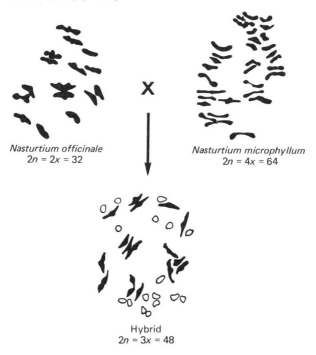

Nasturtium officinale
2*n* = 2*x* = 32

X

Nasturtium microphyllum
2*n* = 4*x* = 64

Hybrid
2*n* = 3*x* = 48

to be entirely possible. In the chapters that follow we shall examine in more detail these two contrasting modes of speciation, and try to assess their relative importance in evolution as a whole.

10

Gradual speciation and hybridisation

In Chapter 9 we presented a simple model of gradual speciation. Two gamodemes derived from a common ancestor and occupying different geographical areas (i.e. allopatric) pass through an ecotypic phase, followed by a period of independent change yielding derivatives of different morphology, which are reproductively isolated from each other. In such cases the existence of isolating mechanisms is revealed if the taxa come to occupy the same area (i.e. become sympatric). To take account of the complexities of different situations likely to be important in nature, however, this simple model must be modified to yield a group of models incorporating a range of different assumptions (Grant, 1971). For example, models may differ in the relative importance they attach to the effects of selective processes on the one hand, and the influence of chance on the other. Thus, one or a small number of individuals may be involved in the establishment of a new gamodeme by long-range dispersal to 'islands' of different sorts, whether they be oceanic islands, isolated mountain peaks, or landlocked lakes. In other model systems, geographical isolation of daughter gamodemes may be achieved by the destruction of land bridges (e.g. the opening of the Irish and North Seas following post-glacial sea-level changes) or by mountain-building processes or continental drift. Such changes may or may not involve severe reductions in numbers at some period in the independent evolution of daughter gamodemes. Models may also be constructed in which geographical isolation is not necessarily totally maintained throughout the evolution of daughter gamodemes. The possibility of rare or even frequent migration between daughter gamodemes could be incorporated. In constructing model systems different assumptions might be made about differentiation. Thus, morphological changes in daughter gamodemes might proceed along with the sort of genetic changes which yield eventual isolating mechanisms or, alternatively, morphological change and 'reproductive isolation' may evolve at different rates. Many different hypothetical systems of gradual speciation might be constructed on the basis of a different balance of factors.

Evidence for gradual speciation

In considering the evidence for various models of gradual speciation, the extended timescale of hundreds of generations presents an immediate difficulty. It seems that the details of the processes of change from a single ancestral gamodeme to two biological species must remain unknown. However, it has been suggested (for example, by Clausen, 1951) that, if groups of different taxa are examined, they may be at different stages in speciation; thus, pairs of gamodemes may be at the ecotypic stage, as, for example, in *Potentilla glandulosa*, whilst others, perhaps given subspecific or specific rank by taxonomists, may be at a later stage in speciation. By examining a range of different types of situation, from ecotypes to island endemics, from local races to vicariads (two similar taxa occupying different geographical areas), a composite picture of gradual speciation might be built up.

To test the ideas of Clausen, it is necessary to discover the degree of reproductive isolation between collections of different taxa – taxa which, on account of their distribution, morphology, etc., are likely to share, either closely or remotely, a common ancestor. Material from different areas is tested under conditions of 'artificial' sympatry by appropriate crossing experiments, followed by studies of the viability and fertility of any resulting hybrids.

In some early studies there are references to interesting results in crossing experiments. For example, Müntzing (1929) demonstrated partial sterility in populations of *Galeopsis tetrahit*, and Babcock (1947), in his classic account of the genus *Crepis*, recorded c. 50% sterility between plants of *C. capillaris* from South Europe and Denmark. More recently botanists have carried out very extensive and elaborate crossing programmes, the results of which are set out in crossing polygons, in which the fertility of F_1 (and sometimes F_2) hybrids is revealed in diagrammatic fashion as a geometric figure. It will be helpful to examine a series of such experiments.

Crosses between populations falling within a single taxonomic species

Clausen (1951) and associates studied the results of crossing different subspecies of *Layia glandulosa*. Evidence of partial genetic barriers is provided by some of these crosses (Fig. 10.1).

Fig. 10.1. Diagram of intraspecific crossings within *Layia glandulosa*.

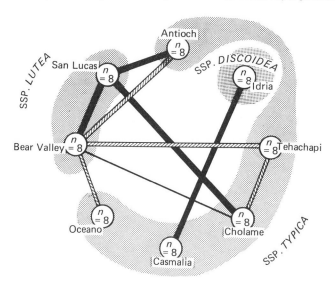

F₁ fully fertile, F₂ vigorous
F₁ fully fertile, F₂ reduced in vigour
F₁ partially sterile, F₂ reduced in vigour
Limits of subspecies

Fig. 10.2. Crossing polygon indicating pollen fertility in F_1 hybrids between the 12 taxa of the *Nigella arvensis* complex recognised in the Aegean area. (From Strid, 1970)

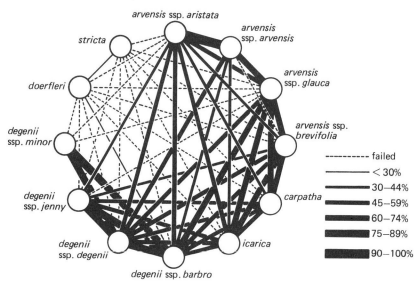

In studying the taxonomically complex species *Nigella arvensis*, Strid (1970) examined the crossing behaviour of 12 taxa drawn from different islands in the Aegean Sea (Fig. 10.2). In this case a complex pattern was revealed. Some taxa (e.g. *N. stricta*) produced hybrids of low fertility in crosses, whilst crosses between other taxa, quite distinct morphologically, were apparently fully fertile. Strid considers that founder effects and genetic drift are highly important in the differentiation of island populations.

Several studies have been made of groups which do not appear to be complex taxonomically. For instance, in studies of the American grass

Fig. 10.3. A summary of hybridisation between populations of *Elymus glaucus* collected along a 75-mile transect in the Sierra Nevada. The diagram shows developmental behaviour and pollen fertility of the hybrids. Snyder concludes from the cytological behaviour of the hybrids at meiosis that much of the sterility is caused by small structural differences in the chromosomes and by specific genes. He also suggests that hybridisation in nature between *Elymus glaucus* and species of the related genera *Agropyron*, *Hordeum* and *Sitanion* might be responsible for much of the variability. (From Snyder, 1950, 1951)

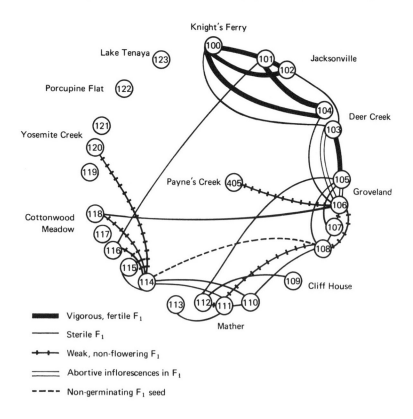

Elymus glaucus, Snyder (1950, 1951) found that, of the F_1 hybrids he was able to produce, the majority had low pollen and seed fertility (Fig. 10.3).

Crosses between and within taxonomic species

The experiments of Vickery (1964) on the *Mimulus* group provide an informative example of crosses between different taxonomic species (Fig. 10.4). Some interspecific hybrids proved relatively infertile, whilst other F_1 hybrids were fully fertile. Some intraspecific crosses included in these experiments gave infertile hybrids, e.g. within *M. laciniatus* and *M. nasutus*. Paradoxically, some interspecific crosses between *M. laciniatus* and *M. nasutus* proved fully fertile.

Fig. 10.4. Crossing polygon indicating fertility of F_1 hybrids between different populations of the *Mimulus guttatus* complex in North America. (From Vickery, 1964)

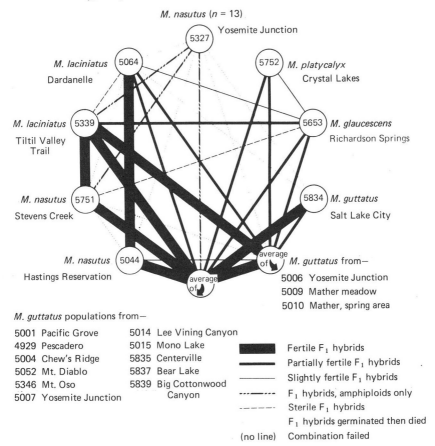

M. nasutus (n = 13)
5327 Yosemite Junction
M. laciniatus 5064 Dardanelle
5752 M. platycalyx Crystal Lakes
M. laciniatus 5339 Tiltil Valley Trail
5653 M. glaucescens Richardson Springs
M. nasutus 5751 Stevens Creek
5834 M. guttatus Salt Lake City
M. nasutus 5044 Hastings Reservation
average of
average of
M. guttatus from—
5006 Yosemite Junction
5009 Mather meadow
5010 Mather, spring area

M. guttatus populations from—

5001 Pacific Grove	5014 Lee Vining Canyon
4929 Pescadero	5015 Mono Lake
5004 Chew's Ridge	5835 Centerville
5052 Mt. Diablo	5837 Bear Lake
5346 Mt. Oso	5839 Big Cottonwood
5007 Yosemite Junction	Canyon

Fertile F_1 hybrids
Partially fertile F_1 hybrids
Slightly fertile F_1 hybrids
F_1 hybrids, amphiploids only
Sterile F_1 hybrids
F_1 hybrids germinated then died
(no line) Combination failed

While crossing experiments of this type provide valuable information on speciation, they require cautious interpretation. First and foremost, only small numbers of plants are used in many crossing programmes and, in the extreme case, a *single* plant is taken as a representative of a population! Crossing experiments are often halted at the F_1 stage, as to raise F_2 families a large amount of garden space may be needed. There are other grounds for caution. Glasshouses are frequently used in breeding experiments. Insect pests and fungal diseases may suddenly reach epidemic proportions during a series of crossing experiments, and success or failure of crossing may be influenced by such pests and diseases. The effects of bagging flowers must also be considered. Geiger (1965) reviews the relevant literature, which indicates that temperature inside pollen/insect-proof bags might be up to 15°C higher than ambient temperatures in the daytime and 1–2°C lower at night. As pollen sterility and other effects may be induced at high temperatures, failure in crossing may be due to these external factors rather than to intrinsic differences. Furthermore, pollen 'fertilities' are commonly estimated, not by direct study of germinability or in a crossing test, but by staining with acetocarmine or other stain. Fully formed grains with nuclear staining are assessed as 'good' pollen, and mis-shapen, undersized, inadequately stained grains are judged to be 'bad'. The reliability of staining as an indicator of fertility is rarely, if ever, put to the test. We discuss pollen staining tests again, in a slightly different context, below. In assessing crosses, the possible complicating factor of genetic incompatibility is often forgotten. Also artificial crossing experiments do not often assess possible prezygotic isolating factors, but concentrate on the results of experiments which often involve crude surgery. Will such experiments reveal what would happen if isolated populations became sympatric in nature and only the normal agencies of pollen transfer were to operate?

A more fundamental problem must also be faced. There will always be uncertainty about the magnitude and sequence of past events. Founder effects, mutations and chromosome and other changes are likely in the history of populations. In reality gradual speciation is dependent upon abrupt events. If these events have small effects and many such changes are 'needed' for speciation, then models of gradual speciation may have some validity. It is clear, however, that the division of speciation into gradual and abrupt modes may be difficult, especially if the experimenter does not study his material cytologically. Such studies might provide clues as to the possibility of significant chromosome differences between isolated gamodemes, differences likely to have had an abrupt origin. We will discuss abrupt speciation further in Chapter 11.

Our brief discussions have revealed many difficulties in the study of gradual speciation. Because of the uncertainties of past history, the formidable problems of studying the crossing behaviour of plants from different populations and the technical difficulties of unravelling chromosome relationships as revealed in meiotic studies of hybrids, gradual speciation as a phenomenon is very little understood. Reflecting on the idea of Clausen (1951) of producing a composite picture of gradual speciation, one is forced to the conclusion that the 'ecotypic phase' is highly complex and that no simple patterns emerge in the study of 'higher groupings' in different experiments. Each group has its own special history, and crossing polygons do not reveal any consistent picture linking the degree of morphological differentiation with the crossing behaviour. Taxonomists studying the different groups have named various subspecies or species according to the degree of morphological distinctness exhibited by plants. In some cases there is a suggestion that different entities recognised as taxonomic species are reproductively isolated from other such groups, but it is abundantly clear that there is no necessary correlation between the presence of sterility barriers and morphological differences. Thus in *Elymus glaucus*, which taxonomists treat as a single species, there would appear to be sterility barriers between different collections. On the other hand, certain taxonomic species of *Mimulus* appear to be fully interfertile. Such findings provide clear evidence that morphological factors and sterility barriers do not necessarily evolve in step.

Natural hybridisation

So far we have been examining artificial hybridisations. Experimental investigation of natural hybridisation between different taxonomic species has also contributed to our understanding of speciation. The genus *Geum*, widespread in the temperate regions of the world, will serve to illustrate a number of important points.

The two most widespread European species are *Geum rivale* and *G. urbanum*. The latter, which has a somewhat 'weedy' tendency to which we shall refer later, has also become widely naturalised in North America. As Fig. 10.5 shows, the flowers of these two species are very different in general appearance, and one of them is clearly adapted to a particular kind of visiting insect. *Geum rivale* is a typical 'bee' flower and species of Humble Bee (*Bombus*) are recorded as the commonest visitors. The purplish colour and the somewhat concealed entrance to the hanging flower are features shown by many 'bee' flowers. Contrast with this the smaller, open, erect, yellow flower of *Geum urbanum*, which shows no specialisation for the visits of particular insects and seems to be frequently self-pollinated.

Over much of Europe these two species are sympatric (Fig. 10.6). They are, however, usually effectively separated by ecological differences. *Geum rivale* often lives in damp shady places in southern and central Europe, and is more or less confined to the upland regions. It is absent from much of the Mediterranean region, but it occurs in Iceland, from which *G. urbanum* is absent. Over most of lowland Europe, however, *G. urbanum* is the common plant, growing particularly in hedgerow and woodland communities affected by man. It has long been known that plants with somewhat intermediate characters sometimes occur in abundance in woods and scrub where the two species meet; such obviously hybrid plants were called *Geum intermedium* by Ehrhart as early as 1791. These hybrids have attracted much attention, mainly because the two parent species look so different, and in some places hybrid swarms are found in which there is a remarkable range of variation. Such a hybrid population formed the basis of a detailed genetic study by Marsden-Jones (1930), who was able to work out to some extent the inheritance along Mendelian lines of several of the characters determining the differences between the two species. This work was followed, and greatly enlarged, by the Polish

Fig. 10.5. (*a*) *Geum rivale*. (*b*) *G. urbanum*. (From Roles, 1960)

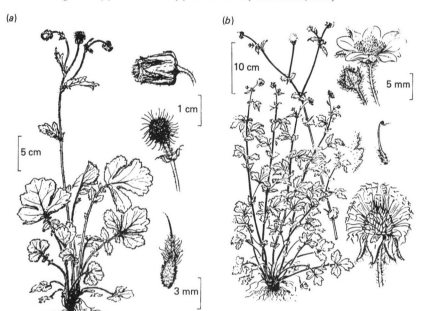

botanist Gajewski, whose study of the genus *Geum*, published in 1957, is the fruit of many years' experimental study and is of outstanding value.

The situation described in detail by Marsden-Jones for a wet wood of Alder (*Alnus glutinosa*) at Bradfield, Berkshire, could be paralleled in a good many places in lowland England. On the other hand, Gajewski, who studied *Geum* mainly in Poland, emphasises that large hybrid populations are rare, and that even where the two species are growing near together, there are often very few intermediate plants. What is the cause of this apparent difference in efficiency of isolation between England and Poland? To answer this we must look at the field and experimental evidence together.

The first relevant point is that artificial F_1 hybrids between the two species can be made, though not with ease, and that such plants are highly fertile, the F_2 showing a range of segregates as might be expected if (as Marsden-Jones' detailed genetic experiments bore out) many genes are involved in the specific differences. Judged purely in terms of the theoretically possible gene flow, therefore, *Geum rivale* and *G. urbanum* would fall within a single biological species. As a matter of fact, Gajewski's work demonstrated that all 25 species in the subgenus *Geum* (to which our two species belong) will hybridise with each other, and that most of these hybrids are at least partially fertile. In practice, however, a

Fig. 10.6. Map showing the distribution in Eurasia of *Geum rivale* and *G. urbanum*. (From Gajewski, 1957)

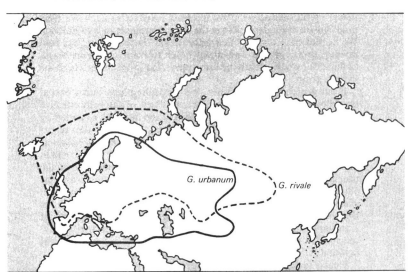

good many species-pairs or species-groups are allopatric and hybridisation does not take place in nature (Fig. 10.7).

We have already seen that ecological preference will normally separate, at least partially, mixed populations of the two species. Moreover, the difference in type and colour of flower would ensure a segregation of insect pollinators. To this difference should be added a rather obvious difference in the times of the beginning of flowering, which in Britain differ by 3 or 4 weeks, *G. rivale* being the earlier. This would mean that, even in mixed populations, seed set on the early flowers would be necessarily 'pure' in the case of *G. rivale*. Such considerations point the way to at least a tentative answer to our question. Gajewski records that he grew seeds taken from plants of each species growing in a mixed stand in a Polish locality where hybrid plants were rare and found that the

Fig. 10.7. Fertility in F₁ hybrids among different hexaploid *Geum* species. Thick lines indicate fertile hybrids; double lines partially fertile hybrids; thin lines sterile hybrids. (From Gajewski, 1957)

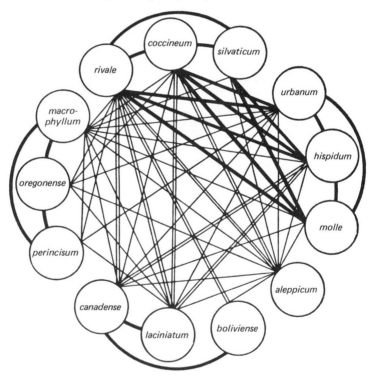

progeny were 'pure' with no detectable sign of hybridisation. Clearly factors such as the preferences of insect visitors, coupled with differences in flowering-time, effectively prevent more than a minimum of gene flow between the species in Poland. What is different about the English conditions? No complete answer can be given but several tentative ones come to mind. The most important difference lies probably in the complex history of man's interference with the vegetation. Gajewski's observations were made, partly at any rate, in the great forest nature reserve of Białowieza, in eastern Poland, where the forest is as little affected by human activity as anywhere in lowland Europe. Here such hybrids as he recorded were single individuals on roadsides and in forest rides, where *Geum urbanum* probably owes its existence to its having accompanied man into these new habitats. *Geum rivale* behaves as the original, native species. In England, on the other hand, most woodland is more or less obviously artificial and the disturbed marginal habitats suitable for *Geum urbanum* have clearly been enormously extended over the centuries by the human activities of drainage, forest clearance, hedgerow planting, etc.

Although this difference in vegetation history may be the most important factor in determining the local frequency of hybrid *Geum* populations, we should bear in mind the possibility that the other isolating factors may also be less effective in some circumstances than others. Is it possible, for example, that the separation in flowering time between the two species is less effective in the relatively mild climate of England than in the more continental one of Poland? We do not have any detailed information on this point, but field investigations of this and many other relevant questions about pollination would clearly be very interesting in the areas where the hybrids grow.

If our general thesis is correct, we are dealing here with a partial breakdown, brought about by man's activities, of naturally effective ecological isolation. This kind of explanation has been extended by Gajewski, admittedly more speculatively, to cover the recent evolutionary history of both species. He pictures *Geum urbanum* as originally evolved in geographical isolation from *Geum rivale*, perhaps in South-east Europe. Certain adaptations, among them the unspecialised type of pollination (and often self-pollination) and the efficient, small, animal-dispersed fruits, made *Geum urbanum* an effective 'weed' of marginal woodland habitats created by man. In this way, the species became sympatric with *G. rivale* over much of Europe. In this new situation, the advantage lies with the 'weedy' species, for most vegetational change brought about by man will favour it rather than its relative.

We have dealt at some length with the *Geum* example, because it provides an excellent paradigm of what we take to be a general phenomenon. Man's activities in the exploitation and management of various areas have led to a variable degree of breakdown of ecological isolation. In areas long settled by man there is hardly a square inch which has not in some way or other been influenced by human activity. Burning, forest clearance, drainage, grazing and other agricultural practices, mining, and all aspects of urban industrialisation from road building to atmospheric pollution – all these activities have contributed to changes in vegetation. Man has 'hybridised the habitats' in the famous phrase of Anderson (1949). Thus the acreage of natural forests, wetlands, mountain communities and coastal vegetation types has gradually diminished. 'Islands' of apparently unchanged vegetation sometimes survive on land less fitted for some form of agriculture or other development, but on inspection these too have been exploited by man for food, fuel (peat, wood), building material (wood, thatching material), sport or in other ways.

Man's activities result not only in complex patterns of breakdown of ecological isolation, but also in changes of geographical isolation. Evidence suggests that by man's deliberate or accidental transport of plants, the 'natural' distribution of some plants has been greatly changed. In particular, weed species have been carried across the world, and plants of horticultural interest and agricultural importance have been widely disseminated.

Thus, in considering the course of gradual speciation, a further complication is beginning to emerge. In some areas man's activities are likely to have influenced, perhaps decisively, not only the past and present distributions of plants, but also the population variation at different sites. Therefore our models of gradual speciation must take account of the fact that in relatively recent times (geologically speaking) man has become the dominant factor in landscape management, greatly influencing patterns of distribution in many areas of the world. The breakdown of geographical and ecological isolation may lead, as in the *Geum* example, to hybridisation, and the consequences of such hybridisation are likely to differ in different circumstances. What consequences might we expect in various circumstances?

The consequences of hybridisation: some theoretical considerations

Consider the case of two gamodemes, A and B, which have become sympatric after a period of allopatry. The gamodemes may have changed genetically in isolation to such an extent that in sympatry they are unable to cross freely, hybridisation being a rare event leading to infertile products (pre- and post-zygotic factors may be important). The two gamodemes are in effect behaving as two biological species. It may be, on the other hand, that the two gamodemes have come to differ in morphology and ecological requirements, but crossing experiments reveal only partial, or no, major barriers to interbreeding. As daughter gamodemes A and B differ in ecological preferences, the fate of any hybrids produced in nature is likely to be influenced decisively by ecological factors.

If no intermediate habitats are found between those preferred by taxa A and B, then hybrids between A and B are likely to be at a selective disadvantage. In comparison with crosses A × A and B × B the hybrids derived from the crosses A × B and B × A may yield fewer (or no) viable offspring. Selection, in favouring the progeny of A × A and B × B, will act against the products of hybridisation. Any partial isolating mechanism minimising A × B and B × A will be subject to selection. The perfecting of partial isolating mechanisms under the impact of selection may then take place, such a change being referred to as the 'Wallace effect' (Grant, 1966) after Alfred Russel Wallace, the co-founder with Darwin of the theory of evolution by natural selection, who speculated on this subject.

If habitats intermediate between those preferred by A and B occur, then in such habitats hybrids between A and B may be 'fitter' (i.e. contribute a greater number of offspring to future generations) than the progeny of A × A and B × B. Any partial isolating factors developed in the once allopatric gamodemes are likely to be overcome and the distinctness of the two gamodemes may be lost, locally or regionally, in a mass of hybrids and backcrosses.

These two models, which are in reality extremes of a spectrum of possibilities, will now be examined in further detail. It is clearly very difficult to study natural populations to discover the validity, or otherwise, of the Wallace effect, but there is a certain amount of circumstantial evidence which supports the model.

The first line of evidence comes from studies of 'experimental' populations. In the Fruit Fly *Drosophila*, studies have shown that intensification of reproductive isolation may result from artificial selection (e.g. Koopman, 1950; Knight, Robertson & Waddington, 1956). A most elegant botanical example was studied by Paterniani (1969), who grew two varieties of Maize (*Zea mays*) together in an experimental garden. To explain Paterniani's experiment some details of the stocks used are necessary. One variety had white, flint cob characters (genetically, *yy SuSu*): the other had yellow, sweet cob characters (*YY susu*). Thus each stock had one dominant and one recessive marker gene. Pollen of the white flint stock had the genotype *y Su*, while pollen of yellow sweet had the genotype *Y su*. Any cross between the two stocks would yield kernels with yellow colour in the white flint parent and flint (often called starchy) kernels in the yellow sweet parent. Because these markers give pronounced effects which are visible in the developing cob, it is possible to determine the degree of hybridisation between varieties without raising progeny, enabling the experiment to be carried out in a relatively short time. Using the experimental stocks, planted in such a way as to maximise the opportunity for hybridisation between varieties, Paterniani investi-

Fig. 10.8. Histograms showing the numbers of days to flowering for the original stocks of two varieties of Maize (*Zea*), and for the stocks after the selection experiment had run for six generations. Maize has separate male and female flowers in the same plant. Hence the separate recording of tassel (♂) and ear (♀). Details in text. (From Paterniani, 1969)

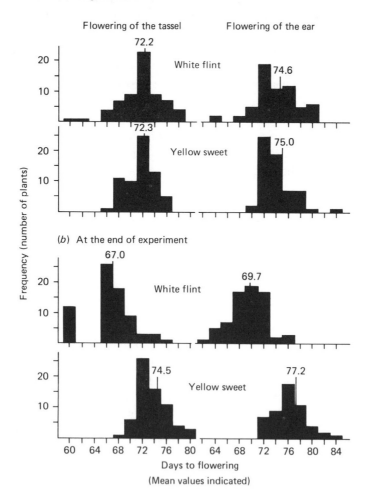

(a) Original stocks

(b) At the end of experiment

Days to flowering
(Mean values indicated)

gated what happened when selection for reproductive isolation was carried out. This was achieved by selecting at each of a number of generations the cobs which showed the *least* hybridisation. From such cobs grains of phenotype white flint and yellow sweet were used to provide seedlings for the next generation. At the start of the experiment the degrees of outcrossing were 35.8% and 46.7% for white flint and yellow sweet stocks, respectively. After six generations of selection the levels of intercrossing were 4.9% for white flint and 3.4% for yellow sweet. The factors involved in this isolation were examined and it was discovered that the number of days from sowing to flowering was probably the most important factor. The original stocks flowered on average at the same time, but by the end of the experiment the mean flowering time for white flint was 5 days earlier, while the mean for yellow sweet was 2 days later. By selecting against hybridisation the mean flowering times of the two stocks had been separated by about 7 days (Fig. 10.8).

This experiment suggests that similar effects on flowering time might be expected in wild populations, where hybrids are at a selective disadvantage. One such example, involving flowering-time differences in grasses found on mine spoil and adjacent pasture, has been published and will be examined in a different context in a later chapter.

The second line of evidence is provided by artificial crossing experiments. If isolating mechanisms are initiated in allopatric gamodemes but perfected in sympatric situations, then it follows that crossing within groups of sympatric and allopatric taxa might be revealing. Such studies have been carried out by Grant (1966), who studied nine species of *Gilia* (Fig. 10.9). Five of the species were sympatric, having overlapping distributions in the foothills and valleys of West California. The other four species were allopatric in maritime habitats in North and South America. As Table 10.1 shows, crossing between sympatric taxa yields fewer seeds per flower than that between allopatric taxa. A similar study of other *Gilia* species gave the same sort of result. However, crosses between the *Gilia splendens* and *G. australis* groups revealed the opposite pattern, with sympatric populations giving greater seed yield when crossed *inter se* than allopatric populations similarly crossed. Thus two of the three situations examined experimentally provided some support for a Wallace effect.

Population studies of wild plants provide a third area of evidence concerning the Wallace effect. A modified population containing *Phlox glaberrima* (mean pollen size 55 µm) and *Phlox pilosa* (mean pollen size 30 µm) was studied by Levin & Kerster (1967). In order to investigate the

Table 10.1. *Comparative crossability of species with different geographical relations in the Leafy-stemmed Gilias (Grant, 1966)*

Geographical relation of parental species	No. of combinations of parental species	Mean number of seeds per flower		Mean of means
Foothill species inter se (*sympatric*)	9	0.0	0.1	0.2
		0.0	0.1	
		0.0	0.4	
		0.0	1.2	
		0.0		
Maritime species inter se (*allopatric*)	5	7.7		18.1
		16.7		
		19.6		
		21.9		
		24.8		

Fig. 10.9. Geographical distribution of the nine species of *Gilia* studied by Grant in America.

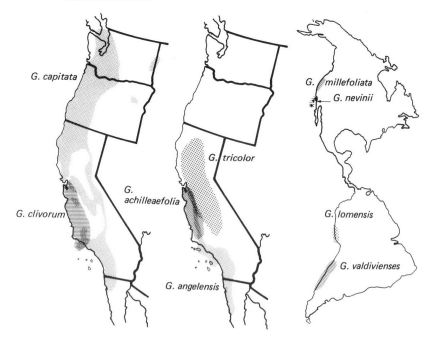

effect of flower colour on interspecific pollinations made by Lepidoptera, transplants of red-flowered variants of *P. pilosa* were added to the naturally occurring white-flowered variants of this species at a site in Cook County, Illinois. Thus two colour variants of *P. pilosa* were available for interspecific pollinations with the red-flowered *P. glaberrima*. Evidence of the grains on a sample of *P. pilosa* stigmas suggested that while 30% of the red flowers of *P. pilosa* received alien pollen, only 12% of the white ones bore such pollen. It would appear that the flower colour differences in the unmodified population of *Phlox glaberrima* (red) and *P. pilosa* (white) might act as aids to pollinator discrimination and thus reduce interspecific pollinations. It is possible to devise models of selection for reproductive isolation between two daughter gamodemes which have come to differ in flower colour, one gamodeme (A) being polymorphic for flower colour, while gamodeme B has only one colour phase, the flower colour of B being one of the colours possible in A. If pollinator activities result in cross-pollination, A × B, between similar colour variants and this leads to less fit progeny, then the balance of colour variation in gamodeme A could change. Ultimately A and B in sympatric situations could come to be characterised by different flower colour. While these cases provide some support for the Wallace effect, many more studies of wild populations are obviously needed.

Introgression and other patterns of hybridisation

We can now turn our attention to what happens where hybrids are at a selective advantage. In 1949 the American geneticist Anderson published a book entitled *Introgressive hybridisation* in which he postulated that in some cases when species hybridise and the environment has been disturbed naturally or by man's activities, gradual infiltration of the germplasm of one species into another might occur, as a consequence of hybridisation and repeated backcrossing. This process, for which the shorter term 'introgression' is now used, Anderson claimed was much more widespread and important in evolution than had been previously thought (Fig. 10.10). Anderson invented several simple methods of displaying variation in hybridising populations; his hybrid index method is illustrated in Fig. 10.11. By devising a suitable scale of numerical values an investigator can make a rapid survey of field collections. In the interpretation of complex populations, interspecific hybridisation is often assumed to be the cause of the pattern. But it would seem better to see the 'hybrid index' method as a means of describing the degree of separation of plants of different morphology, whether or not interspecific hybridisation is actually involved. Furthermore, it is essential to have some clear idea of the variation to be expected in the pure species. To this end samples should be collected from sites where the two 'pure' parents are not in contact with each other. (We may note in parenthesis that the collection of 'pure' parents is often a highly subjective and difficult task, especially in cases where the 'parental' taxa are broadly sympatric.) The hybrid index method yields a scale of variation, with pure parental colonies gaining the highest and lowest scores, and plants of intermediate morphology having intermediate scores.

The method has the disadvantage that the variation of individual plants is effectively 'lost' and in the display of the results, plants with the same hybrid index score may differ phenotypically. Anderson's 'pictorialised scatter diagram' technique, illustrated in Fig. 10.12, provides an attractive method of display which overcomes this problem to some extent. Material from complex and putative pure parental populations is scored for a number of features, and two quantitative characters are used to generate an ordinary scatter diagram. The figures for each plant serve to determine the location of a spot on the diagram. Spots may be of different shape or colour, and are decorated with appropriate arms to show the qualitative and quantitative characteristics of each specimen.

In the excellent example of studies of *Primula vulgaris* and *P. veris* populations, Woodell (1965) explored the variation in 'parental' populations at Marley Wood and Dickleburgh (Fig. 10.12a) and in a complex population at Boarstall Wood (Fig. 10.12b). The results of Woodell's study are also displayed as a hybrid index (Fig. 10.12c and Table 10.2).

Fig. 10.10. Diagram illustrating introgression between two species. Backcrossing of the F_1 hybrid to species B ultimately results in the absorption of some genes from species A into at least some individuals of species B. (From Benson, 1962)

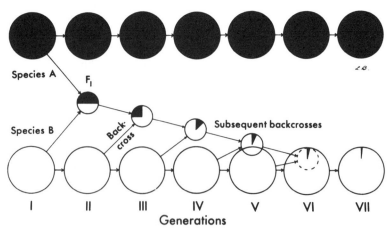

Fig. 10.11. Hybridisation between *Iris fulva* and *I. hexagona* var. *giganticaerulea* in Louisiana, USA, as an example of the use of the hybrid index. First, a list is compiled of the differences between the two taxa in, for example, flower colour. Next, one species is arbitrarily picked to be at the low end of the index scale, and the other at the upper end. Specimens from natural populations are then scored character by character giving the appropriate score as outlined on the following scale:

	Tube colour	Sepal blade colour	Sepal length	Petal shape	Exer- tion of stamens	Stylar append- age	Crest
Like *I. fulva* score	0	0	0	0	0	0	0
Intermediates score	1	1, 2 or 3	1, 2	1	1	1	1
Like var. *giganticaerulea* score	2	4	3	2	2	2	2

Plants exactly like *I. fulva* score 0 for each character, giving a grand total of 0. Total score for plants like var. *giganticaerulea* is 17. Intermediate plants score between 1 and 16. Riley's results shown here are for three populations (sample size 23). Colonies F and G are more or less pure parental species. Colony H1, on the other hand, contains many hybrid plants. In the main these hybrids resembled var. *giganticaerulea* rather than *I. fulva*. (From Riley, 1938.) A number of botanists (e.g. Stebbins, 1950; Grant, 1971) have considered this example to be a clear case of introgressive hybridisation. However, Randolph *et al.* (1967) re-examined the problem of field hybridisation of *Iris* species in Louisiana. Although they established that hybridisation occurred not infrequently and in one case were obliged to interpret a particular population as being a 'new' stable taxon of hybrid origin, they came to the conclusion that their results provided no evidence that introgression had altered significantly the status of the three cross-compatible species of Louisiana Irises as stable taxonomic units.

Hybrid index values

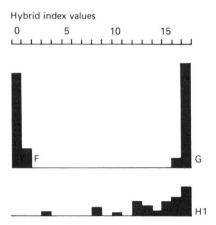

Fig. 10.12. Scatter diagrams and hybrid index histograms representing populations of *Primula vulgaris*, *P. veris* and hybrids in English woodland and meadow sites. (From Woodell, 1965.) (*a*) Scatter diagram of pure populations of *Primula vulgaris* from Marley Wood, Berkshire and *P. veris* from Dickleburgh, Norfolk. (See (*b*) for key to symbols.) (*b*) Scatter diagram of hybrid population from Boarstall Wood, Buckinghamshire. (*c*) Frequency distributions of scores of plants using hybrid index. (For characters employed in calculating indices see Table 10.2.)

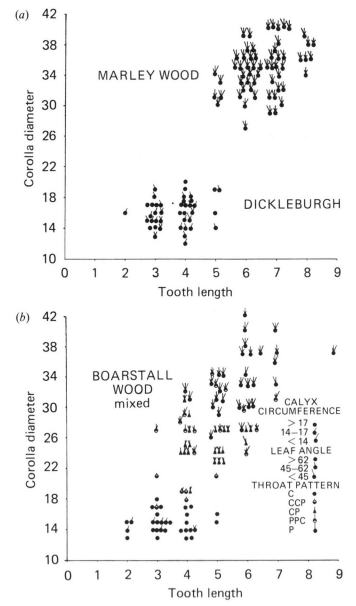

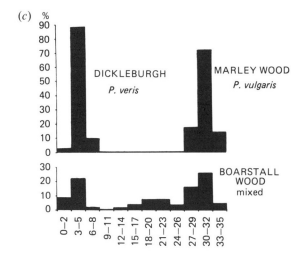

Table 10.2(*a*). *Characters (scored 0–9) used in construction of the hybrid index* (Fig. 10.12)

Character	veris								vulgaris	
	0	1	2	3	4	5	6	7	8	9
1. Corolla diameter (mm)	12–14	15–17	18–20	21–23	24–26	27–29	30–32	33–35	36–38	39–41
2. Calyx circumference (mm)	24–22	21–19	18–16	15–13	12–10					
3. Calyx tooth length (mm)	2–3	3–4	4–5	5–6	6–7	7–8	8–9	9–10		
4. Throat pattern	C	C(P)	CP	(C)P	P					
5. Calyx hair	C	C(P)	CP	(C)P	P					
6. Pedicel hair	C	C(P)	CP	(C)P	P					
7. Leaf hair	C	C(P)	CP	(C)P	P					

Note: In characters 4–7, C = *P. veris;* C(P) = putative backcross to *P. veris;* CP = putative F_1; (C)P = putative backcross to *P. vulgaris;* P = *P. vulgaris.* The minimum score, representing 'pure' *P. veris*, would be 0, that representing 'pure' *P. vulgaris* would be 36.

Table 10.2(*b*). *Pollen fertility of all plants sampled from Marley Wood, Dickleburgh and Boarstall Wood, as judged by staining with acetocarmine*

Species	Locality	No. of plants	Mean of fertility	Individual values
P. vulgaris	Marley Wood	60	99.97	99, 99, 58 at 100
	Boarstall Wood	60	99.98	99, 59 at 100
P. veris	Dickleburgh	42	99.93	98, 99, 40 at 100
	Boarstall Wood	39	99.95	99, 99, 37 at 100
P. veris × vulgaris	Boarstall Wood	21	42.95	9, 17, 20, 24, 31, 33, 33, 35, 38, 40, 45, 46, 50, 50, 51, 60, 60, 62, 63, 65, 70
Putative backcross to *P. veris*	Boarstall Wood	6	88.17	63, 70, 98, 99, 99, 100
Putative backcross to *P. vulgaris*	Boarstall Wood	12	77.00	2, 62, 63, 72, 75, 80, 85, 95, 95, 96, 99, 100

Table 10.2(*c*) overleaf

Table 10.2(*c*). *Variation in hybrid* Primula *populations*

Species	Corolla diameter	Calyx circumference	Calyx tooth length	Leaf length breadth ratio
P. veris thrum	—	—	*	—
P. veris pin	—	—	—	—
P. vulgaris thrum	*	—	*	—
P. vulgaris pin	—	—	—	—

As an increase in variability of the parental taxa is likely to follow introgressive hybridisation, Woodell examined by means of a statistical test whether the two parental taxa were more variable in the hybrid population than the pure populations. In only three cases (marked *) was there statistically significantly greater variation in the hybrid population and Woodell concluded that there was little evidence of introgression.

Fig. 10.13. Polygonal graphs of five quantitative characters of *Viola lactea*, *V. riviniana* and their hybrid. (From Moore, 1959)

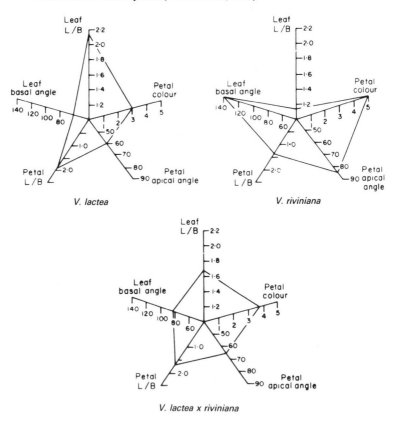

V. lactea

V. riviniana

V. lactea x riviniana

They raise a point of especial interest. In scoring corolla diameter 0 to 9, and calyx tooth length 0 to 7, while giving other characters scores 0 to 4, the characters of corolla diameter and calyx tooth length have been given greater weight in the calculation of the index. The justification of weighting and the ways this might be achieved have received particular attention from various experimentalists: for example, Gay (1960) and Hathaway (1962).

Another method of displaying the variation of individual plants is provided by the polygonal graph method of Hutchinson (1936), later elaborated by Davidson (1947). An excellent example is provided in the study by Moore (1959) of *Viola* hybrids (Fig. 10.13).

Recently, with the advent of more readily available computer facilities, various multivariate methods of studying field collections have been devised. Fig. 10.14, for example, shows the pattern of variation revealed by principal component analysis of populations of *Quercus robur*, *Q. petraea* and a complex population thought to contain hybrids (Rushton, 1978, 1979). It is not our intention to discuss how such analyses are carried out; the point we wish to make is that students of evolution are still

Fig. 10.14. Hybridisation between the two species of Oak (*Quercus robur* and *Q. petraea* in Britain (see Morris & Perring, 1974). This scatter diagram illustrates the use of principal component analysis in the separation of *Q. robur* (dots) from *Q. petraea* (squares), and the intermediate nature of a putatively hybrid population (open circle). (From Rushton, 1978, 1979, where further details can be found.)

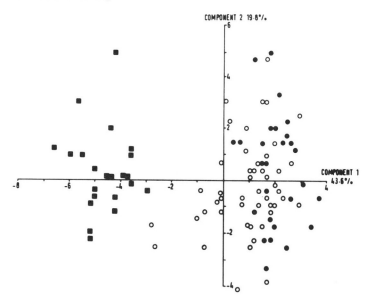

devising techniques for displaying interesting phenotypic variation in populations, the latest methods yielding patterns of dots arranged around axes which are of a highly derived kind. While many botanists are fascinated by multivariate grouping and clustering techniques, others, not necessarily the mathematically incompetent, find this trend in data processing unsatisfactory. It yields diagrams of over-rich complexity from which it is impossible to judge the phenotypic characteristics of any particular specimen. Perhaps the valuable insights offered by complex multivariate techniques need to be combined with displays which reveal the characteristics of the plants under study.

A character which is often scored in field samples, and sometimes in herbarium studies, is pollen stainability, as judged by staining with acetocarmine or other stains. As we have mentioned above, pollen stainability sometimes masquerades as pollen fertility in the report of investigations. Since the relation of pollen stainability to fertility is almost never investigated, it would seem necessary to exercise caution in interpretation. The rationale behind the study of pollen stainability in the context of natural hybridisation is as follows. Two gamodemes, as a consequence of genetic differences developed in their period of allopatric isolation, may produce primary hybrids of low fertility, and, in association with morphological intermediacy, low pollen stainability may therefore be taken as evidence of hybridisation.

Genetic investigations of hybridisation

Interpretation of the variation displayed in Andersonian 'pictorialised scatter diagrams' and hybrid index histograms in terms of F_1, F_2 and backcross derivatives is often attempted by botanists. It would seem important to put these hypotheses to the test. Artificial F_1, F_2 and backcrosses should be produced from pure parental stocks (Baker, 1947, 1951); comparison of synthesised plants with wild ones may then be made, as in the case of *Geum* outlined above. In particular, the range of F_1 variation from different plants from the parental stocks should be examined and the dominance or recessiveness of various alleles should be studied.

Thus, in the case of the *Primula* investigation already discussed, the existence of earlier experimental hybridisation studies makes the interpretation more secure. Artificial F_1 hybrids have been made between *P. veris* and *P. vulgaris*, but only with *P. veris* as female parent, the reciprocal cross giving empty or imperfect seeds. Artificial F_1 hybrids are vigorous with pollen stainability of about 30% of that of parental types (some cytological irregularities have been discovered in meiotic studies of hybrids) and backcrosses have been made (Valentine, 1975*b*, and references to earlier studies cited).

Sometimes, as part of genetic studies, progeny trials are carried out using seed collected in experiments and in the wild (see Heiser, 1949, for examples). As the pollen parent is almost always unknown in seed stocks collected from nature, care must be exercised in interpretation (Baker, 1947).

Chemical studies of hybridisation

Chemical investigations of plant variation have provided important insights at many levels. The reviews by Alston & Turner, 1963*a*, Smith, 1976 and Ferguson, 1980 may be consulted for the history of the development of this important subject. In the present context we may note that the examination of chemical characteristics has proved particularly helpful in interpreting population variation where hybridisation is suspected. Often it is found that taxon A and taxon B differ by more than one chemical character. The chemical constitution of hybrids between A and B will mostly be additive, that is to say, the hybrid A × B will have a spectrum of compounds representing the sum of chemical constituents A and B. Generally so-called secondary plant products (terpenes, alkaloids, phenolics, etc.) are examined in researches.

As an example of the power of resolution of chemical methods we have chosen to discuss the experiments of Fröst & Ising (1968) using the widespread northern species of *Vaccinium*, *V. myrtillus* and *V. vitis-idaea*. Their sterile F_1 hybrid has long been known as *V. × intermedium*. A study of the phenolic compounds of leaf extracts by two-dimensional chromatography was undertaken by Fröst & Ising using Scandinavian material (Fig. 10.15). They discovered:

1. Differences between *V. vitis-idaea* and *V. myrtillus* in two localities. While the variation within *V. vitis-idaea* was small, *V. myrtillus* showed considerable differences between sites.
2. Generally speaking *V. × intermedium* (which may or may not have been produced from the particular individuals of the parents studied chromatographically) had the 'spots' of both parental taxa, but the patterns were not identical at the two sites. As the genotypes of parental stocks may differ, it is obvious that F_1 variation is to be expected, although Ritchie (1955*a, b*) found F_1 plants in British populations to be homogeneous morphologically.

There are many examples of the use of chemical methods in studying complex variation. For example, Alston & Turner (1963*b*) have studied the flavonoids present in population samples of *Baptisia*, in an attempt to resolve patterns of natural hybridisation.

Fig. 10.15. Chromatograms of *Vaccinium*. (*a*) *V. vitis-idaea*. (*b*) *V. myrtillus*. (*c*) The almost completely sterile hybrid, *V.* × *intermedium*. Note the largely additive effect in the hybrid pattern. (From Fröst & Ising, 1968)

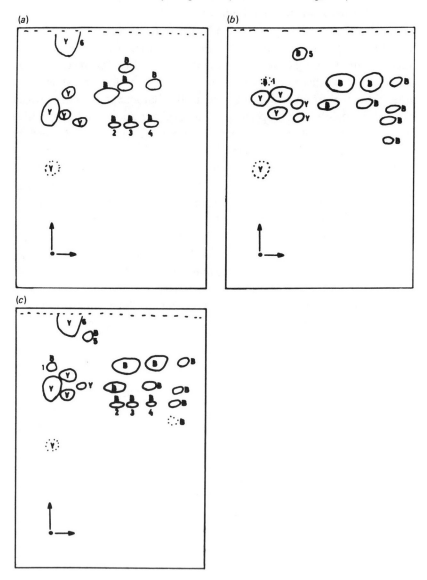

Interpretation of complex patterns of variation as cases of introgression

Our survey of lines of evidence – morphological, genetic, bio-chemical – would suggest that a great deal of information is necessary before an informed interpretation of patterns of variation may be attempted. Thorough studies of natural populations, perhaps over several years, should be undertaken, and some study of the history of the site would seem important, especially in view of the speed with which some wild populations change. An excellent example is provided by Stebbins & Daly (1961), who analysed over several years a hybrid population of *Helianthus* (Fig. 10.16). Riley (see Fig. 10.11) had compared equal numbers of *Iris* plants from three populations. With *Helianthus*, sample size was different in different years and comparisons of histograms of frequency would therefore not be meaningful. It is, however, useful and valid to make a direct comparison of histograms of hybrid index, where the values are expressed as percentage frequency. The hybrid index in this case is based on examination of six characters, scored in such a way that extreme *H. bolanderi*-like plants received a score of 0–2, and *annuus*-like plants score 14–17. The genus *Helianthus* has many advantages for a study of this kind. All the species are self-incompatible annuals with the same chromosome number ($2n = 34$), and the interspecific hybrids, although highly sterile, are not completely so. The study was made on a

Fig. 10.16. Histograms of hybrid index frequencies in a hybrid population of *Helianthus*. (From Stebbins & Daly, 1961)

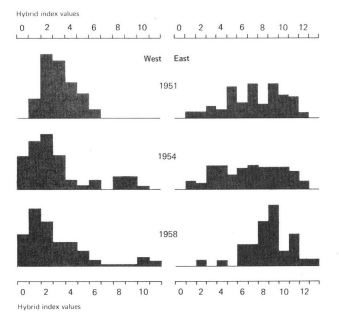

hybrid population of *H. annuus* and *H. bolanderi* in California, first described by Heiser in 1947 as consisting of a few hundred plants, among which were hybrids of at least the F_2 generation. It was possible to date the origin of this hybrid swarm as not earlier than 1942. By 1951 the original population had become divided into a western and eastern half by the invasion of grass species. In the succeeding 8 years during which the population was carefully studied, the western and eastern parts changed independently of each other, in spite of the very short distance (120 m) separating them, and the ease with which pollinating bees could cover such a distance. The western population maintained a high proportion of plants closely resembling *H. bolanderi*; in the eastern population an initial change in the direction of plants resembling *H. bolanderi* was reversed in 1955, so that by 1958 the net change was significantly in the direction of plants resembling *H. annuus*. Intermediate plants, which in 1948 Heiser had found to be highly (though not completely) sterile, were in 1958 significantly more fertile, judged by production of 'good' pollen.

In the interesting discussion of these results, Stebbins & Daly draw, tentatively, two important conclusions. They stress the importance of selection in such hybrid populations, saying that there is good reason to believe that the different behaviour of the western and eastern populations was due to different selection pressures rather than the effects of isolation *per se*. Secondly, they point to the significance of the increased fertility of intermediate plants after 12 years, and say this is susceptible to the general explanation originally advanced by Müntzing (1938) in discussing sterility phenomena in *Galeopsis*, namely that genes responsible for morphological differences can segregate quite independently of those genic and chromosomal differences responsible for F_1 sterility. This fact would mean that recombination of this genetic material can produce 'genotypes containing combinations of alleles derived from the two parental species, but homozygous for either the original or new combinations of chromosome segments'.

The impressive feature of this study is the speed with which the composition of the hybrid can change under pressure of selection, not only in morphology, but also in fertility. With such favourable material, evolutionary processes can certainly be detected in action. A number of other studies of changes in hybrid populations have been published, e.g. *Helianthus divaricatus* and *H. microcephalus* (22-year period) (Heiser, 1979), and two species of *Carduus* (after 5 and 10 years) (Moore & Mulligan, 1964).

If the hypothesis of introgression is to be accepted, there must be evidence of infiltration of germplasm from one species into another. For example, species A, through introgression of alleles or genes from B, may come to show a different pattern of variation from that in sites where A exists alone. Flower colour, leaf-shape and other 'markers' may signal the presence of genes from species B in species A. Alternatively, species A may become more variable in quantitative characteristics, a possibility examined, for instance, by Woodell in *Primula* (Table 10.2).

Given thorough studies it might be possible to test the hypothesis of introgression. Regrettably, however, many botanists studying patterns of variation in field collections alone have stated without reservation that introgression is occurring (or has occurred) in the plants under study. (See reviews by Heiser, 1949, 1973.) If field collections are the sole evidence available, as they must often be, we feel the proper course is to make a provisional statement, making it completely clear how far the situation has been explored, and in particular whether artificial hybrids have or have not been made experimentally.

Another difficulty with the literature on introgression is that botanists are content to infer that the process has occurred, and so do not discuss any other hypotheses even where other explanations are equally plausible. Introgression is essentially a down-grade process in which gamodemes developed in isolation come together with local or regional blurring of pattern. Various other explanations of the variability of taxon A in the direction of taxon B might be devised which do not necessarily involve present or recent introgression. For instance, as taxa A and B are likely to have developed from a common stock, mutations in A may produce effects which are independent of any involvement with B, yet might appear to be introgressants. Taxa A and B may have a history of incomplete isolation. Introgressive hybridisation in times long past may account for the variability of A. Some patterns of variation may be primary (up-grade) situations of great complexity. In some cases, introgressive effects have been claimed over a wide geographical range. For example, Hall (1952) described a case in *Juniperus virginiana*. This North American species has a number of distinct 'races', where it meets *J. horizontalis* to the north, *J. scopulorum* in the west, *J. ashei* in the south-west and *J. barbadense* in the south (Fig. 10.17). Hall considered that these variants resulted from allopatric introgression involving the other species, but this picture has been criticised by Barber & Jackson (1957) who think that other explanations (e.g. ecotypic differentiation) should be considered. This case has continued to intrigue botanists and a large number of chemical studies have been undertaken. While there is some evidence for introgression between *J. virginiana* and *J. scopulorum*, studies of other taxa have not supported this interpretation for the other variants (see Flake, Urbatsch & Turner, 1978, and references cited therein).

Such has been the criticism of the facile use of the concept of introgression (e.g. Barber & Jackson, 1957) that, in the most up-to-date review, Heiser (1973) concludes that many of the classic cases of introgression have not been confirmed by recent researches. He considers that introgression has been important in some cases but the evidence is rarely complete. With present techniques it is difficult to see how very small infiltrations of genetic material could be detected in wild populations and their origin from an introgressive event convincingly deduced. Perhaps molecular biologists will eventually turn their attention to this fascinating subject.

While the interpretation of particular patterns in wild populations in terms of introgression often remains uncertain, there is excellent evidence for the artificial transfer of genetic information in cultivated stocks. Suppose taxon A is known to be resistant to a certain disease organism,

Fig. 10.17. Distribution in the south-eastern United States of *Juniperus virginiana* and other partially sympatric species of *Juniperus*. Hall claimed that 'races' of *J. virginiana* were the result of introgression from the other species, but this interpretation has been challenged. (From Anderson, 1953, modified from Hall.)

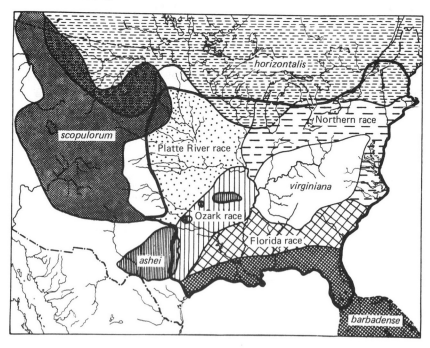

while taxon B is at all times and in all places susceptible to the same pathogen; transfer of disease-resistance genes from A to B might be attempted by crossing A and B, followed by backcrossing AB to taxon B. At every step in the hybridisation programme resistant variants would be selected by infecting all material with the pathogen. It might safely be concluded that transfer of genetic information (in this case a gene confering disease resistance) from A to B had been successful if a disease-resistant strain of variety B could be produced. In a number of plant breeding programmes such controlled introgression for disease resistance has been successfully accomplished (see Knott & Dvořák, 1976). These examples undoubtedly provide the most convincing instances of introgressive hybridisation.

Studies of artificial and natural hybrids have shed a great deal of light on variation and given us a number of insights into gradual speciation. It may be argued that in our account we have introduced too many complexities and uncertainties. Perhaps it is appropriate to close our chapter with a quotation from Lewontin & Birch (1966):

> It is a fundamental difficulty of an historical science like evolution that one can never establish the cause of a past event. It is only possible to show that certain causes are plausible or at most likely, but because each species is a unique historical event we cannot say for certain what its genetic history was.

11

Abrupt speciation

In contrast to the gradual processes whereby two species may diverge under geographical and ecological separation from a single ancestral species, the sudden origin of new species must now be considered. By far the most important kind of abrupt speciation is associated with the general phenomenon of polyploidy; we shall therefore first review this topic, but later refer to other possible modes of abrupt speciation which have been discussed by botanists.

Polyploidy

Botanists have been studying the chromosome numbers of plants for more than 50 years and the results of their work have been published in a multitude of books and scientific papers. In an attempt to make this body of information accessible to biologists, 'chromosome atlases' have been produced (e.g. Tischler, 1950; Darlington & Wylie, 1955; Löve & Löve, 1961). The most up-to-date reference work available for the chromosome numbers of flowering plants worldwide, Bolkhovskikh *et al.* (1969), includes counts published up to the end of 1967. Many new chromosome counts are still being published in an 'Index of plant chromosome numbers' in the volumes of the journal *Taxon* and, recently, as part of the *Flora Europaea* project, a verified list of chromosome numbers of European plants has been published (Moore, 1982).

An examination of the available information makes possible some important generalisations. First, while polyploidy is rare in animals, only being found in a small number of groups (White, 1978), it is a common phenomenon in plants. Calculations of the numbers of polyploid species in flowering plants and the proportion of polyploids in different regions are rendered difficult by differences of opinion between taxonomists as to 'what constitutes a species'. Further, since only a small fraction of the world's flora, mostly in temperate regions, has been studied cytologically, extrapolation to tropical floras is difficult. Also some, perhaps many, chromosome numbers published in the literature are incorrect (Fig. 11.1). When the information for different higher plant groups is examined, a number of patterns emerge. The species in many genera show well-developed series, being simple multiples of a minimum or 'basic number'. For instance, in *Rumex* subgenus *Rumex* there is a series with the base number $x = 10$ which runs from $2n = 2x = 20$ for *Rumex sanguineus*, through $2n = 4x = 40$, in, for example, *R. obtusifolius*, up to $2n = 20x = 200$ in *R. hydrolapathum*. The species of the genus *Chrysanthemum* ($x = 9$): $2n = 18, 36, 54, 72, 90, 198$ and *Solanum* ($x = 12$): $2n = 24, 36, 48, 60, 72, 96, 120, 144$ also provide excellent examples of polyploid series.

In contrast, we find at the opposite extreme of a spectrum of intermediate patterns that all the species of some groups have high chromosome numbers, and it is generally supposed that these plants are of ancient polyploid origin, the lower multiples of the base number having been lost. An extreme case is the fern *Ophioglossum reticulatum* which has $2n = c.1260$. Other species of the genus have $2n = 240, 480, 720$ and 960, so that if a base number of $x = 15$ is assumed, then *Ophioglossum reticulatum* is 84-ploid!

Stebbins (1971) has calculated that some 30–35% of flowering plants are 'straightforward' polyploids fitting into polyploid series. As we shall see later, however, some polyploids do not fit into simple series of multiples of a basic number. Working on the premise that plants with chromosome numbers in excess of $x = 13$ are polyploids, Grant (1971) came to the conclusion that 47% of the angiosperms are polyploids, the figure for dicotyledons being 43% and that for monocotyledons 58%. Recent reviews of the subject have suggested an even higher proportion. Goldblatt (1980) considers that many species with $n = 11$, $n = 10$ and even $n = 9$ have polyploidy in their ancestry, giving a figure of at least 70% for the incidence of polyploidy in the monocotyledons, and Lewis (1980*a*) concludes similarly that perhaps 70–80% of dicotyledonous species may be of polyploid origin. The importance of polyploidy in angiosperms is clearly beyond question.

Studies of lower plants permit the following cautious generalisations (see Lewis, 1980*a*, *b*, *c* for review articles on the incidence of polyploidy in all groups). Polyploidy is apparently rare in fungi and gymnosperms, but is recorded in algae and many bryophytes, and is particularly common in ferns and their allies. Grant (1971) calculates that, on the basis of chromosome numbers in excess of $x = 13$, 95% of fern species are polyploids.

Fig. 11.1. Chromosome numbers in *Myosotis* checked by Merxmüller and Grau. (From Merxmüller, 1970.) Cytologists rarely admit in print that the information in chromosome lists is often wrong. Chromosome numbers, especially in the early literature, may be wrongly counted and are in any case often based on a single cell preparation. Furthermore, no voucher specimen is available in many cases so that verification of the identification is impossible. Such checking is essential, especially in groups where the taxonomy is not straightforward. This survey by Merxmüller and Grau provides a clear warning example against uncritical acceptance of published information.

	Chromosome numbers in TISCHLER (1950)	Correct	Correct chromosom numbers in 1969
M. sparsiflora MIKAN	$2n = 18$	!	$2n = 18$
M. sylvatica EHRH. ex HOFFM	$2n = \boxed{14}$		$2n = 18$
	$2n = 18$	!	
	$2n = \textcircled{24}$ (= *M. alpestris*)		
	$2n = \textcircled{32}$ (= *M. decumbens*)		
M. alpestris F. W. SCHMIDT	$2n = \boxed{14}$		$2n = 24, 48, 72$
	$2n = 24$	!	
M. lithospermifolia HORNEM.	$2n = \textcircled{48}$ (= *M. alpestris*)		$2n = 24$
M. suaveolens W. & K.	$2n = \textcircled{72}$ (= *M. alpestris*)		$2n = 24, 48$
M. decumbens HOST			
ssp. *kerneri* (D.T. & SARNTH.) GRAU	$2n = \boxed{16}$		$2n = 32$
M. rehsteineri WARTM.	$2n = 22$	!	$2n = 22$
M. palustris (L.) NATHH.	$2n = \boxed{64}$		$2n = 66$
	$2n = \boxed{18}$		
	$2n = \boxed{42}$		$2n = 44$
M. laxa LEHM. ssp. *caespitosa*			
(C.F.) SCHULTZ	$2n = c. \boxed{80}$		$2n = 88$
M. stricta LINK	$2n = \boxed{36-40}$		$2n = 48$
M. ramosissima ROCHEL	$2n = 48$	!	$2n = 48$
M. arvensis (L.) HILL	$2n = \boxed{24}$		$2n = 52$
	$2n = \textcircled{48}$ (= *M. ramosissima*)		
	$2n = \boxed{54}$		
M. discolor PERSOON	$2n = c. \boxed{60}$		$2n = 72$

Chromosome numbers in TISCHLER

correct (23.8 %) !   wrong determinations (23.8 %) ◯

wrong counts (28.6 %) ☐   everything wrong (23.8 %) ⬚

Auto- and allopolyploidy

In Chapter 9 we discussed two types of polyploidy, auto- and allopolyploidy. It is instructive at this point to consider their cytology and the light this sheds on problems of ancestry.

Autopolyploidy involves the multiplication of the same chromosome set. Thus a diploid, which has two like chromosome sets (genomes), could give rise to an autotetraploid with four such sets by chromosome doubling. Such a change could be represented symbolically as follows: AA→AAAA. As we saw in Chapter 4, normal sexual reproduction in diploids involves the production of gametes by meiosis, a process in which the homologous pairs of chromosomes become associated together and eventually separate after an exchange of a portion of genetic material in crossing-over. This regular pairing at meiosis is dependent upon there being two, and two only, of each homologous chromosome, forming a bivalent. In the autotetraploid four members of each homologue are present. Evidence suggests that chromosome pairing is only possible between two homologues at any particular point on the chromosomes, but the proximity of four homologues, and the fact that pairing may begin at several different points during the pairing process, results in the association of three or four chromosomes, and single chromosomes (univalents) may remain unpaired. The association of three or four chromosomes leads to chromosome structures known as multivalents, such associations being easily recognised in favourable cytological material. In most cases multivalent production results in a failure of normal separation of chromosomes: for example, bridges of chromosome material may be stretched across the division figures at anaphase I of meiosis as multivalents 'attempt' disjunction. Univalents may be segregated in different (unbalanced) numbers in the two products of the first division of meiosis. It is easy to see how such meiotic irregularity may lead to sterility in gametes. This may be detected on the male side in the production of a high proportion of irregular-sized, mis-shapen pollen grains.

The allopolyploid is the product of the addition of unlike chromosome sets, usually following hybridisation between two species. Thus, two diploid taxa AA and BB may yield an infertile hybrid AB and the production of unreduced gametes from such a plant will give an allotetraploid of formula AABB. In contrast to the autopolyploid situation discussed above, the typical allopolyploid is usually fertile. It is not difficult to see why this should be so. There is no longer any tendency for multivalents to be formed, since each chromosome can pair with its exact partner and no other. This lack of correspondence, which ensures proper pairing in the allopolyploid preventing the association of A with B genomes, is the likely cause of sterility in the primary hybrid AB.

As studies of polyploids progressed in the 1920s and 1930s it seemed possible for a time that the study of meiosis in polyploids would provide an easy way of detecting ancestry. Multivalent associations would indicate autopolyploidy: bivalent pairing would indicate an allopolyploid origin. Such a simple classification was quickly abandoned as mixtures of multivalents, bivalents and univalents were found in the meiosis of many polyploid species (Darlington, 1937).

It is interesting to consider why intermediate meiotic situations may be found. The distinction made in Chapter 9 between autopolyploidy and allopolyploidy, though useful and clear enough in the extreme cases, now seems to be misleading when applied to the evolution of groups of polyploid taxa. The difficulty can be appreciated if we consider what we mean by a hybrid individual with A and B genome sets. We have seen in the previous chapters that ordinary diploid sexual species with some degree of outbreeding are genetically very variable. The genomes of any two individuals of such a species are most unlikely to be identical. It is therefore a conventional oversimplification to represent such an individual as having identical genomes contributed by each parent, and it would be better to write in such cases

$$AA' \xrightarrow{\text{doubling}} AAA'A'$$

to represent the origin of a polyploid derivative. As soon as we do this, we see the nature of the difficulty. Is this situation to be described as autopolyploidy or as allopolyploidy? Clearly the answer hinges on our definition of these terms. If we restrict allopolyploidy to those cases where a *sterile species-hybrid* (represented by AB) gives rise to a fertile polyploid derivative (AABB), then all the other cases where the parents of the diploid belong to the same species would be described as autopolyploid. This is a very unsatisfactory definition, for it obscures the essential similarity in the two situations. A better solution would be to use, as Stebbins (1947) suggested, a third term, 'segmental allopolyploidy', for all the intermediate cases where the parent diploid possesses some measure of chromosomal and genic difference between its genome sets, but where its parents are sufficiently similar to be assigned to the same species.

Experimental studies of polyploids

The oversimple picture of two distinct types of ancestry in polyploids must clearly be discarded. But suppose a botanist wishes to explore the origin of a given polyploid, what experimental procedures might be carried out to provide evidence in particular cases? A number of different ways of studying polyploids have been devised. We will examine each in turn, giving some indication of its history and limitations.

Experiments with deviant plants arising in experimental or cultivated stocks

A classic case of allopolyploidy is provided by *Primula kewensis*, which was discovered amongst seedlings of *P. floribunda* at Kew in 1899. The proposition that *P. kewensis* was a hybrid between *P. floribunda* and *P. verticillata* was put to the test, by making the hybrid experimentally, using *P. floribunda* as female parent. *P. kewensis* was morphologically intermediate between the parental stocks and had the same chromosome number, $2n = 18$ (Digby, 1912; Newton & Pellew, 1929). Although meiosis was regular in the hybrid the plants were sterile, presumably because of genic imbalance. This sterile hybrid was vegetatively propagated and widely distributed as an ornamental garden plant. On three occasions, however, hybrid plants were observed to set good seed (in 1905 at the nurseries of Messrs Veitch, in 1923 at Kew Gardens, and in 1926 at the John Innes Horticultural Institution). In each case the progeny proved to be tetraploid and fertile. Moreover, in one original hybrid plant the investigators discovered that vegetative cells from the fertile stem were tetraploid, the parent plant itself being largely diploid with sterile inflorescences, showing that a sterile hybrid had become fertile by somatic doubling. The fertile *Primula kewensis* behaves as a new species, morphologically similar to the sterile hybrid stocks from which it was derived, but distinct from both parents.

It can readily be seen that, if the primary hybrid and allopolyploid derivatives arise in cultivation, they may be detected and their ancestry investigated. Many other examples of allopolyploids arising in experiment and in cultivation from sterile hybrids are known (e.g. Darlington, 1937; Grant, 1971; Lewis, 1980a) and essentially similar means have been used to deduce ancestry.

Resynthesis of wild polyploids

In the reconstruction of the origin of allopolyploids arising in experiments or in cultivation, the parental stocks may be obvious or the number of candidates limited. As there are often many diploid taxa in a genus, unravelling the ancestry of wild polyploids is altogether a more formidable undertaking. It is very instructive to examine the famous experiments of Müntzing (1930a, b), who studied the origin of the weedy species *Galeopsis tetrahit* ($2n = 4x = 32$). First he made a careful study of six diploid species ($2n = 2x = 16$); from these, *G. pubescens* and *G. speciosa* were selected as closest in morphology to *G. tetrahit*. F_1 hybrids ($2n = 2x = 16$) were produced between *G. pubescens* ($\female$) and *G. speciosa* ($\male$) which proved to be highly, but not absolutely, sterile. After self-fertilisation these F_1 plants yielded strongly variable F_2 progeny, amongst which was a triploid plant ($2n = 3x = 24$) presumably arising from the union of one reduced and one unreduced gamete. This highly sterile triploid was backcrossed to *G. pubescens* and yielded one seed which germinated and grew to give a plant with $2n = 4x = 32$ chromosomes (presumably derived from a cross between an unreduced gamete from the triploid plant and *G. pubescens* pollen). Morphologically this tetraploid derivative was very like *G. tetrahit*. In a study of the 'status' of the experimentally produced tetraploid, Müntzing (1930*a*, *b*) discovered that there was no difficulty in crossing it with wild stocks of *G. tetrahit*. Moreover, fertile offspring were produced.

What may we deduce from these experiments? Is *Galeopsis tetrahit* the allopolyploid derivative of *G. pubescens* × *G. speciosa*? Given our total ignorance of the history of the natural taxon *G. tetrahit*, and the fact that diploid and tetraploid taxa of *Galeopsis* may have evolved in the period after the formation of *G. tetrahit*, it is important to make a cautious deduction. Thus it is highly likely that the ancestors of present-day *G. tetrahit* arose from stocks ancestral to present-day *G. speciosa* and *G. pubescens*.

While the approaches employed by Müntzing to the problem of the resynthesis of a wild polyploid are conceptually sound, the method suffers from the weakness that the investigation is dependent upon a chance event which produces the polyploid derivative from diploid stocks. Also in the *Galeopsis* experiments the one triploid plant produced in the F_2 progeny might have been overlooked, and the *single* seed produced on crossing this plant with *G. pubescens* might have failed to grow. What was needed was a method of generating polyploids at will from diploid stocks.

As long ago as 1904 the Czech botanist Nemeč reported that chromosome doubling in root cells could be induced by treatment with chloral hydrate or other narcotics, but that such changes in the root did not influence 'the germ line'. Blakeslee & Avery (1937), studying the effects of various substances and treatments on stem tissue, reported that the alkaloid colchicine had the property of inducing chromosome doubling. Different methods of application were devised. Seeds could be soaked in dilute colchicine solutions, or the alkaloid could be applied to plants as a lanolin mixture, in agar blocks, by application of drops of solution to bud tissue, or by atomised sprays. The cytological effect of colchicine is illustrated and discussed in Fig. 11.2. The discovery that colchicine treatment could induce polyploidy was a major breakthrough in experimental taxonomic studies and we may now examine an example of its use in an experiment.

In investigations at the famous Iowa Agricultural Experimental Station, Tome & Johnson (1945) studied the tetraploid species *Lotus corniculatus* ($2n = 4x = 24$), which had been introduced into North America

Fig. 11.2. Diagram of the difference between normal mitosis (above) and mitosis that has been changed by the action of colchicine (below). (From Müntzing, 1961.) Note the characteristically widely spaced chromatids in the metaphase after treatment with colchicine. The normal mitosis gives rise to two cells, each with four chromosomes. By the action of colchicine, in which the mechanism of movement of the chromosomes is anaesthetized, one cell with eight chromosomes is formed. If a cell with eight chromosomes is removed from colchicine, a spindle may form at the next nuclear division producing daughter cells which are 'polyploid'. If the cells remain in colchicine, additional C-mitoses may take place with further increments of polyploidy: for example, Onion (*Allium*) roots left in colchicine for 4 days have cells with more than 1000 chromosomes! (Levan, 1938). On removal from colchicine, competition between cells of different number occurs and a new thick root may grow out of diploid cells. Experimental treatments with colchicine are likely to yield mosaics of cells, some polyploid, but others, not actively dividing at the time of treatment, may remain diploid. For further information on colchicine the monograph by Eigsti & Dustin (1955) may be consulted. Of particular interest is the detailed history of studies of the effects of colchicine, especially the question of priority in understanding the significance of chromosome doubling. Biologists intending to use colchicine for experiments should be aware of its carcinogenic properties.

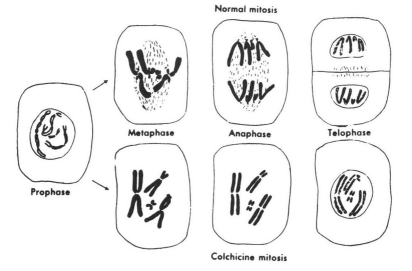

Normal mitosis

Prophase   Metaphase   Anaphase   Telophase

Colchicine mitosis

from Europe in the nineteenth century. They tested the hypothesis that this agriculturally important plant had arisen by autopolyploidy from the diploid *Lotus tenuis* $(2n = 2x = 12)$, which it somewhat resembles. Seedlings of *L. tenuis* were germinated on damp filter paper until the roots were about one inch (2–3 cm) long, the seedlings were then rolled in filter paper and the cotyledons allowed to grow out of the tube so formed. This roll was then suspended above a colchicine solution in such a way that only the cotyledons were immersed in the solution. After washing in water, seedlings were transplanted. As the root system had not been treated with colchicine, any seedlings in which polyploidy had been induced in the cells of the shoot would have diploid roots. The young plants were grown on and, from stem tissue, rooted cuttings were produced and the chromosome number of root-tip cells was examined. Autotetraploid *L. tenuis* was successfully produced by this means. While the autotetraploid resembled *L. corniculatus* in some respects, it differed in a number of ways and, in reciprocal crossing tests, it was not possible to cross the autotetraploid *L. tenuis* with *L. corniculatus*. It would seem that the results do not fit the hypothesis. The authors quite properly point out, however, that *L. corniculatus* may have undergone changes since its formation and that it may not be compatible with an experimentally produced autotetraploid of its true parentage.

It may, however, be an oversimplification to take only the widespread diploid *L. tenuis* into consideration when we are looking for possible ancestors for the common tetraploid species. As the investigation of *Lotus* has proceeded, other diploid taxa have been found (for example, *L. borbasii* in East Central Europe), and there remain yet other taxa which have hardly been investigated. Grant (1965) provides a survey of the taxonomic and cytogenetic complexity of the genus, and Somaroo & Grant (1971) summarise the evidence from crossing experiments.

Studies of karyotypes

In cases where chromosomes are large, an examination of karyotypic differences is often of great value in understanding the origin of particular polyploids. An elegant example, elucidating the evolutionary relationships of common species, is provided by the work of Jones (1958) with the widespread European grasses *Holcus lanatus* and *H. mollis*. *H. lanatus* is uniformly diploid and fertile, with $2n = 2x = 14$. *H. mollis*, on the other hand, contains four cytodemes with $2n = 28, 35, 42$ and 49. In many areas in Britain, *H. mollis* is represented by the sterile pentaploid $2n = 5x = 35$, which reproduces entirely vegetatively. Jones studied the chromosomes of *H. mollis* and *H. lanatus*. In the latter he found a particular chromosome of the basic set of seven which was conveniently recognisable by carrying a 'satellite'. Tetraploid *H. mollis* ($2n = 4x = 28$) also had a pair of satellited chromosomes, but these were much shorter and easily distinguished from those in *H. lanatus*. In pentaploid *H. mollis* a pair of short satellited chromosomes and a single long satellited chromosome can be recognised. In a triploid hybrid between the two species found in the wild, both long and short satellited chromosomes were found. Jones concluded that the only simple sequence of events which would yield a pentaploid of the karyotype as seen would be the prior origin of the triploid hybrid and then a backcross of an unreduced gamete with a normal gamete of the tetraploid parent *H. mollis*. The evidence can be presented diagrammatically (Fig. 11.3).

Note that *H. mollis*, a Linnaean species, is shown by this analysis to consist (at least in Britain) largely of a complex pentaploid hybrid in the parentage of which *H. lanatus* is involved, and that it is not possible to

Fig. 11.3. Karyotype analysis of *Holcus*. (From Jones, 1958)

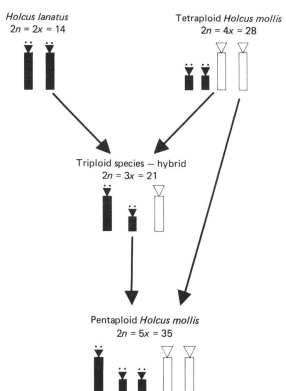

Holcus lanatus
$2n = 2x = 14$

Tetraploid *Holcus mollis*
$2n = 4x = 28$

Triploid species – hybrid
$2n = 3x = 21$

Pentaploid *Holcus mollis*
$2n = 5x = 35$

distinguish morphologically between the various cytodemes of *H. mollis* (Jones & Carroll, 1962).

Another interesting example of the use of karyotypic information concerns the origin of the widespread polyploid weed *Poa annua* ($2n = 4x = 28$). Nannfeldt (1937) suggested that this plant was the allopolyploid derivative of the cross between diploids *P. supina* and *P. infirma*. In contrast, de Litardière (1939) suggested that *P. annua* could be an autopolyploid derived from *P. infirma*. Crosses between *P. infirma* and *P. supina* were made by Tutin (1957) and yielded *P. annua*-like plants. Tutin suggested that during the Pleistocene period *P. supina*, a perennial mountain species, was probably driven down to lower altitudes and came into contact with the ephemeral grass *P. infirma* in the northern Mediterranean region. Crossing between the two taxa could have occurred, especially where mountains are close to the coast, giving rise to the tetraploid plant which is now found worldwide.

Fig. 11.4. Idiograms of the karyotype of *Poa annua*. (From Koshy, 1968.) (Note *P. infirma* is sometimes called *P. exilis*.)

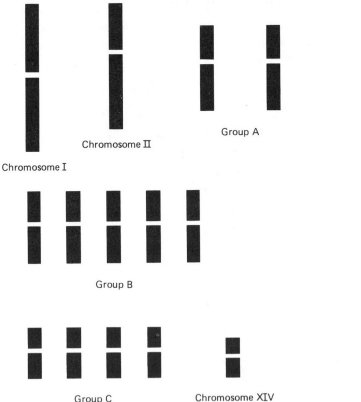

More recently, Koshy (1968) made a detailed study of the karyotype of *Poa annua* and showed that there are three particularly distinctive chromosomes, each present as a pair (Fig. 11.4). Nannfeldt (1937) showed that the karyotypes of *P. supina* and *P. infirma* were rather similar. Koshy draws the following conclusions: *Poa annua* does not have four identical sets of chromosomes as would be required in an autotetraploid, nor does it have the sum of the karyotypes of *P. infirma* and *P. supina*. Either *P. annua* has undergone structural changes since its formation from *P. infirma* and *P. supina*, or it may be derived from either *P. infirma* or *P. supina* forming an allopolyploid derivative with another, as yet unknown, species of *Poa*. Clearly, more studies will be needed to sort out the ancestry of *P. annua*.

Genome analysis

The cytological study of polyploids and their hybrids frequently yields valuable evidence on ancestry. The basic idea of genome analysis may be appreciated by considering an example. If an allopolyploid ($2n = 4x = 28$) of genomic constitution AABB is crossed with a plant thought to be an ancestral diploid $2n = 2x = 14$, a triploid ($2n = 3x = 21$) is formed. Suppose at metaphase I of meiosis seven bivalents and seven univalents are seen in division figures. We could deduce that the diploid and tetraploid shared a genome in common, say genome 'A', by employing the following argument. In the triploid, the A genome from the tetraploid would form bivalents with the A genome originating from the diploid; the 'B' genome would have no pairing partner and remain as univalents. A search could be made for the donor of the B genome by

Fig. 11.5. Genome analysis of a presumed allopolyploid. Meiotic pairing noted in brackets (I: a univalent; II: a bivalent).

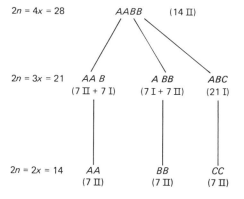

making triploids from crosses between the tetraploid and other diploids. In searching for such a plant, crosses might be made which would yield a triploid in which there was *no pairing* at meiosis, and the parental diploid could be supposed to have a genome different from either A or B, say 'C'. This example is set out diagrammatically in Fig. 11.5.

Genome analysis has been employed in the experimental investigation of many species. A very convenient and familiar example is provided by the common Polypody Fern widespread in many parts of Europe and with related species in North America (the details are provided by Manton, 1950, and Shivas, 1961*a, b*). In central Sweden, where Linnaeus knew the plant, it is not a variable species: to Linnaeus this was *Polypodium vulgare*. This rather narrow-leaved, mainly northern European plant is then *P. vulgare* L. *sens, strict.* If we now look at *Polypodium* in southern Europe, we find a rather obviously different-looking plant, with roughly triangular leaves, which is clearly adapted to a Mediterranean climate of mild, damp winters and hot, dry summers, producing new fronds in the autumn and withering in the summer. This species was called, appropriately, *P. australe* by Fée in 1850. Not only are the two *Polypodium* species separable on general appearance, ecological requirements and geographical distribution, but there are also quite precise characters of the reproductive structures which serve to distinguish them. Further, the cytological situation is clear: *P. australe* is a diploid with $2n = 74$, while *P. vulgare* is a tetraploid with $2n = 148$. In many parts of north-western and western Europe, however, *Polypodium* plants do not divide readily into these two taxa, and a third taxon, somewhat intermediate between the other two, is common. This is the allohexaploid, *P. interjectum* Shivas, which has $2n = 222$ and overlaps in morphology and distribution with both its parent species. It was the presence of the allopolyploid which confused the traditional taxonomy; no clear recognition of the three taxa had been made before Manton and Shivas carried out their experimental and cytological investigations. Typical fronds of the cytodemes are illustrated in Fig. 11.6 and the results of the cytological investigations are set out in Fig. 11.7. The evidence suggests that *P. interjectum* is the allopolyploid derivative (AABBCC) of the cross between *P. australe* and *P. vulgare*. Furthermore, there appears to be no common genome between *P. australe* (CC) and *P. vulgare* (AABB). Genome analyses showed, however, that the European tetraploid has a genome in common with the North American *P. virginianum*, but the source of the other genome is unknown (see Lovis, 1977).

Genome analysis has been used to investigate relationships in a number of genera, e.g. *Viola* (Moore, 1976: Fig. 11.8), and it has also proved to be of enormous value in studying the origins of important crop plants such as the Turnip and its relatives (*Brassica spp.*), Cotton (*Gossypium*), Banana (*Musa*), Tobacco (*Nicotiana tabacum*), Potato (*Solanum tuberosum*) and Grape (*Vitis*). (See Simmonds, 1976, for a review.)

To exemplify the important insights offered into the analyses of polyploids we will discuss some of the experiments investigating the

Fig. 11.6. Cytodemes of *Polypodium vulgare*. (*a*) Diploid ($2n = 2x = 74$) from Cheddar, England. (*b*) The triploid hybrid ($2n = 3x = 111$) between diploid and tetraploid from Roches, Switzerland. (*c*) Tetraploid ($2n = 4x = 148$) from North Wales. (*d*) The pentaploid ($2n = 5x = 185$) hybrid from Bolton Abbey, Yorkshire, England. (*e*) Hexaploid ($2n = 6x = 222$) from Ireland. (*f*) The tetraploid ($2n = 4x = 148$) hybrid between diploid and hexaploid from Istanbul, Turkey. (Fronds × 0.2.) ((*a*)–(*e*) from Manton, 1950; (*f*) from Shivas, 1961*a*.)

ancestry of the hexaploid Bread Wheats (*Triticum aestivum*). After decades of experiments, observations and speculation, there seemed to be general agreement on the origin of Wheat (Riley, 1965: Fig. 11.9). McFadden & Sears (1946) had been successful in crossing tetraploid Wheat ($2n = 4x = 28$) with *Aegilops squarrosa* ($2n = 14$) and showed that when the triploid product was treated with colchicine the synthetic hexaploid so formed resembled certain hexaploid Wheats. Sarkar & Stebbins (1956), after studying the patterns of variation in *Triticum* and *Aegilops*, considered that *A. speltoides* was the most likely donor of the 'B' genome. The logic of the argument used in their studies is of general interest. Given a hybrid, knowing the morphology of one of its parents,

Fig. 11.7. Diagrammatic summary of the results of cytotaxonomic study of *Polypodium* based on the work of Manton (1950) and Shivas (1961*a*, *b*). (*P. australe* Fée is now called *P. cambricum* L.)

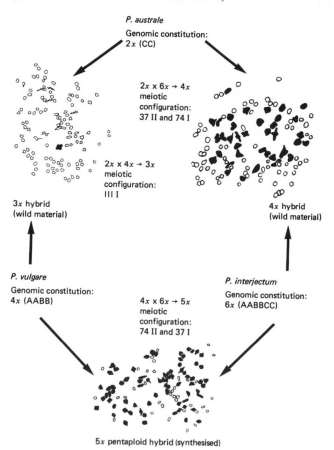

P. australe
Genomic constitution:
2x (CC)

2x x 6x → 4x
meiotic
configuration:
37 II and 74 I

2x x 4x → 3x
meiotic
configuration:
III I

3x hybrid
(wild material)

4x hybrid
(wild material)

P. vulgare
Genomic constitution:
4x (AABB)

4x x 6x → 5x
meiotic
configuration:
74 II and 37 I

P. interjectum
Genomic constitution:
6x (AABBCC)

5x pentaploid hybrid (synthesised)

and considering that most hybrids are intermediate between their parents in quantitative characteristics, then it should be possible to pick out the 'missing' parent from an array of taxa. This technique was called 'the method of extrapolated correlates' by its inventor Anderson (1949) and it has been used in many studies of introgression and other situations involving hybridisation.

In the case of Wheat the variation in diploid and tetraploid *Triticum* was reasonably well known. That an *Aegilops* species was implicated in the ancestry seemed to be likely on cytological grounds. From the diploid species of the genus *Aegilops*, Sarkar & Stebbins deduced that *A. speltoides* was the likely source of the 'B' genome. Early studies of karyotypes supported this idea (Riley, Unrau & Chapman, 1958). The Wheat 'story' displayed in Fig. 11.9 has been widely quoted in the literature, often without giving the evidence. However, recent studies of meiosis (Kimber & Athwal, 1972), seed proteins (Johnson, 1972) and chromosome staining patterns (Gill & Kimber, 1974: Fig. 11.10) have now effectively eliminated *A. speltoides* as the donor of the 'B' set. Furthermore, it is possible that very considerable hybridisation may have taken place at the tetraploid level and that this may have modified the genomes to the point where the tracing of ancestry may be very difficult (Zohary & Feldman, 1962; Pazy & Zohary, 1965). Thus the ancestry of Wheat remains an open question. The whole affair is more a cautionary tale than a 'simple story', reminding us of the complexities likely to be involved in the unravelling of ancestry in polyploids.

Studies of meiosis in hexaploid Wheat offer yet another fascinating insight into the nature and origins of polyploids. Pairing patterns in hybrids between diploid taxa suggested that some homologies between wheat genomes might exist. Indeed, the scheme for the evolution of wheat

Fig. 11.8. Genomic constitutions of species of *Viola* subsect. *Rostratae*, and chromosome pairing in hybrids. Unsuccessful crosses are shown by broken lines. (Based partly on Moore & Harvey, 1961; from Moore, 1976.)

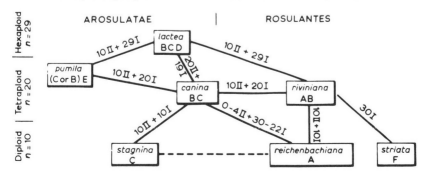

(Fig. 11.9) supposes that the taxa responsible for the formation of Wheat had a common ancestor and it might be supposed that while some chromosome differentiation has taken place, some degree of residual homology remains between the genomes, i.e. the genomes could be said to be homoeologous. On the basis of these considerations *T. aestivum* would be classified as a segmental allohexaploid. When meiosis in hexaploid Wheat is examined, however, pairing is seen to be strictly bivalent in character; pairing behaviour does not appear to reflect the presumed origin!

Experiments by Riley & Chapman were designed to investigate this phenomenon. In hexaploid Wheat (and many other polyploids) plants may survive and reproduce in the absence of a full chromosome complement and plants with $2n = 40$, with a pair of chromosomes missing, can be produced experimentally. With $2n = 6x = 42$, 21 such types – known as nullisomics – may be produced. Riley & Chapman (1958) discovered that if a certain pair of chromosomes was absent (chromosome 5 of genome set 'B') then multivalent pairing occurred at meiosis. Studies of the other 20 nullisomics revealed strict bivalent pairing. They deduced that chromosome '5B' carried a gene (or genes) which enforced bivalent pairing. In the presence of 5B, homologies were overruled and bivalents were produced: when 5B was absent, multivalents were produced reflecting the segmental allopolyploid nature of *T. aestivum*. It seems possible that such 'diploidis-

Fig. 11.9. Ancestry of the Bread Wheats (*Triticum aestivum*). (From Riley, 1965.) (Note: in some of the literature on Wheat, *Aegilops squarrosa* is called *Triticum tauschii*.)

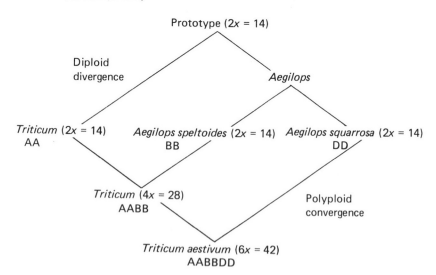

ing' mechanisms occur also in Oats ($2n = 6x = 42$; Rajhathy & Thomas, 1972), *Festuca arundinacea* ($2n = 6x = 42$; Jauhar, 1975) and perhaps in Cotton and Tobacco (Kimber, 1961; Riley & Law, 1965). These findings, of special concern to plant breeders, are also of general interest. In the

Fig. 11.10. Chromosome staining patterns in the three genomes (A, B and D) which make up the chromosome complement of Wheat (*Triticum*). (From Gill & Kimber, 1974.) *A. speltoides* does not have the same pattern of giemsa staining as the 'B' genome in Wheat.

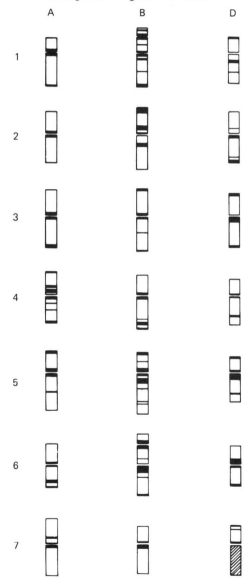

past it had been supposed that pairing behaviour was a rather mechanical affair, dictated by chromosome homologies alone. Such a view forms the rationale behind genome analysis. Now that the situation in wheat and other plants is beginning to be revealed in its complexity, the basis of genome analysis no longer seems secure. How can we ever rule out the possibility that some degree of genetic control is being exercised in the pairing behaviour of polyploids? How would such a control affect pairing in triploids? The experiments necessary to demonstrate the presence of a diploidising genetic system are time-consuming and costly. They may not be technically feasible in plants with small rather undifferentiated chromosomes, since identification of different nullisomics is a necessary part of the analysis.

Another complexity must be raised at this point. It has been known for some time that certain polyploids producing some multivalent associations at meiosis are nevertheless fertile: e.g. *Agrostis canina, Arrhenatherum elatius, Dactylis glomerata* and *Tradescantia virginiana*. After studying *Anthoxanthum odoratum* $(2n = 4x = 20)$, Jones (1964) has argued that multivalent associations need not necessarily indicate a chaotic meiosis. It is possible to devise models of multivalent formation and separation which will yield balanced chromosome products, such systems being under genetic control. It is envisaged therefore that in segmental allopolyploids genetic control of meiotic behaviour may arise, taking the form either of a diploidising mechanism or of a system of co-ordinated multivalent formation and separation, both mechanisms yielding fertile offspring.

Given the possibility of genetic control of meiotic behaviour, the uncritical use of genome analysis must be avoided (see de Wet & Harlan, 1972). Perhaps the method is best used in association with other experimental approaches, such as karyotype analysis and crossing experiments.

Chemical studies

A study of the variation in chemical constituents, especially secondary plant products, is often an important element in the investigation of plant variation, and many botanists have found such studies of great value in investigating ancestry. A particularly elegant example is provided by studies of North American *Asplenium* species (Fig. 11.11).

Further examples can be used to illustrate the historical development of the subject. In many experiments in the early 1960s, no attempt was made to identify the secondary chemical compounds separated by various techniques and deductions were made on the basis of patterns. For example, Stebbins *et al.* (1963) found chromatographic evidence in support of the view that the tetraploid *Viola quercetorum* ($2n = 4x = 24$) is a polyploid hybrid derivative of the cross between the diploid ($2n = 2x = 12$) taxa *V. purpurea* subsp. *purpurea* and *V. aurea* subsp. *mohavensis*. Furthermore, in many of these classic studies a single individual served to represent a population or a taxon. Later investigations suggested, however, that it was important to consider variation within as well as between taxa. A fine example of such an approach is provided by the studies of Bose & Fröst (1967), who investigated, by thin-layer chromatography, the variation in phenolic compounds in *Galeopsis pubescens*, *G. speciosa* and their presumed polyploid derivative *G. tetrahit*. Not only did they study wild populations of *G. tetrahit*, but they also examined descendants of plants from Müntzing's famous resynthesis experiments which we mentioned earlier.

While many chemotaxonomists still study 'patterns' of variation in unknown substances in extracts of plants, more recent studies of variation, many of which concern 'ancestry', have involved an attempt to identify the chemical compounds. For instance, the Red Horse-Chestnut (*Aesculus* × *carnea*: $2n = 4x = 80$), which originated sometime before 1818 in Europe (Li, 1956), has long been considered to be the allopolyploid derivative from the European *A. hippocastanum* and the introduced North American species *A. pavia* (both taxa are diploid: $2n = 2x = 40$).

Fig. 11.11. Diagrammatic representation of two-dimensional chromatograms of three species of *Asplenium* and their hybrids. The flavonoid pattern for *A.* × *kentuckiense* appears to combine the profiles of three diploid species. This evidence adds weight to the hypothesis that *A.* × *kentuckiense* is the trigenomic allopolyploid derivative from the hybridisations outlined in the diagram. (*a*) *A. rhizophyllum*. (*b*) *A. montanum*. (*c*) *A. platyneuron*. (*d*) Bigenomic allopolyploid *A. rhizophyllum* × *A. montanum*. (*e*) Trigenomic allopolyploid *A.* × *kentuckiense* (*A. rhizophyllum* × *A. montanum* × *A. platyneuron*). (From Smith & Levin, in Heywood, 1976.) Recently a much more detailed study of the flavonoids in various *Asplenium* taxa has been undertaken which confirms these earlier findings (Harborne, Williams & Smith, 1973).

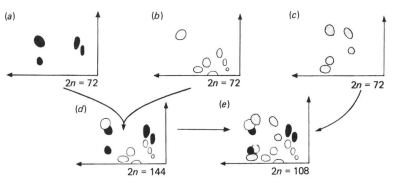

Recent chromatographic studies of the phenolic compounds (Hsiao & Li, 1973) offer support for this hypothesis. The studies of *Aesculus* also point to another recent development in chemical taxonomy. Instead of relying on the evidence provided by extracts of one organ, Hsiao & Li studied the different patterns produced from extracts of leaves and flowers.

A comparison of early chemotaxonomic literature with the volumes of material currently being published in this field reveals the enormous advances made in the techniques of separation and identification of chemical compounds in plants. A recent study of Wheat may serve as an illustration of the remarkable potential of modern investigations, in this case revealing the likely direction of the cross in the production of the hexaploid Wheat. The experiments involved the study of fraction I protein, which is located in the chloroplasts and forms the main soluble protein found in leaves. It has been discovered that a large portion of the fraction I protein molecule is specified by chloroplast DNA, and thus shows maternal inheritance in crosses. As well as large sub-units specified by the maternal parent, there are smaller sub-units specified by both maternal and paternal parents. Thus, by using appropriate techniques on extracts of leaves it is possible to study the parentage of hybrids. For instance, we have seen that hexaploid Wheat (AABBDD) is likely to have been produced by the cross *Triticum dicoccum* (AABB) × *Aegilops squarrosa* (DD). By studying the fraction I proteins, Chew, Gray & Wildman (1975) agreed with this proposal and also concluded that in the production of *Triticum aestivum* the cross must have been *T. dicoccum* (♀) × *A. squarrosa* (♂). For a discussion of the methods of analysis and results of a number of studies in such genera as *Nicotiana* and *Avena* see Gray (1980).

More and more biologists with the necessary training in experimental molecular biology and biochemistry are considering aspects of micro-evolution and there can be no doubt that many more insights into ancestry and relationships will be forthcoming in the next few years.

Reconstruction of the ancestry of polyploids

The careful application of the techniques outlined above may often lead to remarkable insights into the history of particular species. We would, however, stress the need for caution in interpreting evidence. Observations of morphology and experimental studies may provide evidence with which to judge the validity of particular hypotheses, but the evidence is almost always incomplete, and it is regrettable that the cautious interpretations found in scientific papers often undergo such extreme condensation and metamorphosis that a 'good short story', shorn of every vestige of doubt, is all that appears in text books. A good example is provided by the studies of the famous polyploid '*Spartina townsendii*', a new biological species which arose in historical times. Huskins (1930) provides the classic account of this species, but the paper contains many ambiguities and the situation has been re-examined by Marchant (1963, 1967, 1968) who discovered the inaccuracy of Huskins' chromosome counts. Taking all the evidence at present available it is likely that the fertile allopolyploid was produced on Southampton Water in southern Britain, when the North American species *S. alterniflora*, $2n = 62$ (introduced in the nineteenth century) hybridised with the native *S. maritima*, $2n = 60$. However, populations of so-called *S. townsendii* consist either of a sterile hybrid, or the fertile allopolyploid or mixtures of the two. *S.* × *townsendii* is the correct name for the sterile hybrid ($2n = 62$) first collected in 1870. The fertile allopolyploid ($2n = 120, 122, 124$), first noted in 1892, is now called *S. anglica*. The story of *Spartina* is curiously incomplete. Repeated attempts to produce the artificial hybrid between the supposed parents have been unsuccessful. Furthermore, the attempted experimental production of polyploid derivatives from *S. alterniflora*, *S. maritima* and *S.* × *townsendii* by treatment with colchicine has also failed (Marchant, 1968).

Continuing our theme about the need for a cautious interpretation of experiments designed to study the ancestry of polyploids, we should note that many botanists subconsciously assume that a particular polyploid is formed only once. In evolution, if related diploid taxa regularly come into contact, allopolyploid derivatives may be produced repeatedly, sometimes polytopically (i.e. in different sites). The clearest case yet published of the polytopic origin of allopolyploid species in nature is that of *Tragopogon* investigated by Ownbey (1950). Three European species of this Composite genus occur in North America as weeds of roadsides and disturbed ground: *T. dubius*, *T. pratensis* and *T. porrifolius*. All these are diploid species with $2n = 12$ and highly sterile F_1 hybrids between all the pairs are known in Europe. In the area in North America where Ownbey studied them, he found it very easy to detect these hybrids by their failure to set good heads of seed. In four separate localities, however, he found small groups of plants which had the intermediate characters of the hybrids but which were nevertheless quite highly fertile; these proved to be tetraploid with $2n = 24$ and on their morphology it was easy to show that two of them must have arisen from the cross *T. dubius* × *porrifolius* and the other two from *T. dubius* × *pratensis*. Since these fertile allopolyploids are both morphologically distinct and genetically isolated from their parent species, Ownbey described them as new species: *T. mirus* and *T. miscellus*. Recent studies of karyotypes (Brehm & Ownbey, 1965; Ownbey &

238

Fig. 11.12(a)

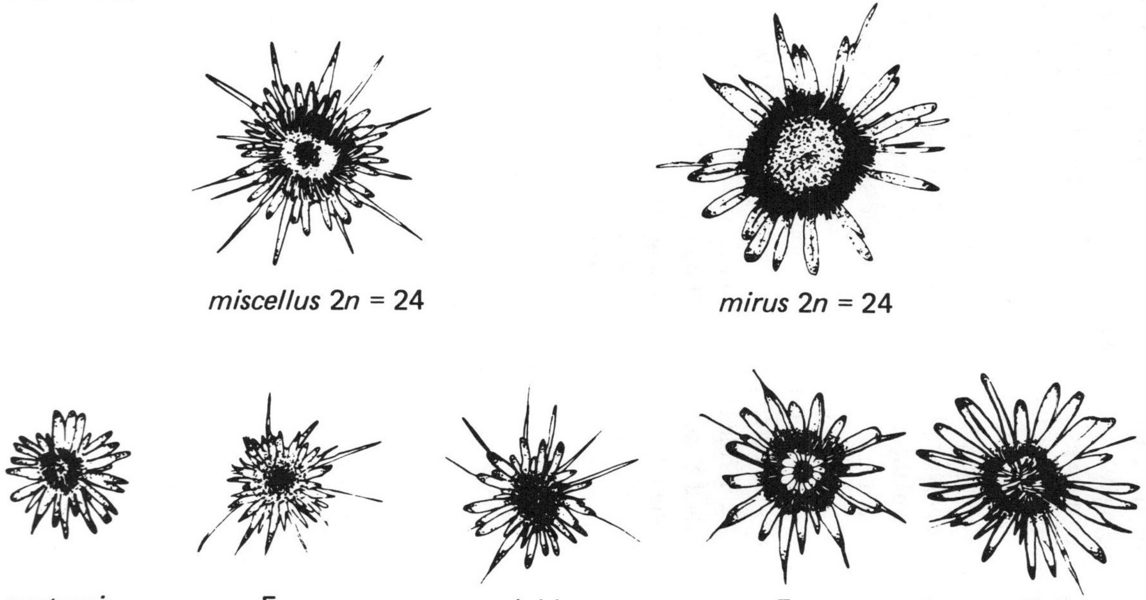

mirus 2n = 24

miscellus 2n = 24

porrifolius
2n = 12

F₁
2n = 12

dubius
2n = 12

F₁
2n = 12

pratensis
2n = 12

McCollum, 1953, 1954) and biochemical characteristics (Belzer & Ownbey, 1971; Roose & Gottlieb, 1976) support Ownbey's ideas about
ancestry. Cytological study has also supported the independent origin of
the separate populations of *T. mirus* from the parent species (Ownbey &
McCollum, 1954). This elegant study is still not complete, for it lacks the
artificial production of the fertile allopolyploids from the species-hybrids;
if and when this is successfully done, it will be a really convincing
demonstration of polyploid evolution in nature (Fig. 11.12). Linnaeus,

Fig. 11.12. *Tragopogon* species in the United States. (*a*, facing page) Flowering
heads of three diploid species of *Tragopogon* (bottom row); *T. pratensis* (left),
T. dubius (centre) and *T. porrifolius* (right), and of F₁ hybrids which led to the
polyploid hybrid species *T. miscellus* (top row, left) and *T. mirus* (top row,
right). (From Ownbey, 1950.) (*b*) Occurrence, generalised, of the European
diploid species *Tragopogon dubius* as a naturalised weed in the western United
States, and localities where its hybrid polyploids with *T. pratensis* (*T. miscellus*)
and with *T. porrifolius* (*T. mirus*) have been found. The diploid species *T.
pratensis* is extensively naturalised throughout the area of *T. dubius*, while *T.
porrifolius* occurs chiefly in the western part of the area. (From Stebbins, 1971,
based on unpublished data of Ownbey.) Both *T. mirus* and *T. miscellus* have
recently been discovered in Arizona (Brown & Schaack, 1972).

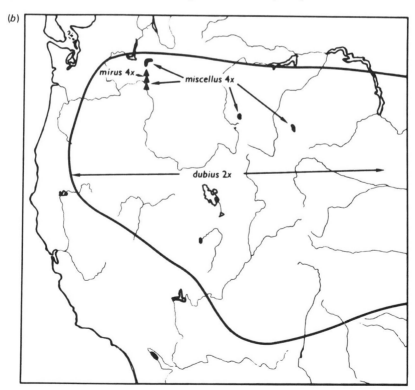

whose work on hybrid *Tragopogon* (described in Chapter 2) gained him the Imperial Academy of Science's prize in St Petersburg in 1760, would have been particularly pleased by this example! Clausen (1966) gives an interesting review of the investigation of these hybrids over two centuries.

Ideas about ancestry must also take into account changes which may occur after the formation of a polyploid derivative. First, gene flow may be considered. The simple picture of an allopolyploid species arising suddenly and achieving at one bound fertility and genetic isolation is unlikely to be the whole story in the complex polyploid evolution of plants. In particular the genetic isolation of allopolyploids must be questioned. Jones & Borrill (1961) report very interesting work on the important pasture grass *Dactylis*, which illustrates the problem. The common *Dactylis glomerata* is a variable tetraploid with $2n = 4x = 28$. Its origin is clearly allopolyploid and there are several European diploid taxa which could be involved, most of which have been recognised taxonomically at least as varietally distinct from *D. glomerata*. Using *D. glomerata* and one of these diploids, *D. glomerata* subsp. *aschersoniana*, a triploid hybrid can be produced which is male-sterile, but partially fertile as female parent. Backcrosses of the triploid with subsp. *aschersoniana* were less successful than with the tetraploid *D. glomerata*. Female gametes produced by the triploid ranged in chromosome number from $n = 7$ to $n = 23$; in general those with $n = 14$ could function well in the backcross to the tetraploid. Thus hybrid tetraploids (or near-tetraploids) could arise relatively easily and moreover showed a high fertility equal to that of wild *D. glomerata*. There seems to be little doubt, therefore, that there is effective gene flow in this case from diploid into tetraploid.

Jones & Borrill ask the important question: 'How likely is this gene flow in natural populations?'. Obviously there is only fragmentary evidence here, but they quote the statistics of Zohary & Nur (1959) who reported that a deliberate search in an area in Israel where both diploid and tetraploid populations occur resulted in seven triploid plants in a total of 4000 examined. On this basis we can say that the event might be rare but not insignificant.

Levin (1978*b*) discusses a number of examples of gene flow in polyploids. Although this is usually a unilateral occurrence from diploids via triploid to tetraploids, he presents the evidence from natural and experimental situations, e.g. in *Solanum*, *Viola* and *Papaver*, in which gene flow occurred in the reverse direction from tetraploid via triploid to diploid.

In considering the possibilities of gene flow between diploid and tetraploid plants, models can be constructed which do not assume the involvement of triploid plants. Studying experimentally the variation of

Betula, Johnson (1945) suggested that gene flow might occur directly between diploids and tetraploids via unreduced gametes of *B. pendula* ($2n = 28$) which could fuse with the normal ($n = 28$) gametes of *B. pubescens* ($2n = 56$) to give hybrid plants of tetraploid chromosome number. Elkington (1968) has also considered this possibility for species of *Betula*.

Not only is there evidence in particular cases of gene flow between diploid and allopolyploid, but there is also the important possibility of new hybridisation at the polyploid level. A case of particular interest was described by Fagerlind as early as 1937 in the common European genus *Galium*. The white-flowered *Galium mollugo* and the yellow-flowered *G. verum* are both represented in South-east Europe by diploid cytodemes ($2n = 2x = 22$). These diploids are completely intersterile. In central and northern Europe, however, the common representatives of both species are tetraploid ($2n = 4x = 44$) and the hybrid between them shows almost normal meiosis and some degree of fertility. We are in this case forced to conclude that an effective sterility evolved at the earlier diploid level has been broken down in the tetraploid. Such cases open up possibilities of 'reticulate' evolution in polyploids which could be extremely difficult to elucidate.

Gene flow has important implications for the investigation of the origin of particular polyploids. Other changes are possible after the initial formation of polyploids but we know very little about such changes in nature. In experimentally produced polyploids there is a good deal of evidence that fertility may increase as 'raw' polyploids are subjected to artificial selection (Stebbins, 1950; de Wet, 1980). Natural polyploids are likely to change with time, through the action of mutation and selection, and as a consequence of chance events. The reconstruction of the ancestry of very old polyploids would therefore seem to be very difficult on this account; indeed, it will often be impossible, because the diploid stocks ancestral to the polyploid have become extinct.

Before we leave the subject of the ancestry of polyploids, we should look closely at the ways in which polyploids are produced. In an important review Harlan & de Wet (1975) have pointed out that, in many accounts of polyploidy, this is left as a shadowy area or it is said that polyploids arise by 'hybridisation followed by chromosome doubling'. This rather ambiguous assertion could imply either that somatic doubling of chromosomes occurs in the primary diploid hybrid, giving rise to the polyploid derivative, or that unreduced gametes are involved. Harlan & de Wet (1975) and de Wet (1980) review the copious literature on the subject and conclude that in very few cases does somatic doubling seem to be implicated (as in *Primula kewensis*) and that unreduced gametes are of supreme importance. This means that, although *Spartina anglica* could have arisen directly from its presumed diploid parents, it is much more likely that the polyploid derivatives, which have three chromosome numbers, $2n = 120$, 122 and 124, had their origin in meiotic 'events' rather than by somatic doubling and clonal propagation. Harlan & de Wet (1975) also review the literature on the frequency with which polyploid individuals arise. Some Maize stocks, for example, produce more than 3% of unreduced eggs, but tetraploid Maize does not compete well with diploid Maize under field conditions. It seems safe to conclude that, while polyploids are constantly being generated in nature and in cultivated stocks, few survive to produce new species.

A consideration of the behaviour of gametes in diploids and polyploids brings us to yet another fascinating possibility. While there is a clear mechanism for increasing chromosome numbers, there is no obvious way of descending a polyploid series. However, Raven & Thompson (1964) drew attention to the possible significance of the occasional phenomenon of 'polyhaploidy' – the derivation of functional diploids from polyploids via unfertilised gametes – and Ornduff (1970) and de Wet (1971) re-examined the possible evolutionary implications in the light of a few described cases of polyhaploidy. By employing anther culture techniques, haploid plants have been experimentally produced in several genera (see for example Kasha, 1974; Sunderland, 1980) and future research may greatly increase our understanding of polyhaploids.

Properties of polyploids

The widespread occurrence of polyploids provokes us to consider their properties and to assess the advantages and disadvantages of the polyploid condition. Except in strict autotetraploids, polyploids combine the genomes of two or more parental taxa, the parental contributions being genetically different. Polytopic origins of polyploids and gene flow between and within different polyploid levels may further increase the variability of polyploids, and genetic recombination will release a great range of variation. Thus the ecological amplitude and geographical distribution of polyploids may exceed that of parental diploid taxa. For example, diploids in *Achillea millefolium* have a rather restricted distribution in southern Europe at the present time, in contrast to the more widespread tetraploid and hexaploid plants (Ehrendorfer, 1959). The enormous variability which may be released in the recombination of polyploids may be seen as a potential advantage in certain environments.

It is also possible to view polyploidy as a means of conserving variation. In diploid organisms of heterozygous genotype *Aa* (where *A* is dominant to *a*), the allele *a* is sheltered from the immediate effects of selection and may survive in the population even if *aa* is deleterious. However, at meiosis gametes containing either *A* or *a* are produced and selection may act on *a* at this stage. The diploid state offers some 'shelter' for recessive alleles: polyploids, with their several genomes giving multiple representation at the *A* locus, have a greater capacity to store variation. This is particularly important in considering the effects of repeated self-fertilisation. Many polyploids are self-compatible; the self-incompatibility systems present in many parental stocks no longer operate at the polyploid level. This breakdown of self-incompatibility may be a very important advantage in certain polyploids where long-distance dispersal of a single self-fertile individual may be sufficient to found a new population (Baker, 1955). We have seen in Chapter 6 how repeated self-fertilisation may lead to homozygous derivatives and inbreeding depression. In comparison with the diploid, in a polyploid the march to homozygosity with inbreeding is not quite so rapid. Thus, in a diploid plant of genotype *Aa*, selfing produces an F_2 in which the progeny are distributed in the familiar Mendelian ratio 1*AA*:2*Aa*:1*aa* and 50% of the progeny are homozygous. In a tetraploid plant of genotype *AAaa*, in which the alleles are located near the centromere on different chromosomes and where the four homologous chromosomes separate at random in pairs, the F_2 ratio on selfing is 1/36*AAAA*:8/36*AAAa*:18/36*AAaa*:8/36*Aaaa*:1/36*aaaa*. Segregation follows Mendelian principles but with a different ratio, yielding only 1/18 homozygous derivatives; 94.5% are heterozygotes of various genotypes.

Models may be constructed in which selfing proceeds for several generations and where the genotype frequencies are not influenced by selection. Thus to reduce the percentage of heterozygotes to less than 1% from a population initially wholly heterozygous (*Aa* in the diploid and *AAaa* in the autotetraploid) will take seven generations for the diploid but 27 generations for the tetraploid. About 46 generations would be needed to achieve less than 1% heterozygosity in the autohexaploid (Parsons, 1959). It is clear that polyploids are a more efficient store of variation than diploids.

The 'Achilles' heel' of many polyploids, however, remains defective meiosis and the consequent sterility. From our discussion of polyploidy it is easy to see which classes are likely to show defective meiosis: those which have odd numbers of genomes (e.g. 3*x*, 5*x*), those which lack genomic homology and show multivalent and univalent formation, and those with abnormal, defective segregants because of genomic incompatibility. Many such polyploids are known to reproduce apomictically, whilst the related diploid taxa are sexual. Thus, within the variable species *Ranunculus ficaria*, sterile variants with bulbils (vegetative apomixis) are triploid $(2n = 3x = 24)$ or tetraploid $(2n = 4x = 32)$, whilst diploid plants $(2n = 2x = 16)$ set seed by normal sexual means (Taylor & Markham, 1978). In the case of *Sorbus* (Fig. 11.13), three widespread and variable species in Europe are diploid and sexual, whilst other more restricted taxa, some of which have leaf-shape and other characters intermediate between two of the diploid species, are triploid or tetraploid and reproduce apomictically. These apomictic taxa can very plausibly be derived by hybridisation and polyploidy from the diploids. In *Alchemilla*, very few of the hundreds of microspecies distinguished in Europe show any trace of sexuality and most have high polyploid chromosome numbers; they look like ancient polyploids derived from sexual ancestors which are now extinct (Walters, 1972).

The delimitation of taxonomic species in polyploid groups

We shall deal with these issues in Chapter 12, but we might note at this point that polyploids are formed by the addition of like or unlike genomes and that the degree of morphological distinction between presumed ancestral diploids and derived polyploid may be considerable or very slight. All will depend upon the degree of morphological difference between the diploid taxa contributing genomes to the polyploid in question. In experimental polyploids, an increase in chromosome number yields larger nuclei and cells with larger diameters and volumes. The members of a polyploid series may differ in mean pollen and stomatal cell size and measurement of samples of such cells may be a reliable way to separate plants of different chromosome numbers. In using this method, e.g. on herbarium specimens, it is important to investigate cell sizes in plants whose chromosome number has been determined. While the study of cell sizes has been helpful in distinguishing cytodemes in some polyploid groups, other groups do not show so-called 'gigas' effects in the plants with higher chromosome numbers (Stebbins, 1950; Davis & Heywood, 1963; Lewis, 1980*a*).

One interesting property of polyploids is that plants which have additional or too few chromosomes may survive – the so-called aneuploid condition, as we saw above. Aneuploids have been found in many cultivated and wild plants and often it is not clear in wild material whether such plants are merely an ephemeral component of populations. In some species both polyploids and aneuploids derived from them are found: for example, in the extreme case of *Claytonia virginica* there are diploids with $2n = 12$, 14 and 16 and polyploids with the numbers $2n = 17$ to 37 inclusive, 40, 42, 44, 46, 48, 50, 72, 81, 85, 86, 87, 91, 93, 94, 96, 98,

Fig. 11.13. Sexual and apomictic species in *Sorbus*. A typical leaf of each is shown (×0.33). The arrows indicate possible origins of the apomicts from the sexual species.

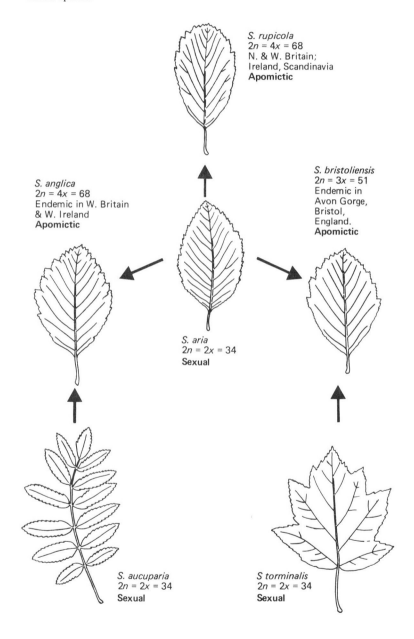

S. rupicola
2n = 4x = 68
N. & W. Britain;
Ireland, Scandinavia
Apomictic

S. anglica
2n = 4x = 68
Endemic in W. Britain
& W. Ireland
Apomictic

S. bristoliensis
2n = 3x = 51
Endemic in
Avon Gorge,
Bristol,
England.
Apomictic

S. aria
2n = 2x = 34
Sexual

S. aucuparia
2n = 2x = 34
Sexual

S torminalis
2n = 2x = 34
Sexual

102, 103, 104, 105, 110, 121, 173, 177 and 191 (Lewis, 1976). Some of the variation in chromosome number may be the result of aneuploidy, but it is possible that allopolyploidy has occurred repeatedly in this group, in which there is said to be relatively little taxonomic variation. Another taxonomic species with a spectacular array of cytodemes is *Cardamine pratensis*, with $2n = 16, 24, 28, 30, 32, 33-37, 38, 40-46, 48, 52-55, 56, 57, 58, 59-63, 64, 67-71, 72, 73-96$ (see Lövkvist, 1956).

Polyploidy and aneuploidy involve differences in respect of whole chromosome sets. In most plant groups a definite centromere is present in each chromosome. Centromeres are involved with the spindle fibres in chromosome disjunction and proper separation of chromosome arms lacking centromeres or chromosome fragments is not possible. In certain groups, however, notably the Juncaceae (e.g. *Luzula* and *Juncus*) and Cyperaceae (e.g. *Carex* and *Scirpus*), centromeric activity is not localised: the plants are said to have diffuse centromeres. In this case fragmentation of the chromosomes may occur; such plants may be viable, and a whole series of chromosome numbers may be generated thereby (a condition known as agmatoploidy).

In the large genus *Carex* there is an exceptionally long series of chromosome numbers from $n = 6$ to $n = 56$, with every number represented between 12 and 43. Whilst some elements in this series could be due to normal polyploidy, both aneuploidy and agmatoploidy are probably also involved (Davies, 1956; see also White, 1978). Similarly, in the genus *Luzula* there is not only evidence of orthodox polyploid series but Nordenskiöld (1949, 1951, 1956, 1961) has described in the *Luzula campestris* and *L. spicata* groups how diploid, tetraploid and octoploid cytodemes have chromosomes in descending order of size, a situation which is interpreted as being due to chromosome fragmentation, rather than multiplication of chromosome sets. While recent studies confirm the notion of the importance of fragmentation, this cannot be the whole story, for plants with the same chromosome numbers have different DNA contents (Barlow & Nevin, 1976).

Before we leave the subject of polyploidy, we might consider some of the now classical experimental investigations of species in the light of our present knowledge. The case of *Leucanthemum vulgare* (*Chrysanthemum leucanthemum*), for example, which in Chapter 3 we used to illustrate the biometricians' interest in 'local races', is now known to be complicated by the widespread occurrence of cytodemes which are to some extent morphologically separable on a number of quantitative characters (Favarger & Villard, 1965). In a rather similar way we now know that the kind of difference which Burkill found between Cambridge and Yorkshire populations of *Ranunculus ficaria* (Chapter 3, Table 3.8) is to be found between diploid and tetraploid cytodemes of this common and variable species. Turning to the work of Turesson, we find again that part at least of the variability which he was able to detect in common European species (such as *Achillea millefolium*, *Caltha palustris* and *Dactylis glomerata sens. lat.*) is certainly attributable to the occurrence within these Linnaean species of more than one cytodeme. This does not, of course, in any way cast doubt upon his demonstration of ecotypic differentiation; it merely emphasises that species recognised on grounds of morphology are often highly complex entities when studied experimentally. Finally, there is the case of *Erophila verna*, the common variable annual weed which, as we saw in Chapter 2, Jordan studied in such detail. A number of cytologists have studied this group, discovering the following chromosome numbers: $2n = 14, 24, 28, 30, 32, 34, 36, 38, 39, 40, 52, 54, 58, 64, 94$ (see Winge, 1940).

Abrupt speciation at the diploid level

So far in this chapter we have discussed abrupt speciation by means of polyploidy. In doing so we have introduced the idea of a basic number of chromosomes and multiplication of this number in the formation of polyploids. The basic numbers found in genera are different and sometimes a genus has more than one basic number. In accounting for different basic numbers in the next few pages, we will show in outline how one number may arise from another and how this process represents a mode of speciation quite different from polyploidy. First, a model of how chromosome changes may occur is presented and then a number of experimental studies are considered. For more complete accounts, see Jones (1978) and White (1978).

We have already discussed the role of the centromere in nuclear division. With the exception of certain diffuse centromere types the chromosomes of plants have a defined centromere which, together with the spindle fibres, ensures proper chromosome disjunction at meiosis and mitosis. Centromeres cannot arise *de novo*, but are formed from pre-existing centromeres. We have also seen that, in many polyploids, plants with fewer or more chromosomes than the normal complement may be viable. Tolerance of aneuploidy may owe its origin to homologies between the 'different' genomes and in polyploids some duplication of genetic material is likely which cushions the plant against chromosome loss. Loss or gain of chromosomes may, however, be much more damaging in diploids. Loss of a chromosome implies that the plant may have a portion of the normal DNA missing from its genotype and such plants are likely to be inviable. Given the sensitivity of diploids to chromosome loss (and perhaps gain also) changes in the base number requiring, as they do, losses or gains, seem difficult to explain.

A simple model suggests a possible mechanism however. As we saw in Chapter 6, chromosomes are composed of genetically active materials – euchromatin, which appears to be essential for the viability of the organism, and heterochromatin, the role of which is less clear but which is probably less essential. The model supposes that chromosomes composed of heterochromatin may be lost from the genome, with perhaps some effect on viability and fertility, but without the catastrophic effect of the loss of a chromosome composed largely of euchromatin. Thus given a plant with, say, $n = 4$ chromosomes, reciprocal translocations between two of the chromosomes may produce a situation where one of the chromosomes becomes entirely heterochromatic. In gamete formation the small, derived heterochromatic chromosome may be lost from the chromosome complement or, by misdivision of the chromosomes, may be present more than once. A possible mechanism for the generation of plants with new base numbers is now at hand. Clearly there are enormous hurdles to surmount before gametes with reduced or increased chromosome numbers can give rise to a population with a different basic number. Many aneuploid individuals must fail to survive, but evidence suggests that changes in basic number have occurred in the evolution of plants and constitute an important class of abrupt changes (Fig. 11.14).

Fig. 11.14. Diagram of translocation, illustrating the possible mode of origin of a new basic chromosome number ($x = 3$) from $x = 4$ by loss of a small heterochromatic new chromosome.

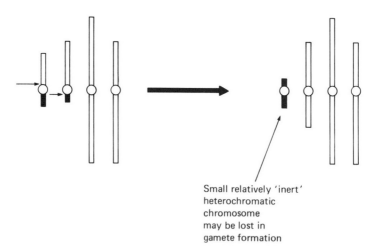

Small relatively 'inert'
heterochromatic
chromosome
may be lost in
gamete formation

An excellent example of an aneuploid change is provided by the studies of Kyhos (1965) on three species of the genus *Chaenactis* (Compositae). He studied the yellow-flowered *C. glabriuscula* ($2n = 12$), a western North American plant of mesic habitats, and two white-flowered Californian desert species, *C. stevioides* and *C. fremontii* (both $2n = 10$). By examining meiosis in the hybrids, both artificial and natural, between these three species, Kyhos was able to make a careful study of chromosome associations. He deduced that *C. stevioides* and *C. fremontii* had been independently produced from *C. glabriuscula* by processes involving chromosome translocations and loss (for interpretations of the changes see Fig. 11.15).

A number of other examples of changes in chromosome number have been reported in the literature, e.g. *Crepis fuliginosa* ($n = 3$) derived from *C. neglecta* ($n = 4$) or its near ancestor (Tobgy, 1943); *Crepis kotschyana* ($n = 4$) derived from an ancestor like *C. foetida* ($n = 5$) (Sherman, 1946); and *Haplopappus gracilis* ($n = 2$) from plants with $n = 4$ (Jackson, 1962, 1965).

For many years Lewis and associates have studied the evolutionary relationships between a number of diploids in the genus *Clarkia* (for a review of these studies, see Lewis, 1973). The results of their work, which are based on analysis of meiosis in hybrids, are set out in Fig. 11.16. In some cases the evidence suggests an aneuploid origin of derived taxa with a reduction in chromosome number. However, as in the case of the origin

Fig. 11.15. Chromosome structure of *Chaenactis* showing the possible origin of related species by chromosome translocation and loss. (After Kyhos, 1965, from Moore, 1976.)

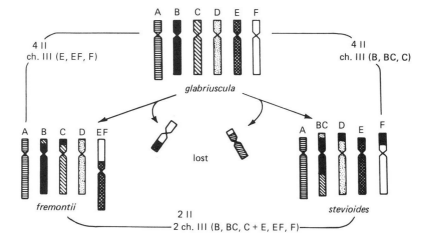

of *C. lingulata* from *C. biloba*, an increase in chromosome number may sometimes be involved. Speciation does not always involve a change in chromosome number. For example, analysis of the chromosome associations of the highly sterile artificial hybrid between *C. franciscana* and its presumed ancestral species *C. rubicunda* (both with $n = 7$) showed that the two species differed in respect of at least three translocations and four inversions. Evidence from this study and a number of others suggests that the repatterning of chromosomes is important in plant speciation. Botanists have, however, made less progress in this area than zoologists and some important insights have been obtained in studies of animals. (See White, 1978, for a thorough review of the subject.)

The idea of chromosome repatterning has an interesting history. For instance, Goldschmidt (1940, 1955) favoured models of speciation which involved wholesale repatterning of the karyotype, the expression of genes in their new positions being modified by neighbouring loci. The end point of these changes has sometimes been referred to as 'a hopeful monster'. However, research in various organisms does not support the idea of the tens or hundreds of changes and the ideas of Goldschmidt fell into disrepute. Recent studies suggest that perhaps one or a few chromosome translocations or inversions may be sufficient to produce a post-zygotic isolating factor between parental and derived taxa.

While we are beginning to see the importance of aneuploidy and

Fig. 11.16. Diagram showing the relationship of species of *Clarkia*. Haploid chromosome numbers are indicated. Predominantly self-pollinating species are underlined; all others are normally outcrossed. (From Lewis, 1973)

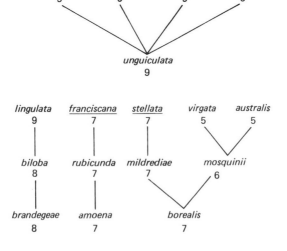

repatterning as modes of abrupt speciation, we know almost nothing about chromosome changes at the population level. It is easy to construct models to account for the origin of plants with different chromosome numbers and repatterned chromosomes, but such variants are likely to be less well adapted than parental stocks. How do these variants, produced in the first place as deviant individuals, come to form populations of a new species? Lewis (1973) has speculated on the likely steps involved in the evolution of species of *Clarkia*, calling the rapid process 'saltational speciation' due to 'catastrophic selection'. He envisages the following:

1. An exceptional drought reduces a normally outcrossing population to very few plants or eliminates it.
2. The few survivors of founders that re-establish the population undergo self-pollination.
3. Extensive chromosomal rearrangements occur; structural heterozygotes are partly sterile.
4. Chance formation of a chromosomally monomorphic population [occurs] from homozygous combinations of rearranged chromosomes.
5. The 'neospecies' is genetically isolated from the parent species by hybrid sterility and because it is self-fertilizing.

The above outline is quoted from White (1978), who has condensed and slightly modified the detailed account of Lewis (1973).

Whether the process of saltational speciation as envisaged in *Clarkia* by Lewis is a general phenomenon has yet to be determined. Many more studies are necessary before either gradual or abrupt speciation is understood. Progress has been slow because of the formidable difficulties of interpreting chromosome associations in hybrid plants, as analysis is only possible in diploids with a small number of large chromosomes. The unravelling of repatterning processes and aneuploid events in polyploids, with their high chromosome numbers and complex meioses, seems impossible using present cytological techniques. As molecular biologists have now begun to study DNA sequences in closely related taxa, however, (see for example Dover & Flavell, 1982), it is likely that our understanding of the processes of speciation will be quickly enlarged.

12

Taxonomy and biosystematics

The success of biosystematics

At this stage in our survey of knowledge of the variation shown by higher plants we can attempt to assess what success the experimental taxonomists have had in some 50 years of effort either to 'improve' the plant classifications we actually use, or (inevitably a more nebulous assessment) to clarify ideas on the basic taxonomic categories of *genus* and *species*. It might seem at first sight not too difficult to answer the former question: most published studies which are called 'experimental taxonomic' have concentrated on relatively small groups (taxonomically speaking, usually a selection of species belonging to a single genus), and it is not difficult to assess in such cases how far 'experimental' data have reinforced or contradicted the existing taxonomic framework. Let us therefore attempt such an assessment.

The first, and perhaps quite unexpected, complication concerns the use of the adjective 'experimental'. Much modern study of the variation of plants involves the use of sophisticated techniques or apparatus, but remains in the strict sense *observational*, not experimental. Good examples are provided by chemotaxonomy and cytotaxonomy, both of which have provided a large amount of new information which has significance for the study of variation. Stace (1980) clearly points to this disadvantage of the term 'experimental taxonomy' in his introductory discussion of the scope of taxonomy and recommends the term 'biosystematics' for studies 'largely concerned with genetical, cytological and ecological aspects of taxonomy (which) must involve studies in the field and experimental garden'. It is, presumably, such studies we wish to assess. We should note, however, that some influential writers have not even agreed with this relatively modest practical aim. Thus Darlington, in his important work *Chromosome botany* (2nd edn, 1963) says:

we can now turn to the question of how we are to graft the Linnean system on to the chromosome system or *vice versa*: it is the great question for both the practical and the fundamental aspect of classification ... Can we add the chromosomes as an appendix to the otherwise sound work of museum classification? The answer is that we cannot. We must make clear at once that the two systems are unavoidably in conflict.

Having made this clear statement, Darlington immediately follows it with an example, apparently commended to his readers, which does exactly what he says is impossible! He cites the remarkable study of Babcock (1947) on the genus *Crepis*, pointing out, very reasonably (if in flat contradiction to his previous statement), that Babcock's taxonomic framework for *Crepis*, which he had of course taken from the existing literature, *had* been modified in the light of cytological and genetic examination.

Pace Darlington, therefore, we can proceed to generalise about biosystematic studies and their effect on taxonomic arrangements, and might hazard the following conclusions:

1. In many cases where some test of reproductive isolation has been made, plant species already recognised on their morphology do also show a measure of genetic isolation. Such examples 'fit', or at least do not contradict, a biological species definition and reinforce the experience of naturalists that most species are 'real' in nature and intermediate individuals or populations scarce or absent. Many examples could be cited: perhaps the familiar one of *Primula* will suffice (see Valentine, 1966, and references cited earlier).

2. The largest single cause of complication in higher plants is undoubtedly polyploidy. In most genera where common species are clearly polyploid, there is some difficulty in deciding whether to treat morphologically similar taxa (which are already in many cases recognised at least at varietal level) as species or not. A typical case is that concerning the fern *Polypodium* (Shivas, 1961*a*, *b*), which we discussed in the previous chapter. Decisions as to whether the appropriate level for a taxon is species, subspecies or variety are seen in such cases to be subjective; in practice, the decision often turns on whether a reasonably convenient minor morphological or anatomical character can be found reliably to distinguish the taxon in question.

3. Linked to the phenomenon of polyploidy is that of hybridisation. In spectacular and well-described cases such as *Geum*, described in Chapter 10, where breakdown of geographical or ecological

isolation has preceded large-scale species-hybridisation, there is no doubt about the phenomenon, but the extent to which past hybridisation is responsible for present taxonomic difficulty is not easy to assess. It is, however, on any assessment a more widespread and significant phenomenon amongst higher plants than amongst higher animals, as Grant (1957), Raven (1976) and others have made clear.

4. The elucidation of the relation between 'critical' taxonomy and apomixis, whilst not much affecting taxonomic practice, must be counted as a success for biosystematic studies. We no longer argue about the 'reality' of the numerous 'microspecies' of *Alchemilla, Hieracium, Sorbus* and *Taraxacum*, but are prepared to treat them as special cases needing peculiar treatment and possessing microevolutionary implications. Perhaps, in a negative sense, biosystematic information about apomixis influences taxonomic procedure in cases where the peculiar variation patterns are shown to be correlated with partial or incomplete apomixis. In such cases any attempts to distinguish taxa within species on the basis of local variants are likely to be suppressed. The study of Smith (1971) on species of *Potentilla* illustrates well this kind of situation.

5. The relation between habitual self-pollination and taxonomic difficulty, often referred to and postulated on theoretical grounds, is quite obscure. It is true that we have largely abandoned the Jordanian 'species' of *Capsella* and *Erophila*, and are content to dismiss the critical taxonomy of these genera as being the reflection of local, ephemeral, self-pollinated variants; but it is by no means clear why other equally common, predominantly self-pollinated weed species such as *Stellaria media* and *Senecio vulgaris* show quite a different pattern. Here the result of the impact of biosystematics on taxonomy has been to leave us puzzled. We revert to this topic, considering recent experiments with self-pollinating species, in Chapter 14.

Summing up, we can say that a good many sexual species behave, so far as we can test, as 'biological species', but that many others do not. In these latter cases, some adjustments of the existing taxonomy may have been made to accommodate the new, uncomfortable data, but always in such a way as to retain a practical framework of morphologically based taxa. This should not surprise us, for classification is and must remain a practical activity. The impact of experimental studies has been mainly in showing why certain groups are taxonomically difficult, rather than in provoking wholesale changes in taxonomic practice.

The species concept

We can now turn to our *ideas* of genus and species. How have these stood up to testing by experiment? We are in real difficulty here, for the literature is both extensive and ever-increasing. The 'species concept' in biology, as we saw in Chapter 2, has engaged the attention of most of the famous naturalists in earlier centuries, from Ray through Linnaeus to Darwin himself, and the argument continues today. Oversimplifying greatly, we could say that both Ray and Linnaeus thought that the genus and the species were 'God-given', a part of the natural order which could be named and described; Darwin, impressed by change and variation, saw both as mere convenient abstractions; and the majority of biologists today see the genus as an abstraction but the species as 'real'.

The comment of many logicians and philosophers on this kind of controversy tends to be impatient; they dismiss arguments about the reality of taxa as being conducted naïvely and in the wrong framework of conceptual thought, and point to the universality of the problem. Perhaps we can re-frame the question along the following lines. When we look at nature, are the 'units' we recognise and name already there to be recognised, or have we 'made' them in the process of looking? The 'naïve realist' view is that they *are* all out there; but it does not need much training in the philosophy of science to see how unsatisfactory this view must be. Recognising even an individual tree or dog is a complex mental process which cannot be independent of our previous experience. How much more so must the recognition of genera and species be 'subjective' in this sense. We *learn* to recognise both genera and species (and, for that matter, families as well); even if we wanted to, we could not begin without the accumulated wisdom and experience of our ancestors.

In recent years the growth of interest in different 'folk taxonomies' – the classifications made, apparently quite independently, by primitive peoples isolated from the main cultural influences – has stimulated this question about the reality of taxa. If it can be shown that the biological units which are named and classified by a particular tribe in an obscure language are roughly equivalent to those recognised in modern biological taxonomy, the case for the 'reality' of taxa is strengthened. Gould (1979) surveys the results of a number of studies of this kind and is particularly impressed by the work of Berlin, Breedlove & Raven (1974, and earlier references cited) who investigated and analysed the plant classifications of the Tzeltal Indians of Mexico. These authors initially interpreted their data as showing relatively poor correspondence between folk names and Linnaean species but, after a more extensive study which recognised unsatisfactory procedures in their earlier work, Berlin reversed his earlier view and agreed that 'there is a growing body of evidence that suggests that the fundamental taxa recognised in folk systematics correspond fairly closely with scientifically known species'. Gould sides with Mayr in quoting the latter: 'species are the product of evolution and not of the human mind'. We have clearly not heard the last of this kind of argument. It may be that the correspondence between folk names and 'Linnaean species' is especially close with higher animals, less close with angiosperms and relatively poor with 'lower' organisms, whether animals or plants. Common sense dictates that 'any fool can recognise a tiger', but it does not follow that we should recognise, say, the several species of *Hypericum* from each other without explicit training. A difficulty hinted at, but dismissed perhaps too easily, by Gould concerns the nature of any inherent limitations and prejudices which are present in the human mind and which therefore we all have in common. Do we construct similar hierarchical classifications apparently independently because, to use the modern computer jargon, our brains are all programmed in the same manner?

Retreating from this complex, speculative field we might now consider another aspect of modern controversy over the nature of species. Most of the relevant literature is written by zoologists and their examples are largely drawn from the animal world. For obvious reasons the strongest arguments have been about Man and the higher animals to which we are related. Equating the taxonomic with the biological species is easier with vertebrate examples, where polyploidy is unknown and populations consist of sexually differentiated well-defined individuals. The contrast between botany and zoology in these respects is well made by Grant (1957). It is not therefore surprising to find that Mayr (e.g. 1957) has consistently championed a biological species concept and has been followed by many zoologists, whilst botanical authors are forced to retain a healthy scepticism. Stace (1980) is perhaps typical. He writes 'there have been many attempts to define a species, none totally successful', and proceeds to list four criteria which are used either singly or together by 'most taxonomists'. Only one of these is concerned with interbreeding and even this has to be qualified at the outset with the phrase 'in sexual taxa'. Like most botanists, Stace recognises it as 'a fact of life' that our species will be 'equivalent only by designation and must therefore be regarded to a considerable degree as convenient categories to which a name can be attached'. He goes further: 'it is not realistic to consider species which are well differentiated on phenetic, genetic and distributional grounds as ideal or normal, and those whose taxonomic recognition poses great difficulty as non-ideal, abnormal or atypical'. This is a clear rejection of all attempts to apply a biological species concept in the taxonomic context. For a well-documented review of the controversy, which comes down uncompromisingly against any single species definition, we recommend a paper by Levin (1979). Heywood (1980) expresses a similar view.

Not all botanists would agree with Stace, Levin and Heywood. Löve has over many years argued for a biological species concept to apply to all organisms, and has frequently acted upon his principles in, for example, elevating to the rank of species variants which differ cytologically even if their morphological differences are negligible. Even Löve, however, is obliged to recognise that apomictic taxa cannot be dealt with in this way. He says (1962) 'to classify them on the basis of reproductive isolation would lead to a confusion even greater than that created by the morphological-chorological method of study of these groups, since every individual is reproductively isolated from all its relatives'. Our own views are close to those of Stace. Neither in theory nor in practice can we adopt as our definition of species any single criterion or even group of criteria. The taxonomic process provides us with a hierarchical system of categories by means of which we can name, and therefore discuss, our material. It is not reasonable to assume that taxa must be equivalent; nothing in nature looks simple and there is no reason why we should expect it to be simple. Of course we tend to look for simplifying hypotheses which enable us to understand what were previously independent phenomena, but we should not complain when the phenomena remain diverse.

A detailed criticism of the biological species concept in terms of its practical value has been provided by Sokal & Crovello (1970). Taking Mayr's definition (1940): 'groups of actually or potentially interbreeding natural populations, which are reproductively isolated from other such groups', they consider each term, and force attention on the impracticability of actually applying the concept in any concrete case. At the outset, setting the limits of the groups to be investigated presents difficulties. How many taxa do we involve in the experimental study? In the case of *Elymus* investigated by Snyder (1951), as we saw in Chapter 10, hybrids between *E. glaucus* and species belonging to other genera such as *Agropyron*, *Hordeum* and *Sitanion* may well have influenced the variation pattern. We have no way of telling, by looking at the taxonomic information before an experimental study is begun, where we should place the limits. The next phrase – 'actually or potentially interbreeding natural populations' – presents even more obvious difficulties. A laborious series of crossing experiments may begin to provide information on interbreeding under some particular experimental conditions; but what can we make of 'actual or potential'? It is clear why Mayr qualified his definition, because the test of crossability in cultivation produces information which may be wholly irrelevant to the microevolution of the populations under study in the wild; but 'potential interbreeding' is an unworkable criterion. Finally, what is 'reproductive isolation'? Is, for example, a level of 50% pollen stainability of F_1 hybrids a sufficient indication of an effective barrier to crossing, or should we always require information about fertility in the F_2 generation? Clearly, subjective decisions enter into the assessment in almost all concrete experimental investigations of this nature. It is not at all surprising that some botanists who have produced elaborate 'crossing polygons' summarising their results have not tried to recognise 'biological species' in their material. Theorists may continue to employ the concept in their model-building, but present-day botanists studying wild populations increasingly ignore the idea and turn their attention, as we shall see in the following chapters, to the close study of patterns and processes in the variation of local populations. Indeed some students of evolution follow Raven (1976) in considering that the biological species concept has clearly outlived its usefulness and might with profit to current research be assigned to history.

Biosystematics and phylogeny

So far in this chapter we have largely confined our discussion to the *species*, its recognition and definition. It is now time to enlarge our canvas and consider what success biosystematics has had in the problem of understanding evolution. Our starting point can be the general agreement with Darwin's thesis that the 'natural classification' of organisms reflects, however imperfectly, the course of evolution. This was not, of course, a new idea when formulated by Darwin, but the general acceptance of biological evolution after the publication of the *Origin* caused all biological classification to take on an overtly evolutionary flavour. In its extreme form this 'take-over' has produced the quite erroneous equation between 'natural' and 'evolutionary' or 'phylogenetic', an equation which, as we saw in Chapter 1, can be easily refuted on both logical and historical grounds.

Modern discussion about the relation between classification and evolution centres mainly around two separate, but obviously related, areas, which concern on the one hand the nature of the evolutionary process and on the other the relation between phenotypic resemblance (in its widest sense) and evolutionary history. As Darwin saw, the species we recognise are varying both in space and in time, and an understanding of the causes of that variation is a major concern of biologists today. Such studies, at or below the level of taxonomic species, provide the main material for this book; they are often called 'microevolutionary' to distinguish them from studies of biological evolution as a whole, and they differ from phylogenetic studies on a broader scale in one very important respect, namely that they are actually or potentially testable by experiment.

From our present position we can look back on aspects of Darwinian controversy and dismiss them as superseded history. We may not know much about the nature of variation, but we are so far in advance of Darwin that relatively clear ideas are available which we can test by field observations and experiments on wild plants and animals. The concept of a reproductively isolated, freely interbreeding population or *gamodeme*, constantly being tested by selection, is the most important of these ideas, and we have considered how far such populations can or cannot be equated with a taxonomic species. We shall return to this question later, but for the present let us consider whether such 'microevolutionary' pictures can be extrapolated to cover evolution as a whole. The 'economy of hypotheses' principle might incline us to agree with Stebbins (1966) that we need not assume in evolution any processes other than those which we can discern and to some extent investigate experimentally at the level of species. In Chapter 15 we look at aspects of this broader problem; here we can at least say that gradual and abrupt speciation, as already explained, provide some basis for understanding the evolutionary process. We should not, however, delude ourselves into thinking that the few, incomplete investigations of particular species-groups yet available to us can possibly provide an adequate base for extrapolation. Indeed, cases such as that of *Triticum* (Wheat), where unusually detailed investigations have been carried out and have revealed an unexpected degree of complexity, should warn us yet again that evolutionary truth is unlikely to be simple.

Classification and phylogeny

Historically it is clear that arguments about the relation between biological classification and the known or inferred facts of evolution received a great impetus from the new ideas of the 'New Systematics', and in particular from the claim that biosystematics could gradually refine and improve existing classifications until they were 'perfect'. This is the so-called 'omega-taxonomy' concept, expressed by Turrill in the volume of essays edited by Huxley (1940) as follows:

> Classification is a *sine qua non* of any biological research, and there has been no system of classification proposed that could replace the system conveniently known as alpha or orthodox taxonomy, though the desirability of subsidiary classifications for special purposes is not

questioned. On the other hand, no taxonomist would say that the existing system, or any large part of it, is complete or perfect. Further, it is becoming more and more obvious that recent discoveries in cytology, ecology, and genetics have often a bearing on taxonomy. There is, indeed, a reciprocal advantageous reaction between them. The taxonomist has to be prepared to use the constructive criticisms of his colleagues and to incorporate into his system relevant data supplied by them. He may thus be able gradually to develop the existing system and progress from the present relative beginning towards an ideal perfected system which is his goal. Daring the reproaches of his biological colleagues, the taxonomist maintains that his subject is the alpha and has the potentiality of becoming the omega of a very considerable part at least of biological knowledge.

Turrill's vision of a gradually evolving system of natural classification has proved an attractive idea to many taxonomists. Stace (1980), for example, states that an 'omega-taxonomy' 'is the distant goal at which taxonomists should aim'. However, serious criticism of the idea has come from two main directions. The first concerns the purpose (or purposes) of biological classification and takes a historical perspective. Generations of biologists have shaped a natural classification for the organisms they study, a classification based upon many different attributes, and this 'broad map' of variation is essential for biological science. 'Correct' and relatively stable names are necessary as a means of communication and information retrieval, and frequent changes seriously impair the usefulness of the system. These pragmatic considerations were stressed in particular in the work of Gilmour, who pointed out (Gilmour & Gregor, 1939; Gilmour & Heslop-Harrison, 1954) that biosystematists and others interested in the units of microevolutionary change could devise and operate separate 'special-purpose' classifications to reveal the patterns that interested them. For this purpose Gilmour and his colleagues devised the '-deme' terminology (see Glossary for details). The limited acceptance of this special terminology (for an assessment, see Briggs & Block, 1981) should not obscure the very important contribution by Gilmour and his associates. Their insistence that all classifications must be judged in terms of their usefulness and that the same material can be the subject of different classifications for different purposes remains of crucial importance; in such thinking there can be no place for a single, 'correct' omega-taxonomy.

The second, quite separate, line of attack came from the rise of computers in the 1950s and 1960s and the impact of mathematics, in its widest sense, on taxonomy. Computers provided entirely new means of information storage and retrieval and turned the interest of both mathematicians and philosophers towards the problems of classification. The most influential book was undoubtedly that by Sokal & Sneath (1963; see also Sneath & Sokal, 1973), which established 'numerical taxonomy' as an important branch of classification theory and practice. This new school of numerical taxonomy, influenced by the ideas of Gilmour, was very sceptical of the widespread post-Darwinian interpretation of natural classification as evolutionary, and pointed out that an objectively satisfactory taxonomic procedure based upon all available characters of the phenotype would produce, not just a single natural classification, as was envisaged in Turrill's concept of an 'omega-taxonomy', but any number of 'natural classifications' according to the procedure adopted. Such natural classifications came to be called *phenetic* and were consciously contrasted with other, overtly 'evolutionary' classifications, in the production of which the procedures were less neutral and less easily definable. For a strong attack on the 'biological species' concept from advocates of the 'phenetic species' see Sokal & Crovallo (1970).

Numerical taxonomy can be defined (Heywood, 1976) as 'the numerical evaluation of the similarity between groups of organisms and the ordering of these groups into higher-ranking taxa on the basis of these similarities'. Its growth is one aspect of a general tendency towards a quantitative, mathematical approach to biology which aims to render more objective and more repeatable the methods and conclusions of what is still a largely descriptive science; it can be seen as a logical development of biometrics. Many examples of the use of numerical taxonomy might be cited; we recommend one by Edmonds (1978) on the variation in the cosmopolitan group of common 'weeds' related to the familiar Black Nightshade, *Solanum nigrum*, in which it is possible to see something of the advantages and also the limitations of numerical methods. Perhaps the most useful contribution from numerical taxonomy in the future will be in similar cases where the variation within a single species or species-complex is being studied in detail and where a microevolutionary interpretation can be made most plausibly. This view is shared by Jardine & Sibson (1971) in their remarkable work *Mathematical taxonomy*, a book highly recommended for the mathematically competent who are interested in our subject; they give as their considered opinion that 'applications in taxonomy which are not aimed directly at the creation of new taxa probably represent the most valuable applications of methods of automatic classification'.

Although numerical taxonomy in the strict sense has its place, it seems likely that modern data processing and information retrieval systems will have an even greater impact on taxonomic procedure in the long term simply as aids to taxonomy rather than as makers of new or changed classifications. Taxonomy is the handmaid of biology; efficient systems for recording, storing and retrieving information will increasingly use modern techniques, so that the herbaria and museums of the future will come to house computerised data banks. Automatic identification (see Pankhurst, 1975) is also likely to play some part in these future developments, if only because, as in other walks of life, machines can and do take over routine tasks formerly done by trained workers. Such developments will take our subject well outside the competence, or even the taste, of the field biologist, although he may well remain interested in the end-products.

One of the most important effects of the impact of mathematics on biological taxonomy seems to have been in clarifying the thought processes which underlie our use of terms such as 'relationship' and 'similarity'. The procedures of numerical taxonomy are necessarily defined and repeatable; they produce measures of similarity in terms of the characters of the phenotype which are used in the calculations, and any classification which arises as a product is a *natural, phenetic* classification. Arguments as to what such classifications can or cannot tell us about the course of evolution inevitably introduce a measure of interpretation and speculation; there is nothing wrong with such argument, but it can and should be clearly stated that the production of a natural classification has its own justification as providing a predictive information system for the use of the science as a whole.

Controversy about the phylogenetic interpretation of classification, which has never completely died down, has been renewed in zoological circles in recent years, principally centring round the work of Hennig. A short account of Hennig's views, together with a useful appreciation of phenetic and phylogenetic systems of plant classification, is given by Stace (1980); for a more detailed, and strongly partisan, account, Eldredge & Cracraft (1980) could be consulted – although the warning should be given that these authors seem to be almost wholly ignorant of any recent botanical contribution to the fields they are discussing. It is perhaps significant that this fierce zoological controversy is mainly about *methods* of achieving the goal of phylogenetic classification, rather than about the idea itself. Botanical writers continue to be sceptical about the idea.

Looking to the future, we note that interest in phylogeny and its relation to taxonomy has greatly increased in recent years with the development of chemotaxonomy, which deals with the chemical variation of plants. Smith (1976) notes that much of this interest 'lies in the possibility that certain kinds of chemical evidence may be a reliable guide to phylogenetic relationships of living species. This possibility has perhaps been the greatest single reason for the current interest and involvement of chemists and biochemists in this field of biology.' Whilst it is undeniable that biochemists and molecular biologists will provide increasingly important insights into the probable course of evolution, we must hope that the pragmatic arguments for a general-purpose taxonomy with special-purpose classifications are appreciated in this rapidly developing field, to which they so obviously apply.

We shall return in Chapter 15 to some assessment of the state of knowledge about the evolution of the plant kingdom, approaching the question not in terms of 'phylogenetic classification' but from a consideration of all lines of evidence.

The future of experimental studies

In recent years there has been a shift in emphasis in much research which could broadly be called biosystematic away from a concern with definition of categories and towards an examination of the detailed patterns of variation and the processes underlying these patterns. Taxonomy in the ordinary sense plays little or no part in many of these studies – although it is important to remember that without any taxonomic framework it would be impossible to define the area of study in the first place. Many workers still prefer to describe their field of research as genecology, whilst others might consider that the same studies fall within the areas of population ecology or population genetics. To some extent this change is the inevitable consequence of the accumulation of knowledge; as more individual cases are studied and described, and the broad outlines of variation emerge, attention turns more to the finer detail and particularly to the origins of variation. To some extent also, one must confess, it is a question of fashion; the same study could be called 'biosystematic' or 'genecological', and the choice of a title including 'ecology' in it is likely to be more favoured today because of the extraordinary publicity which the term has received in the last decade. Whatever weight we should give to these two factors, it is very clear that plant population biology is a rapidly growing field of considerable significance. The success of Harper's excellent text book (1977), which incorporates so much pioneer work of his own and that of his pupils, is both a stimulus and a measure of growth in the subject. The next two chapters of our book are devoted to these themes.

This 'liberation' of experimental studies from taxonomy makes the words of Heslop-Harrison (1955) now seem quite prophetic. In a paper entitled 'The conflict of categories' he wrote as follows:

However, currently developing among genecologists and evolutionists there is a belief that an 'experimental taxonomy' as such may not be required at all. As Gregor has suggested, genecological classifications tend to be summaries of experimentally or cytologically determined facts about natural plant populations which bear upon their origin, structure and properties. To express this information, it may not be necessary to resort to a classificatory approach and certainly no nomenclatural system is required. It may even be that the carrying over of taxonomic concepts seen in the tendency to define 'types' and 'species' is actually an impediment to genecological research. An example of this may perhaps be seen in the early history of the ecotype, for there is no doubt that here the overstressing of the 'type' aspect long tended to conceal the existence of ecologically conditioned clinal variation. The problems of genecology are the problems of the interrelationship and interaction of organism with organism, and organism and population with the secular environment, all in the continuum of time. Taxonomic typification has little to do with such a study, for the recognition of stages is but a poor substitute for the investigation of processes.

There is, of course, no danger that the new studies can part company with taxonomy. Neither Darlington, who had no kind word to say for 'the dead Linnaean taxonomy', nor Harper, who clearly avoids getting involved in awkward taxonomic controversy, could write a book without using Linnaean binomials on practically every page. In this sense, taxonomy is at the root of what they study. What they, and many others, obviously feel is that their research interests do not need an apparatus of categories, whether these are formal, hierarchical taxonomic categories or special classifications of the biosystematic type. Nothing in this should concern us, except a tendency to take the orthodox taxonomy for granted: it is potentially as dangerous for the science if those in the forefront of research ignore taxonomy as it is if they are obsessed or imprisoned by it.

13

Experimental methods in genecology

In the last three chapters we have discussed some of the findings of experimental taxonomists and examined the impact of such studies on the theory and practice of taxonomy, and in Chapter 8 we reviewed the early history of genecology, concentrating on the quasi-taxonomic aspects of the subject. In the next two chapters recent advances in genecology (often now called ecological genetics) will be discussed. First, in this chapter, we will show how many botanists contributed to a refining of genecological experiments, and in Chapter 14 we shall examine the results of experimental approaches designed to study the detailed patterns and processes in populations.

A study of many of the famous early experiments on ecological genetics reveals that the authors of those experiments rarely stated how they were delimiting the populations under study, and that almost without exception nothing was said about the sampling strategies used in field collections. Furthermore, full details of cultivation techniques were not given and, if statistical tests had been applied, then only the simplest analyses were reported.

Our account follows a logical order, detailing the sequence of important issues to be faced in genecological experimentation. As populations of plants are most often sampled at the start of an experiment in ecological genetics, we will first give some consideration to the meanings of the word 'population' and discuss population structure. A sampling strategy appropriate to the experiment must be devised, followed by cultivation or other experiments, the results of which might usefully be tested with an appropriate statistical test. In the final phase of the experiment the balance of evidence may be examined to see if the results provide grounds for accepting or rejecting the hypothesis. In this short book we cannot provide a thorough treatment of all aspects of experimentation; we can, however, point to several key issues in genecological studies, emphasising the notion of the designed experiment, with sampling, experimental procedures, and statistical analyses framed to test an explicitly formulated hypothesis.

The term 'population'

The word 'population', like so many other familiar terms, seems to present little difficulty until we have to define it. In statistics the concept of population is an abstraction signifying a theoretically large assemblage of individuals from which a particular group under consideration is a sample. Most biological uses of the term, however, imply the total of organisms belonging to a particular taxonomic group (or 'taxon') which are found in a particular place at a particular time. This is, of course, the common-sense usage. As we have briefly mentioned in earlier chapters, the population unit which is of outstanding significance in the study of variation is quite different, being the gamodeme, or local group of individuals of a particular taxon within which free exchange of genes is occurring in nature. Population geneticists have coined other names – Mendelian or panmictic populations – for groups essentially similar to gamodemes, but these units often represent 'model systems' in which various assumptions are made about selection, mating, etc. Such model systems (see Spiess, 1977; Merrell, 1981; Wallace, 1981) provide a valuable framework for the botanist interested in field ecology, but what is actually involved in delimiting gamodemes in the field?

Gamodemes in theory and practice

In order to see which plants are interbreeding it is necessary to study patterns of gene flow. First, the source and ultimate destination of fertilising pollen must be known. Given that some plants may be immature in a particular year, observations may be required over a number of years. For instance, populations of biennial plants often have first-year non-flowering rosette individuals along with flowering and fruiting second-year plants. At first sight it would appear that there are two separate gene pools in biennial plants. However, studies (e.g. of *Senecio jacobaea*: see Harper, 1977) have shown that second-year plants, prevented from flowering by insect or other damage, may flower in their third or later years. The idea that several years' study may be necessary to delimit gamodemes is strengthened by the realisation that gene flow also occurs following the dispersal of seeds and their eventual reproductive success around or at some distance from parental plants.

While it is easy to state what information is required for the delimitation of gamodemes, it must be admitted that little or nothing is known about the local interbreeding groups of plants in nature. This fundamental unit is extraordinarily difficult to study with present techniques. The problems to be faced in studying the reproduction and dispersal of plants and their progeny over a number of generations are probably insoluble. The best that can be achieved at present is an *estimation* of the size of gamodemes from information about pollen movements and seed dispersal.

Because the limits of gamodemes are impossible to define experimentally, it is not clear how estimates based on pollen and seed movement relate to the actual gamodemes. In recognition of this fact, the statistically determined entities are often referred to as 'neighbourhoods', being defined as areas of a colony in which mating is assumed to be random. While information on local interbreeding groups is sparse, we shall see below that experiments with agricultural plants have shed a good deal of light on actual and potential gene flow in plants.

Experimental studies of gene flow

As a prelude to our discussion on the methods and findings of botanists studying gamodemes, we note an important change of outlook which has occurred in the last decade. Early ideas on gamodemes in nature were not clearly formulated but, according to Ehrlich & Raven (1969), many biologists (e.g. Merrell, 1962; Mayr, 1963) thought that they were often large. This view was supported by the following observations of naturalists and others:

1. Pollen, present in enormous quantities in air at the appropriate times of year, is widely dispersed (even being detected, for example, 50 km out to sea in the Gulf of Bothnia: Hesselman, 1919).
2. The various mechanisms for ensuring insect and wind pollination seemed to be widely effective.
3. Fruits and seeds are well adapted to wide dispersal in nature.

These general impressions led naturally to the view that the members of a widespread species may be united in the same gene pool by gene flow. The species was viewed as the important evolutionary unit, gene flow amongst its members making it a breeding unit. Speciation was to be seen as the breakdown of the cohesive gene pool of the species by isolating mechanisms.

This view, that gene flow is widespread and gamodemes therefore large, has been subject to radical reappraisal in the last decade. It is now contended that gene flow in nature is much more restricted than previously thought, in the distribution of both pollen grains and seeds. Thus gamodeme sizes may be small and the 'species' may now be visualised as comprising a multitude of small gamodemes rather than a single or a few large gamodemes (Ehrlich & Raven, 1969; Levin, 1978a).

Information on gene flow comes from various lines of study as is made clear in the impressive review of the subject by Levin & Kerster (1974).

Pollen movement

Wind pollination. By sampling the pollen content of the air, using sticky slides or pollen traps, it has frequently been discovered that from an isolated source pollen is distributed in a leptokurtic fashion, much pollen falling relatively close to the parent plant and only a small portion travelling some distance from the parent (see, for instance, the studies of Wang, Perry & Johnson, 1960, on *Pinus elliottii*). When plotted on log/log scales the 'curves' of relative pollen concentration with distance are approximately linear (Fig. 13.1).

In interpreting the various studies it is important to note the marked effects of wind speed and direction, as well as effects of height of presentation of pollen, pollen grain sizes and the influence of environmental heterogeneity in biotic and topographical factors. It is also important to study the actual pollination mechanism in the field. Fig. 13.2 shows

Fig. 13.1. Change in relative pollen concentration with distance for some wind-pollinated plants. (From Levin & Kerster, 1974)

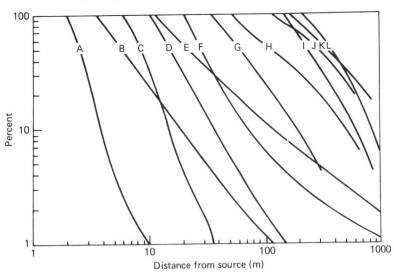

that some plants thought on grounds of morphology to be wind-pollinated are in fact visited by insect pollinators.

Animal pollination. In recent years the study of plant–animal relationships has proved a major growth area in biology. While many plants are apparently well adapted for insect (or bird) pollination, it does not follow that those plants are all effectively pollinated. Entomologists are now inclined to study insect behaviour in terms of cost-benefit analysis. Foraging for food (pollen or nectar) is an energy-consuming activity and 'optimal foraging' may be undertaken by insects, that is to say they visit the flowers offering the best return for the energy expended in searching. The flowers of a particular species under study must now be viewed as an array of 'floral offerings' presented to a wide variety of insects, and many factors such as the degree of faithfulness of an insect to particular plant

Fig. 13.2. Insect pollination of supposedly wind-pollinated plants: sketch of a field experiment with *Plantago lanceolata*. (From Stelleman, 1978)

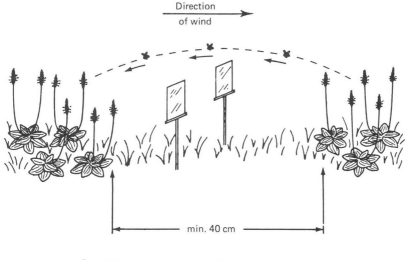

Direction of wind

min. 40 cm

	Potential receptor spikes	Sticky test slides	Donor spikes Pollen coated with dye
Result of experiment	Pollen coated with dye on inflorescence	No pollen with dye	Syrphid flies visit. Dye-coated pollen found on these insects
Deduction	As pollen coated with dye was found on insects and not on sticky test slides, pollen with dye on receptor spikes is most likely to have been carried there by hover (syrphid) flies and not by the wind.		

NOTE: For the sake of clarity the rosette of leaves is drawn as if they are adpressed to the ground (as in *Plantago major* and *P. media*).

species, as well as patch size, spatial pattern and density effects, site heterogeneity and seasonal changes must be taken into account. Given such complexities, the only course of action open to the student of evolution is to study each potential gamodeme separately. Insects and birds must be observed to see which plants are actually visited. As an aid to such studies, pollen may be stained (Simpson, 1954; Sindu & Singh, 1961) or made radioactive (Schlising & Turpin, 1971), so that patterns of dispersal may be discovered. Sometimes intraspecific differences in size of pollen grains have enabled observations of gene flow to be undertaken (e.g. Richards & Ibrahim, 1978). As with wind dispersal of pollen, a leptokurtic distribution of pollen is likely as a consequence of the activities of pollen vectors.

Seed dispersal

Wind dispersal. The distribution of propagules may be studied by catching seeds or fruits on sticky tapes or traps set out around individual plants, or by looking for seeds or seedlings on the soil surface. Sometimes the distribution of marker genes from a carrier parent may be revealing. (See, for example, Bannister, 1965, for a study of the distribution of various markers in plants of progeny found in a study of *Pinus radiata*.) Dispersal of medium to large seeds probably conforms to a leptokurtic distribution, though the 'tail' may be very long. Risking a generalisation, Levin & Kerster consider that even in species with fruits and seeds apparently well adapted with wings or plumes, most fruits or seeds may travel relatively short distances from the parent plant. Much remains to be discovered about wind dispersal; experimental approaches should be encouraged to displace the simple notion that if a plant has a plumed fruit or seed, *ipso facto* its progeny must be widely scattered over large areas at each generation.

Animal dispersal. Current interest in population dynamics and food resources has stimulated research into the relationships of animals to plant propagules. Very little is known, however, about the primary distribution of propagules by animals or the losses due to predation (see Harper, 1977). The phenomenon of secondary dispersal, which may complicate seed flow in many taxa, has received some experimental attention. For instance, in the genus *Viola* primary dispersal is by means of explosive release and wind, but each seed has a protein-rich elaiosome on its surface, and ants locate *Viola* seeds and carry them back to their nests, giving a different ultimate pattern of seed dispersal (Beattie, 1978).

Agricultural experiments

In the past, horticultural and agricultural researches have often been ignored by botanists. This neglect of part of the literature on plants is to be deplored, for some of the most interesting information on gene flow comes from studies of crop plants (Levin & Kerster, 1974) and forest trees (e.g. Wright, 1953). The development of superior cultivars by plant breeding requires the production of large quantities of 'pure' seed. There has therefore been a good deal of interest in minimum distance required to prevent crossing between two different cultivars. Table 13.1 shows the isolation requirement for a number of crop species. It is clear that only a relatively small distance is required in many cases to prevent contamination. In these cases we are dealing with monocultures in the 'bulking up' of seed for agricultural purposes, but other experiments with mixtures of cultivars have provided important insights. Thus, the incidence of hybridisation between mixtures of two crops depends upon planting arrangements in the field: for an example see Fig. 13.3. The degree of crossing also depends on the presence and disposition of taller plants. For example, in an experiment studying gene flow in Cotton (*Gossypium*),

Fig. 13.3. Block planting layout to show the relation between the pattern of planting and the production of hybrid seed. ● = Kale: X = Cabbage (both cultivars of *Brassica oleracea*). Distance between plants = 50 cm. (After Nieuwhof, 1963, from Levin & Kerster, 1974.)

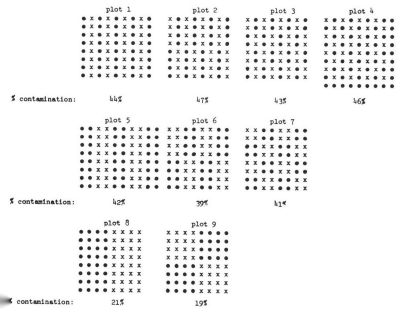

	plot 1	plot 2	plot 3	plot 4
% contamination:	44%	47%	43%	46%

	plot 5	plot 6	plot 7
% contamination:	42%	39%	41%

	plot 8	plot 9
% contamination:	21%	19%

Table 13.1. *Isolation requirements for seed crops (after Kernick, 1961, from Levin & Kerster, 1974)*

Species	Breeding system[a]	Pollination agent[b]	Isolation requirement (m)
Gossypium spp.	S	I	200
Linum usitatissimum	S	I	100–300
Camellia sinensis	S	I	800–3000
Lactuca sativa	S	I	30–60
Avena sativa	S	W	180
Hordeum vulgare	S	W	180
Oryza sativa	S	W	15–30
Sorghum vulgare	S	W	190–270
Triticum aestivum	S	W	1.5–3.0
Cajanus cajan	SC	I	180–360
Citrullus vulgaris	SC	I	900
Apium graveolens	SC	I	1100
Carthamus tinctorius	SC	I	180–270
Papaver somniferum	SC	I	360
Vicia faba	SC	I	90–180
Pastinaca sativa	SC	I	500
Voandzeia subterranea	SC	I	180–360
Nicotiana tabacum	SC	I	400
Coffea arabica	SC	IW	500
Hevea brasiliensis	C	I	2000
Helianthus annuus	C	I	800
Brassica campestris	C	I	900
Daucus carota	C	I	900
Lycopersicum esculentum	C	I	30–60
Anethum graveolens	C	I	300
Brassica oleracea	C	I	600
Allium cepa	C	I	900
Raphanus sativus	C	I	270–300
Brassica oleracea	C	I	970
Carica papaya	C	I	100–1600
Zea mays	C	W	180
Chenopodium ambrosoides	C	W	180–360
Cannabis sativa	C	W	500
Secale cereale	C	W	180
Beta vulgaris	C	IW	3200

[a]S, species which are predominantly self-fertilising; SC, species in which self- and cross-fertilisation are of similar importance; C, species in which there is predominant or exclusive cross-fertilisation.
[b]I, insect pollination; W, wind pollination.

which is insect-pollinated, the degree of intercrossing of two cultivars of Cotton was very much influenced by growing barriers of Maize (*Zea*), c. 2–3 m high, between the rows of plants (Pope, Simpson & Duncan, 1944). In the agricultural landscape the presence of hedgerows and plantations has been known to act as a barrier to crossing (Jensen & Bogh, 1941; Jones & Brooks, 1952).

'Neighbourhoods' in wild populations
From information on pollen flow (and sometimes seed dispersal) calculations of neighbourhood sizes have been made for a number of species (see Table 13.2). A discussion of the equations used is beyond the scope of this book, but full details are presented in Levin & Kerster (1974)

Table 13.2. *Neighbourhood sizes in various species*
(a) *Cases where calculations are based upon pollen and seed dispersal*

	Pollination	Seed dispersal	Area	Numbers of plants
Phlox pilosa (Levin & Kerster, 1968)	Lepidoptera	explosive	11–21 m^2	75–282
Liatris aspera (Levin & Kerster, 1969)	bees	wind	4.5–29 m^2	30–191
Primula veris (Richards & Ibrahim, 1978)	insects	wind	—	1–562

(b) *Cases where calculations are based on pollen dispersal only. Results given below assume no seed dispersal: if seed and pollen dispersal distance are equal, then neighbourhood sizes are doubled*

Lithospermum caroliniense (Kerster & Levin, 1968)	bees[a]		4–21 m^2	2.2–5.4
Forest trees (data from Wright, 1953)				
Pseudotsuga taxifolia	wind		0.5 acres	26
Cedrus atlantica	wind		8.28 acres	207
Fraxinus americana	wind		0.9 acres	9
F. pennsylvanica var. lanceolata	wind		0.9 acres	22
Pinus cembroides var. edulis	wind		0.44 acres	11
Ulmus americana	wind		26.2 acres	262

[a]also cleistogamous flowers.

and Levin (1978*a*). It is clear that these equations, first formulated for bisexual mobile animal populations, are not ideally suited to the study of hermaphrodite, sessile and often clonal plants. Several difficulties arise in the calculation of neighbourhoods. To start with, the assumption of random mating in the wild may be seriously wrong. Some degree of self-fertilisation is possible, and indeed likely, in many species. Pollinator behaviour may also produce non-random (so-called assortative) mating. For example, in plants polymorphic for flower colour, each variant may tend to be visited by a different insect pollinator (see for instance Kay, 1978, on pollination in *Raphanus raphanistrum*, which has white or yellow flowers which tend to be visited by different insects). Furthermore, technical difficulties of watching pollinators and studying seed dispersal might encourage the experimenter to choose a relatively 'simple' area to study. It is unclear whether the results in Table 13.2 are typical of the species in question.

Many of the experiments and observations, of which those detailed above are but a sample, are consistent with the view that gene flow is quite restricted in plants, and gamodemes are therefore small. We feel, however, that many more studies are required, and we should not be seduced into accepting a dogmatic generalisation about gamodeme size. Clearly the gamodemes of some plants with dust-like seed or small spores (e.g. orchids, ferns, fungi) may be very large.

Sampling populations

The avowed, but often unstated, aim of genecological studies is to investigate experimentally the patterns and processes within and between gamodemes. If the basic unit – the gamodeme – is impossible to define, the genecologist is faced with enormous difficulties in sampling. Quite frequently the problem is ignored and 'population samples' are collected without mentioning the difficulties. As we shall see in Chapter 14 many recent studies of small-scale pattern have been attempted. Whilst the extent of neighbourhoods has not often been investigated, there is sometimes circumstantial evidence of gene flow between distinct 'subsites' of heterogeneous areas studied as systems. Thus some informed inferences are sometimes possible, but in many cases biologists must make the best use of samples drawn from sites whose underlying patterns of breeding behaviour are at best imperfectly understood.

In genecology – as in any other subject – it is not often realised that the pattern of sampling will to a very large extent determine the outcome of an experiment. We have already made this point in relation to Turesson's ecotype concept in Chapter 8: it is equally true for any experiment. Much time and effort may be spent in growing and measuring plants and analysing results, but very little attention may have been given to sampling strategies; indeed, the word 'strategy' may be entirely inappropriate for samples of seed snatched at brief roadside stops on car journeys or obtained from Botanic Garden seed lists.

If statistical analysis is to be performed on the results then ideally a random sample of plants must be collected. Ward (1974) has described a simple way in which two people may collect such a sample. Having decided on the area to be sampled – a most difficult problem, as we have seen above – a count is made of the number of individuals in the area (or subsection of an area if the plant population is very large). A decision is then made on the size of the sample, say 25 plants out of 250. Using a table of random numbers (as found, for instance, in Fisher & Yates, 1963) or numbers 'drawn out of a hat', 25 numbers within the range 1–250 are 'selected' and placed in ascending order: say 5, 8, 14, 27, etc. On traversing the sample area again, one person calls out the number of each individual, 1–250, while the second person labels the individuals to be sampled, the 5th, 8th, etc., as determined by the random numbers. This random sample is then used for experimental investigation. Other methods of random sampling are discussed by Yates, 1960, Cochran, 1963, Greig-Smith, 1964 and Green, 1979.

There are some theoretical and practical difficulties to be faced in undertaking such a sampling procedure, which will now be considered.

If the experimenter is studying apparent hybridisation, a random sample might not include all the 'interesting' plants of an area. A deliberate sampling of the plants of the area might be more appropriate in such circumstances. Should the study involve the investigation of variation across a vegetational discontinuity, e.g. woodland to grassland, it might be more informative to collect plants from a transect (sampling at, say, metre intervals) across the ecotone rather than collect a random sample. All will depend on the hypothesis being tested.

There are many habitats where the collections of random samples is very difficult (e.g. tropical rain forests, aquatic and wetland habitats, cliffs). However, where the collection of a random sample is a practical possibility it should be seriously considered.

Since populations often contain individuals at all stages of growth and development from seeds and seedlings to adult plants, a truly random sample should perhaps contain individuals in several different age classes. In practice, a subset of the population is often sampled. The following might usefully be distinguished:

    1. 'Individuals' present as ungerminated seed in the soil ('seed bank').

2. Seedlings, a transitory stage in many habitats, but more important in some plant communities; for instance, in tropical rain forest many tree species growing in deep shade have a long seedling stage: only if disturbance in the canopy causes greater illumination of the ground flora do the seedling trees develop into adults.
3. Immature individuals.
4. Mature individuals.
5. Seeds attached to 4.
6. Diseased and damaged plants. Sometimes, as in the case of the 'choke' disease of grasses caused by fungus, the plants are vegetatively vigorous but the fungal infestation suppresses the formation of inflorescences (Bradshaw, 1959c).

Subsets 4 and 5 are most commonly sampled by experimenters. Different subsets may reveal quite different spectra of variation; we shall see examples in Chapter 14 when we consider attempts to study the effects of natural selection by comparing the variation of different subsets in cultivation.

One of the biggest difficulties in sampling populations concerns the definition of an individual. In open vegetation it is usually possible to define individuals in annual plants and to see patches of individual perennial plants. In closed swards, however, the problem is more difficult. Sometimes the presence of 'marker' genes (e.g. leaf marks in *Trifolium repens:* Davies, 1963) might reveal the extent of particular individuals: but such markers are rare. Theoretically it might be possible to trace root systems in an attempt to establish the extent of individuals, but the practical difficulties are enormous. Furthermore, in some plants, e.g. certain forest trees, root-grafts occur which unite the root systems of several different individuals (Graham & Bormann, 1966; Böhm, 1979).

The problem of defining the individual is further complicated by clone formation, in which the vegetative continuity of an individual breaks down, producing a clonal patch of several individuals of identical genotype (Harper, 1978). Evidence for clonal populations is usually circumstantial, but direct evidence is available in the case of certain self-incompatible species which are very variable morphologically. Variability has been studied in garden trials of population samples, and the material classified into different individuals on the basis of morphology, phenology, susceptibility to pests and diseases, etc. The behaviour of different plants in crossing experiments is then studied. Crosses between dissimilar-looking plants may yield a 'full seed-set', from which we may infer that the plants have different S alleles and are different genotypically. Conversely, crosses between plants which are morphologically indistinguishable may yield little (or no) seed, and can be thought to share the same *S* alleles, and to be of the same genotype or 'isoclonal'. Some caution is necessary in interpreting experiments of this type as the method depends upon a thorough knowledge of the type of incompatibility mechanism involved, a requirement almost never satisfied with wild species. On present evidence it would seem that extensive clones, probably of great age in some instances, occur in many habitats (see Table 13.3).

Why is knowledge of the extent of individual genotypes important in sampling? Suppose we collect two population samples A and B. Fortui-

Table 13.3. *Some examples of studies of clones (Lines of evidence:*
F. = field observations; C. = cultivation trials; H. = hybridisations;
I. = electrophoretic studies of isozymes)

C.	*Anemone nemorosa* (von Bothmer, *et al.*, 1971): large number of clonal patches, of limited size, in Swedish habitats.
C.	*Arum maculatum* (Prime, 1955): small clonal patches, in Britain.
C.H.	*Eichhornia crassipes* (Barrett, 1980*a*,*b*): deliberate introduction of this plant (native of tropical America) has been followed by clonal propagation on such a scale that this plant is now a serious weed of waterways etc. in many parts of the world.
F.	*Erica* (Webb, 1954): extensive clones as judged from plants of distinctive phenotype.
C.H.	*Festuca rubra* (Harberd, 1961): evidence of many genetically different individuals in a study of an area of South Scotland. One particular variant occurred at points *c.* 220 m apart. If this area was achieved by radial growth then the clone must be *c.* 400–1000 years old. However, perhaps the present distribution has been achieved by dispersal of fragments by animals or other causes, or as a consequence of vivipary, which has been recorded in this species (Smith, 1965). Widespread clones also found in *Festuca ovina* (Harberd, 1962).
F.I.	*Larrea tridentata* (Sternberg, 1976; Vasek, 1980): extensive clonal patches, visible on aerial photographs. By radiocarbon dating oldest clone may be 11 700 years old. Isozyme studies reveal that parts of apparent clones are indeed isoclonal.
C.H.	*Lysimachia nummularia* (Dahlgren, 1922): self-sterile clones found in many parts of North and Central Europe; sexual reproduction only occurs when genetically distinct individuals are found in same site.
F.C.	*Petasites hybridus* (Valentine, 1939): in many parts of Britain this dioecious species is represented only by male plants, which form clonal patches.
F.	*Ulmus* spp. (Rackham, 1975): by studying in British woodlands patterns of morphological variation together with incidence of fungal diseases and timing of coming into leaf and leaf fall, evidence of very extensive clonal patches was discovered.
F.	*Vaccinium* spp. (Darrow & Camp, 1945): extensive clonal patches as judged by patterns of morphological markers.

tously, sample A could consist of 25 pieces of a widespread clone, whilst sample B could consist of material of 25 genetically different individuals. A comparison of the two 'populations' in a cultivation trial is likely to show that they are different, but interpreting this difference as a real population difference could be misleading. Perhaps population A *is* largely composed of the clonally propagated individuals of one genotype, while B is variable; on the other hand, populations A and B might both be variable, and the multiple sampling of one clone in population A might be merely the consequence of poor sampling technique. Harberd and others who have made a special study of the problem (Harberd, 1957, 1958; Wilkins, 1959, 1960; Ward, 1974) recommend that spaced samples be collected from populations. From all the evidence available the probable maximum extent of clonal patches is estimated. Sampling at points separated by distances greater than this estimated clonal patch size is then carried out. Widely spaced samples are to be recommended to counteract another problem which arises on studying plant populations. The lepto-kurtic distribution of fruits and seeds must often result in 'family groups of close relatives', perhaps involving several generations (see, for example, Linhart *et al.*, 1981), and distorted comparisons can arise if a sample containing a group of closely related plants is matched against a set whose members are totally unrelated. It must be noted, however, that wide spacing of samples is somewhat at odds with the present trend of studying small systems in detail. Such studies as those of Smith (1965, 1972) reveal that, while some clonal patches are extensive, there is enormous variation within sites. There seems to be no easy solution to the problems raised by clonal propagation: the experimentalist must make the best judgment possible in each situation in relation to the hypothesis under considera-tion.

Another question to be resolved before sampling is undertaken con-cerns the number of sites and samples within sites. Suppose we study a single site with two different soil types, A and B. Patterns of variation may be revealed in samples drawn from the two subsites A and B, and at the end of an experiment some differences related to soil type may be found in plants originating from the two subsites. The experimenter must then decide whether the differences are ecotypic or whether they owe their origin to random variation. With one A/B comparison it is difficult to rule out random events (Wilkins, 1959). A more penetrating study of the patterns of variation might be made by studying several areas where subsites of type A and B are juxtaposed. Furthermore, in collecting from the wild a bulk seed sample may be made to represent each of the subsites A and B, or the seed from a random collection of mature individuals may be separately collected and packeted at each subsite. Family lines may then be grown, patterns of variation within lines offering some insights into the breeding system of the plants under study. This type of sampling – a hierarchical or nested pattern – has much to recommend it, allowing not only a number of A/B comparisons to be made, but also providing some information on variation within subsites. For instance, the plants under study might be obligate apomicts; while 'seed parents' might differ, family lines might be invariable. The cultivation of plants from bulked seed samples in this circumstance would fail to reveal an important strand in the variation pattern.

Cultivation experiments

A study of variation usually requires cultivation of plants. This is true, not only of field collections brought into a common environment to investigate the nature of variation patterns, but also of many sophisticated genetic and physiological studies. In many cases the experimenter wishes to grow material from diverse sources under the same conditions. Thus, if population samples are collected in the wild, and if there are interesting phenotypic differences between populations, a Turessonian cultivation experiment might be carried out, to see if differences between populations persist in cultivation.

At first sight a requirement to grow material 'under the same conditions' appears to present little difficulty. A moment's reflection, however, is sufficient to remind the reader of the variation in soil fertility, drainage, pests and diseases within even the most uniform experimental plot in garden or field. The notion that glasshouses provide a uniform environment is quickly dispelled by studying investigations of yields of vegetable crops on benches in different parts of experimental glasshouses (see, for example, the little-known experiments of Lawrence, 1950).

In designing genecological experiments, the botanist has much to learn from the agricultural scientist. Farmers wish to grow high-yielding varieties of crop plants and since the middle of the last century research workers have struggled to perfect experiments designed to study yield. In this short book we cannot provide a complete review of this interesting subject and will confine our attentions to a few important general issues. Notable advances in the design of field experiments came with the work of Fisher, who studied the famous long-term Broadbalk Wheat experiment at Rothamsted Research Station in South England (Fig. 13.4). A recent book on the life of R.A. Fisher (Box, 1978) provides a useful historical review of field experimentation and explores in detail Fisher's many contributions to the subject.

The basic ideas behind the design of cultivation trials are as follows:

1. Experiments must be designed with sufficient replications of the varieties, populations, treatments, etc. Thus in a simple experiment on yield in, say, spring Wheat several plots of each variety must be grown.

2. Soil fertility, and other edaphic factors, often vary across garden plots and fields, but it is commonly found that adjacent sites have similar fertility, etc. Thus Fisher recommended that the ground available be divided into uniform blocks (not necessarily square). Each block should contain a full complement of the material under study. Within blocks the small plots of each variety should be *randomly* arranged. In early experiments in agriculture and forestry it was hoped that, by careful husbandry, varieties could be given the same conditions. But a critical approach to experimentation suggests that this is a forlorn hope: it is impossible to ignore the variability induced by environmental factors. With a proper layout of experiments differences between blocks can be

Fig. 13.4. Layout of the famous Broadbalk field experiment at Rothamsted, Herts, England studied by R.A. Fisher. Experimental crops of Winter Wheat have been grown continuously in these plots since 1843. Photograph taken in 1954. Copyright Rothamsted Experimental Station.

measured to give an estimate of the random element of variation introduced into the experiment.

3. Another important factor in the design of field experiments is the effect of position. If plants are growing in blocks, those in the centre of the block will be surrounded by neighbouring individuals; in contrast, plants on the margins of blocks are likely to be adjacent to bare soil and subject to very different amounts of root and shoot competition. Thus it is recommended that 'guard rows' of similar plants be planted around the blocks, to provide uniform conditions for the experimental material. Guard rows, usually of the same species as the plants under study, are discarded at final harvesting of the experiment.

It is clear that these ideas may with profit be incorporated into the design of genecological experiments. (It is most interesting that, at an early date, advanced field trial techniques were employed in the famous genecological experiments of Gregor and his associates in studies of variation in *Plantago maritima* at the Scottish Plant Breeding Station (Gregor, 1930, 1939; Gregor, Davey & Lang, 1936; Gregor & Lang, 1950).)

In a simple genecological experiment each individual, say of plants A, B, C and D, may be clonally propagated, the experimental garden may be divided into small blocks and a ramet of each individual A, B, C and D planted in a weed-free plot surrounded by guard rows of the same species. The position of each ramet within blocks is determined by random numbers.

The fundamental ideas influencing the layout of simple field trials may also be incorporated into the design of more complex genecological experiments such as population trials, family lines and experiments involving populations given various treatments. Several excellent books with fully worked examples of various designs are now available for the biologist. Especially suitable for beginners are: Salmon & Hanson (1964); Heath (1970); Bishop (1971); and Parker (1973). More advanced treatment will be found in: Campbell (1974); Ridgman (1975); Snedecor & Cochran (1980); and Sokal & Rohlf (1981).

Studies of agricultural crops have resulted in other important insights into the design of field experiments. At first sight it would seem reasonable to suppose that repeated experiments with the same varieties (or genotypes) would 'give the same results'. In practice there are considerable differences from year to year in the results of experiments estimating yield in cultivated stocks. The principal causes of variability are differences in weather, and changes in the incidence and severity of various pests and diseases (which are themselves probably correlated with past or present weather conditions). Thus in designing genecological experiments the following factors must be taken into account:

1. The pretreatment of seeds and seedlings prior to the experiment is very important. There will be differences in the speed of development of plants between those sown as seed and those set out in the field as young plants. The timing of the experiment in relation to such seasonal factors as cold periods may also be crucial. Thus some plants will not flower unless subjected to cold treatments, and spring and autumn sowing will yield different results.

2. The treatment of plants during the experiment has a profound effect upon their growth and performance. The experimenter must decide whether to water plants in dry weather, apply fertilisers, etc.

3. The incidence of pests and diseases causes considerable problems. In particular, experiments in glasshouses often turn into a struggle to control various insect and fungal pests, and the liberal use of pesticides may be the only means of 'preserving' the experiment. It is important to realise that 'spot-treatments' of badly infected individual plants may seriously affect the random-ised design of the experiment. In the design of garden and field trials, on the other hand, the decision is often taken to allow non-catastrophic invasions of pests and diseases to take their toll of the experimental material. In this way it may be possible to see if any individuals or populations are resistant to fungal or insect attack. Studies of the effects of non-fatal pests and diseases may add a further dimension to our knowledge of population varia-tion.

4. The length of an experiment may be crucial if the investigation involves material dug up from the wild and transplanted into a garden for, as Turesson (1961) discovered, an extended period of adjustment may be necessary before plants may be said to have outgrown the effects of their original habitats. Indeed, it may be difficult to convince a sceptic that a complete adjustment is ever made, especially in the case of woody plants.

5. Agricultural experiments are often designed to be left until final harvest when estimates of yield are made on fruiting material, and in other cases the experiment is so constructed as to permit regular intermediate harvests at selected periods between sowing and final harvest. Such experiments may be poor models for

experiments in the ecological genetics of plants, in which a great deal of information may be gathered by 'non-destructive scoring' of the plants over weeks or months. For instance, given adequate spacing between plants, plant height at different times could be measured, and the timing of flowering and fruiting could be studied. Also samples of leaves could be removed for study, provided that all the material in the experiment is treated alike. Thus a good deal of quantitative information might be obtained by repeated scoring of an experiment. Sometimes it is unnecessary to make measurements: the stages of development or incidence of damage by pests may be recorded by classifying the material into a small number of 'character states'.

The designed experiment

So far we have discussed a number of important factors in the design of genecological experiments. For both the experimenter and the botanist who wishes to interpret the scientific literature it is crucial to take proper account of the problems and possibilities of sampling and cultivation. These are elements in a larger canvas however. Many authors have stressed that genecologists should aim at a *designed experiment* in which hypothesis, sampling, cultivation, analysis and interpretation all take their proper places.

The generation of germinal ideas is a mysterious process. Armed with a knowledge of the literature, provoked by the observations and comments of others, the botanist notices something of interest in the patterns of variation. From this initial interest an idea emerges for an experiment. The process by which ideas occur to experimenters is not to be seen as a mechanical process but as a creative act much as is required for practice of the arts. Next the experimenter formulates a hypothesis leading to an experimental investigation, the results of which are used to consider whether the hypothesis is confirmed or rejected. As part of the investigation the results may be subjected to statistical tests.

The best way to appreciate the different elements in the designed experiment is to study an example. We have chosen to present the results of a simple study on *Plantago major* (Warwick & Briggs, 1979). Our account should be seen as a simplified (certainly oversimplified) introduction to a central concern of science, namely how to devise, execute and interpret experiments. We hope that biologists reading our account will be encouraged to study the many excellent introductory books (which we have noted above) on the design and statistical analysis of experiments.

An experiment to study the variation in Plantago major *growing on droves (grassy tracks) at Wicken Fen Nature Reserve, Cambridgeshire, England*

Many thousands of visitors visit the famous Wicken Fen Nature Reserve each year and the droves (grassy tracks) which cross the fen are subject to severe trampling pressure. *Plantago major* occurs in the heavily trampled areas (as a small prostrate plant) and also in the adjacent grassy sward (in which it is a larger, erect plant). Although no observations or experiments have been carried out, it is almost certain that *P. major* plants in both habitats are within a common gamodeme.

Ecotypic differentiation has been reported in *P. major* (Turesson, 1925; Groot & Boschuizen, 1970; Mølgaard, 1976) and as we shall see in Chapter 14 there is evidence from a number of genecological studies which suggests that differentiation might occur over short distances, despite gene flow. Therefore the possibility exists that dwarf prostrate variants might be selected on the pathway, while taller plants would be at a premium in the adjacent grassy sward. Thus we could formulate the hypothesis that samples taken from the wild might retain their distinctness in cultivation. As the differences involved are those of size, our hypothesis is not very precise in its present form. We cannot make any definite prediction as to the degree of difference to be retained; indeed as we are dealing with quantitative differences it is not at first sight clear how one can make a prediction as to the degree of difference which 'needs' to be retained in order to accept the hypothesis. So far our hypothesis is too vague. However, a precise hypothesis is possible in this case, namely that on cultivation we expect *no* difference between groups of *Plantago* after cultivation. Such a hypothesis is known as a 'null hypothesis'. The concept of the null hypothesis is widely used in biology and such a hypothesis, that zero difference is expected between two sample groups, should always be formulated as part of a designed experiment, for a precise initial hypothesis is likely to lead to a well-designed investigation.

Unbiassed samples, 10 from the trampled area and 10 from adjacent grassy swards, were collected in the autumn of 1974. *P. major* is not a clonally propagating species (although it may be cloned in gardens: Marsden-Jones & Turrill, 1945), but spaced samples were taken at least 10 m apart. Plant material was potted up in John Innes No. 1 compost and the pots, which were randomly arranged, were plunged to the rims in the sand of an outdoor plunge bed. Spacing between pots was very generous and guard rows were not necessary.

In order to allow us to examine the null hypothesis, a statistical test is necessary to enable us to compare the two groups of samples. The test should allow us to compare the variation between and within groups. Clearly variation between groups (from trampled path *versus* adjacent grassy sward) is only likely to be significant if it can be shown to be significantly greater than variation within groups. We shall use for our test the analysis of variance technique, which works by estimating the significance of variation between groups by comparing it with variation within groups. The variation in some measurable trait of 20 plants of *Plantago major* is, by this test, partitioned in such a way as to enable us to see the variation due to subsites at Wicken, while at the same time giving us an estimate of the variation within groups.

The steps in the analysis of variance are a simple extension of those used in Chapter 3. To recapitulate, we showed that:

$$\text{variance } (s^2) = \frac{\sum (x - \bar{x})^2}{n - 1}$$

The sum of the squares of the deviations from the mean could be calculated by subtraction of each value from the mean, squaring the difference and summing the resulting squared deviations. Alternatively we suggested that, if a calculating machine is available, the sum of (deviations from mean)^2 (sum of squares) could more readily be calculated by employing the formula

$$\text{sum of squares} = \sum x^2 - \frac{\left(\sum x\right)^2}{n}$$

Where $\dfrac{\left(\sum x\right)^2}{n}$ is known as the Correction Factor or Term, C.

We may now examine (Table 13.4) the steps in the calculation of simple analysis of variance on the *Plantago major* experiment. The *null hypothesis* is that there is no difference in leaf length between plants grown from trampled drove and from grassy sward.

Table 13.4. Plantago major: *length of longest leaf (cm) in plants after c. 10 months cultivation in the Botanic Garden, Cambridge*

Wicken Fen: trampled areas on droves		Wicken Fen: grassy swards adjacent to droves
	30.5	33.3
	33.4	28.0
	25.5	21.9
	34.2	26.0
	27.4	24.0
	26.5	28.4
	31.5	32.2
	29.3	27.0
	24.8	26.3
	28.0	26.0
Mean	29.110	27.310
Total	291.100	273.100

Grand total = 564.200

$$\text{Correction factor} = \frac{564.200^2}{20} = 15916.082$$

Sum of squares (total) = $(30.5^2 + 33.4^2 + 25.5^2 \dots 26.0^2) - C$
$= 16133.080 - 15916.082 = 216.998$

Having calculated the total sum of squares we now calculate the variations between and within groups.

Between groups is estimated by

$$\frac{291.100^2}{10} + \frac{273.100^2}{10} - C = 15932.282 - 15916.082 = 16.200$$

Within groups is estimated by subtracting 16.200 from the total sum of squares. Within groups sum of squares = $216.998 - 16.200 = 200.798$

Subdivision of the sums of squares into its two parts has been accomplished and the degrees of freedom (19 in all: one less than the number of observations) may now be determined for each component. Between groups: 2 groups, therefore 1 degree of freedom. Within groups: 10 observations per group, each loses 1 degree of freedom, total 18.

The analysis of variance may now be set out in a table showing the sources of variation, the divisions of degrees of freedom and sum of squares. Mean squares (variances) are now calculated. The between-groups mean square gives the variance of the two groups about the grand mean, while the within-groups variance gives the variance of individual values about the two sample means.

Source of variation	Degrees of freedom	Sum of squares	Mean square (variance)	Variance ratio (F)	Probability
Between groups	1	16.200	16.200	1.452	> 0.05
Within groups	18	200.798	11.155		
Total	19	216.998			

If the null hypothesis, *viz.* that there is no difference between the two groups of *Plantago major* plants, is to be confirmed then there should be little or no difference between the variances between and within groups. To estimate the relative size of these two variances we calculate the variance ratio (the *F* value – in honour of R.A. Fisher who developed analysis of variance). If, however, there is a real difference between groups, we would expect variation between groups to exceed that of the variance within groups. Tables of probabilities appropriate to different values of *F* are available (see Appendix). In the case of the *Plantago* experiment it is clear that a good deal of the variation is *within* groups and that the difference between groups is small. The mean values are very similar. Indeed, leaf length of plants grown from the small plants of the trampled area slightly exceeds that for the samples from the tall sward. The null hypothesis, that there is no statistically significant difference between the two groups of plants, is supported by our results. On the strength of present evidence, we have no reason to suppose that 'ecotypic' differentiation has occurred in the trampled and tall sward subsites.

The *Plantago major* investigation was part of a more extensive study of this species in various grasslands (Warwick & Briggs, 1979, 1980*b*). Table 13.5 sets out another comparison. Small phenotypes were found not only in trampled areas (as on the droves at Wicken), but also in closely mown lawns. Samples of plants from the Botanic Garden lawn in Cambridge and from Wicken droves (trampled areas) were compared.

The variation between groups in this case is statistically significantly greater than the variation within groups. Therefore, the null hypothesis, namely that samples do not differ in leaf length, receives no support from the experiment. There would appear to be a real difference in leaf length between the two samples. In Warwick & Briggs (1979, 1980*b*) details are given of the highly distinctive plants of *Plantago major* discovered in the lawns of Cambridge colleges and gardens.

Table 13.5. Plantago major: *the effect of* c. *10 months cultivation on samples of small phenotype from Wicken droves and Botanic Garden lawns: leaf length of longest leaf (cm)*

Wicken: trampled areas on droves		Botanic Garden lawns
	30.5	11.8
	33.4	20.7
	25.5	8.9
	34.2	22.6
	27.4	24.0
	26.5	14.1
	31.5	13.1
	29.3	16.0
	24.8	12.5
	28.0	12.0
Mean	29.110	15.570
Total	291.100	155.700

Grand total = 446.800

$$\text{Correction factor} = \frac{446.800^2}{20} = 9981.512$$

$$\begin{aligned}\text{Sum of squares (total)} &= 30.5^2 + 33.4^2 + 25.5^2 \ldots 12.0^2 - C \\ &= 11228.860 - 9981.512 = 1247.348\end{aligned}$$

$$\text{Between groups sum of squares} = \frac{291.100^2}{10} + \frac{155.700^2}{10} - C$$
$$= 916.658$$

Within groups sum of squares = 1247.348 − 916.658 = 330.690

Source of variation	Degrees of freedom	Sum of squares	Mean square (variance)	Variance ratio (F)	Probability
Between groups	1	916.658	916.658	49.895	<0.001
Within groups	18	330.690	18.372		
Total	19	1247.348			

Our examples of analysis of variance are of a very simple kind, with 'division' of the variation into two parts. Much more elaborate experiments may be devised and the 'overall variation' discovered in experiments may be divided into many parts estimating, where appropriate, the variation due to blocks, population differences, family lines within populations, interacting factors, random events, etc. By looking at the relative magnitude of different segments of the variation very considerable insights into population variation may be obtained.

Analysis of variance is a most elegant technique, which must, however, be used with care. It should only be employed in analyses where the results are 'normally distributed' and in which the variances of the contributing population samples, treatment values, etc., are equal or approximately so. Various tests have been devised to study the 'properties' of arrays of figures to see if they are appropriate for analysis of variance (e.g. for details of Bartlett's test see Salmon & Hanson, 1964; Sokal & Rohlf, 1981). Sometimes it is possible to 'transform' the results to produce equality of variances. For instance, the unsatisfactory raw data may be converted to square roots or to logarithms. If the results cannot be satisfactorily transformed then other statistical tests – so called non-parametric tests– – may be applied (Sokal & Rohlf, 1981). Such tests do not make any assumptions that the figures from the experiment are normally distributed or have equal variances. Non-parametric tests should be more widely used in biology, for the results of many experiments and observations show enormous departures from normality.

The interpretation of experiments

The final phase of an experiment involves interpretation of the results. However, whatever the results of particular experiments, there are usually grounds for a cautious interpretation of genecological studies, for the following reasons:

1. However many plants are grown, or studied in experiments, the size of samples which can conveniently be grown is often minute relative to the size of wild or semi-natural populations. For instance, according to the estimates of Barling (1955), populations of *Ranunculus bulbosus* may reach 257 000 per acre in the English Cotswolds, and continuous populations in adjacent fields of pasture were estimated to contain 14 000 000 plants. Such figures are by no means exceptional.

2. Many experiments are carried out in conditions remote from those in nature. For example, studies of metal tolerance in plants involve measurements of root growth in very simple culture

solutions (see Chapter 14). In the wild, plants grow in soils where conditions are quite different. The attempt to simplify situations in order to study individual factors is clearly justified, but the investigator must not make too facile an extrapolation from simple laboratory tests to the natural situation.

3. As in the case of metal tolerance many experimentalists 'isolate' individual factors of presumed importance and make special studies of the tolerances of population samples. A fascination with the study of critical or limiting factors should not blind the student of evolution to the fact that the concept of a factor is an abstraction. Often particular factors are chosen for study largely because the means to control or vary them in precise ways are available in laboratories. It is often forgotten that plants respond to their environments as a functioning whole. Realising this difficulty, a number of botanists are becoming interested in the experimental studies of the adaptive significance of variation in plants by carrying out experiments in the field. The garden trial with its weed-free, spaced plants is not entirely satisfactory as a means of studying 'adaptation', for the competitive interaction between plants is absent. Thus there has been a revival of interest in the reciprocal clone-transplant experiment, in which cloned material of diverse origin is transplanted into swards subject to different treatments. By close mapping and labelling of plants the survival and growth of transplants may be studied (see Chapter 14). Care in the layout and recording of such experiments may overcome the difficulties which, as we saw in Chapter 6, cast doubt on the historic studies of Bonnier & Clements. By studying the way plants behave in such experiments, the experimentalist may have a very direct insight into the responses of plants to 'whole' environments. There is obviously a place for both types of study – tolerance tests and reciprocal transplant investigations – in the repertoire of techniques available to the genecologist.

4. A final problem facing the experimentalist is that of deciding the nature of the underlying patterns of variation under study. Even after long and complex experiments it may not be possible to conclude with certainty that residual variation in, say, a garden trial, is 'genetic' in origin: breeding experiments are necessary to see if characteristics are transmitted by seed.

In order to study the question 'What genetic changes occur in ecotypic differentiation?' it is necessary, as Clausen & Hiesey (1958) discovered, to carry out many lengthy genetical experiments. As we saw in Chapter 8 there was a good deal of difficulty in sorting out the genetics of quantitative characters in their experiment. For many purposes a simpler method of deciding whether two plants differ genetically could be most valuable, and recently with the study of enzyme polymorphism new insights into population variation in plants are promised.

Enzyme polymorphism

In Chapter 6 we gave a brief account of some ideas in molecular biology. The nucleotide sequences in DNA provide the fundamental information specifying the genotype, but in observations and experiments the botanist must deal with phenotypes which are interaction products of genotype and environment. Even with a rigorous control of external conditions, studies of the phenotype only allow revelation of the underlying genotype in specially contrived experiments. In a very real sense the phenotype is remote from the genotype, from the DNA code. A method of 'direct' access to the genotype would be most interesting, but DNA sequences cannot be routinely studied in plants. Theoretically the structural differences in protein might reveal 'the genotype' as they are only a few steps removed from the code, but there is as yet no simple way of studying the details of protein structure in plants. Sophisticated techniques will allow insights into DNA and protein structure in special cases, but the methods involved are too complicated and expensive for genecological use.

Proteins contain different combinations of amino acids which differ in their molecular size and electrochemistry. The protein molecules therefore differ in mass and net charge, and it has proved possible to study differences in protein structure, particularly enzymes, by their physicochemical properties, using the technique known as gel electrophoresis. Synthetic polymer, starch or agar gels are prepared. A homogenate of plant or animal material is inserted into a sample pocket in the gel and an electric field is generated across the gel (Fig. 13.5). The molecules of protein migrate in accordance with their mass and net charge. Enzymes, which are the most commonly studied proteins, are present in minute quantities in the gel. In order to locate them, for they are invisible, a substrate is added together with a dye which produces a colour reaction at the site of enzymic breakdown of the substrate.

If several plant homogenates from individuals of the same taxon are run in parallel across the gel and many such experiments are carried out, an interesting fact emerges. Many different patterns of banding are found for each class of enzyme. Biochemists have concluded, as a result of critical studies of selected enzymes, that each enzyme is polymorphic, that is, it consists of a family of related structures which 'catalyse' the same general reactions, but the members of each family – often called isoenzymes, or isozymes – differ in amino-acid composition, as a consequence of variation at the DNA level in the locus (or loci) coding for the enzyme. Mutations in DNA result in changes in the detailed pattern of amino acids in the enzyme molecule and 'substitution' of one amino acid for another may yield an enzyme molecule with different properties of mobility under electrophoresis.

Not only do banding patterns reveal polymorphism but, because the phenomena of dominance and recessiveness do not apply in many cases at the molecular level, both homozygotes and heterozygotes may be detected in population studies. Homozygotes have single staining bands whilst, in contrast, heterozygotes have two or three bands. The supposition that a plant is heterozygous may therefore be put to the test. The detection of apparent heterozygosity may be made on a single leaf of a plant, which may be relatively unharmed by its removal. Selfing of the plant may be readily accomplished in self-compatible species and the segregation patterns appropriate to a heterozygote may be looked for in the seedling progeny. From initial testing of the parent to studies of offspring may

Fig. 13.5. Diagram of a vertical-slab gel-electrophoresis apparatus. (From Lewontin, 1974)

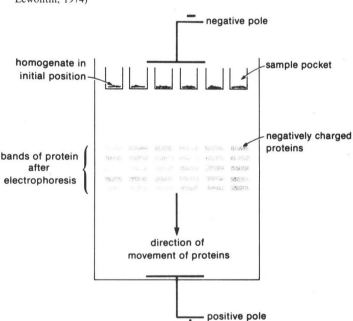

take only a few weeks, and information on many enzyme systems for large samples may be assembled. Some details of the results of experiments on enzyme polymorphisms will be given in Chapter 14.

Considering this recent development in plant genecology, biochemists and geneticists have stressed that the method must be applied with great care, and the following precautions are necessary:

1. Material of exactly the same stage of development should be compared, as different genes are 'switched on' at different stages in a plant's life history.
2. Experimentalists should be careful to minimise the effects of secondary plant products (gums, latexes, etc.), which might interfere with gel electrophoresis and seriously distort the results obtained.
3. Care should also be taken to test the deduction that genetic differences underlie patterns of enzyme polymorphism.
4. It should be noted that not all DNA changes leading to amino-acid changes in enzymes will result in differences in electrophoretic properties (e.g. Boyer, 1972; Garrick, Bricker & Garrick, 1974); thus, the method probably seriously underestimates the level of genetic variation in the material under study.

In this chapter we have discussed some of the important issues to be faced in studying variation in populations. After reading our survey, it is possible that some will charge us with making the subject too difficult, by raising too many problems. Our answer to the charge is that if the subject is to develop and progress then the full complexity of natural systems must be appreciated. Complexity is not a figment of our imagination. How genecologists have reacted to the complexity of nature in the design of experiments to study pattern and process in populations will be the subject of our next chapter.

14

Recent advances in genecology

The masterly review of genecology by Heslop-Harrison (1964) describes and evaluates progress made in the subject from the 1920s to the early 1960s. Heslop-Harrison states that three basic propositions governed early work:

1. Wide-ranging plant species show spatial variation in morphological and physiological characteristics.
2. Much of this infraspecific variation can be correlated with habitat differences.
3. To the extent that ecologically correlated variation is not simply due to plastic response to environment, it is attributable to the action of natural selection in moulding locally adapted populations from the pool of genetical variation available to the species as a whole.

The direction of genecological research

The work of Turesson and others provided plenty of evidence that patterns of variation are habitat-correlated and genetically based, and there was a great deal of discussion of the 'reality' of ecotypes. In the same period, as we saw in Chapter 13, other genecologists were concerned to improve the design of garden trials and other experiments. In contrast, the third proposition, namely that pattern is the result of the action of natural selection, received little experimental investigation.

In attempting a short review of contemporary developments in genecology, we note that botanists are still interested in patterns of variation between and within populations and the means used to investigate such patterns, but a number of other issues have emerged to enrich genecological researches.

The role of natural selection

Genecologists are now fascinated by the possibility of studying natural selection in populations, and theoretical and experimental studies

294

of populations are providing a picture of what is involved in natural selection. Processes which Darwin grouped together under this blanket term are now classified into three main types (Fig. 14.1).

The first type of selection – so-called stabilising or normalising selection – is a process tending to produce conformity and stability. At first sight it is surprising to find selection acting in this way, as many people equate natural selection with change. Imagine, however, a gamodeme which is well adapted to its environment. The environment does, of course, change climatically and biotically with the seasons, but we assume that it is not changing directionally and fundamentally. By sexual reproduction an array of phenotypes is produced, a sample of which may exhibit a typical normal distribution for some character; most of the individuals in the array depart little from the mean, but some segregants in each tail of the distribution are markedly distinct from the mean. Stabilising selection has the effect of eliminating individuals which depart significantly from the mean, giving a bimodal distribution of eliminants.

The second type of selection – directional selection – may be postulated where the environment is changing in a particular direction, as might, for instance, occur with the onset of a glacial epoch. This is directional selection. Here selection will produce a change in the mean values for significant phenotypic characteristics. Again, as a consequence of selection, a portion of the young is eliminated.

The third and final type of selection is called disruptive selection. In this case individuals of different genotype may be at a selective advantage in different places within the total area occupied by a gamodeme, giving a pattern of genetic polymorphism in a mosaic or patchy environment.

These models of natural selection provide guidelines for the design of experiments aiming at detecting selection in action in populations. In all cases selection involves the maturation and reproduction of only a portion of the population, and adults will be less variable (in significant phenotypic traits) than the young from which they develop. Thus, if the variation in young and adult is compared, it should be possible to detect selection at work. This idea is of course not a new one. As we saw in Chapter 5, as long ago as the 1890s the zoologist Weldon studied variation in young and adults in crabs.

Considering the different model systems we might reasonably decide that disruptive selection might be most easily detected in the field. Suppose that in a study area there are two extremely dissimilar habitats 'a' and 'b', and that a common species found in both habitats is genetically polymorphic, with morph A in habitat a and morph B in habitat b. If these sites are contiguous, crossing between A and B may

Fig. 14.1. The three types of natural selection. (From Hardin, 1966)

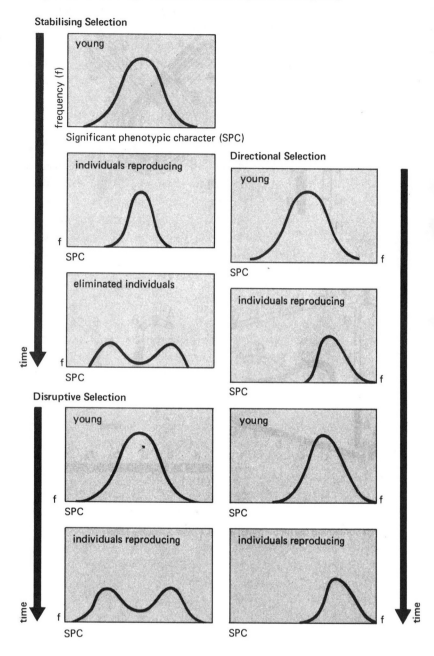

occur at each generation, giving a wide spectrum of variation from A to B. Given the extreme difference between habitats a and b, it is likely that disruptive selection will occur, plants of the appropriate morph maturing and reproducing in each habitat, with the elimination of unfit 'young' individuals at both sites. Whilst early studies often involved sampling over a wide geographical range, more recent work had begun to examine small-scale differentiation patterns in greater detail. For instance, Bradshaw (1959*a*, *b*, *c*, 1960) investigated *Agrostis capillaris* (*A. tenuis*) over a relatively small area of Central Wales, and Cook (1962) studied variation in Californian Poppy (*Eschscholzia californica*) over a wide area, together with a survey of sites along a 35-mile transect. The logical end-point of the trend has now been reached in studies of polymorphic populations in mosaic or patchy environments. The comparison of plants likely to be in different gamodemes, which dominated many early studies, is still an active area of research, but as we shall see below genecologists are now studying local populations of plants likely to be in the same gamodeme, concentrating their attentions in particular on patterns and processes found at the junction of two extremely dissimilar habitats.

The study of widely spaced sites often involves the examination of gross differences in phenotype, say as large as those found in Mendel's stocks of peas – tall (*c*. 2 m or more) *versus* dwarf (*c*. 0.5 m). Almost any garden trial, however crude, would hardly fail to reveal such differences in height. However, with an interest in studying small sites as systems has come a fascination with small-scale (often quantitative) variation. As we saw in Chapter 13, the study of this type of variation requires good sampling techniques, combined with proper attention to design, layout, scoring and analysis of experiment.

Historical ecology and man-made habitats

Evidence for patterns and processes within populations is likely to be more complete if some notion of the history of study sites is obtained. In the last decade many ecologists have become interested in historical ecology, using evidence from different sources to allow the botanist to see present-day vegetation in its historical context. To illustrate such studies, Fig. 14.2 presents a synopsis of the history of Buff Wood, a small wood owned by Cambridge University Botanic Garden. The following genecological investigations have been carried out there: population variation in *Crataegus* investigated by Bradshaw (1953, 1971, 1975); patterns of hybridisation between *Primula vulgaris* and *Primula elatior* studied by Valentine and associates (Valentine, 1975*b*, and references cited therein); and clones of Elm (*Ulmus*) studied by Rackham (1975, 1980).

Fig. 14.2. Historical ecology of Buff Wood, Cambridgeshire, England. This small wood of *c.* 16 ha is situated on a calcareous boulder clay plateau west of Cambridge. Its history is complex; there is an original core of presumed ancient woodland surrounded by areas which have been allowed to revert to woodland from agricultural use or after the decline of human settlement at various periods from 1350 onwards. (*a*) General plan showing historical evidence of earthworks and ridge-and-furrow ploughing. (*b*) Detail of the distribution of elms (*Ulmus*) in the wood. Each letter represents a separate elm invasion. Clones of *Ulmus procera* and *procera* × *minor* are shown by different kinds of double-hatching; a single tree by x. *U. glabra* and hybrids are shown by single-hatching; a single tree (N) by a square. *U. minor* clones are shown by various kinds of stippling; small clones are left blank; a single tree (O) is shown by a triangle. (From Rackham, 1975, 1980)

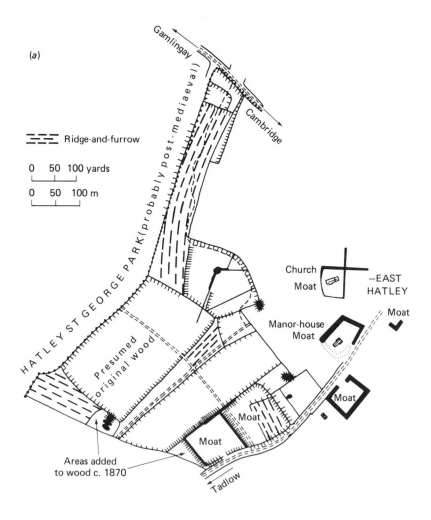

Space is not available to review these researches here but we are concerned to make the general point that each study takes a proper account of historical factors and is enriched thereby. The genecologist neglects the history of his sites at his peril. As we shall see later in this chapter, a knowledge of the history of study areas has provided a framework in which the speed of evolutionary change may be judged.

A study of the examples cited in Heslop-Harrison's review gives the impression that in times past many genecologists chose to study more or less natural communities. As we shall see, however, many insights have come from agricultural experiments. Furthermore, it has recently been

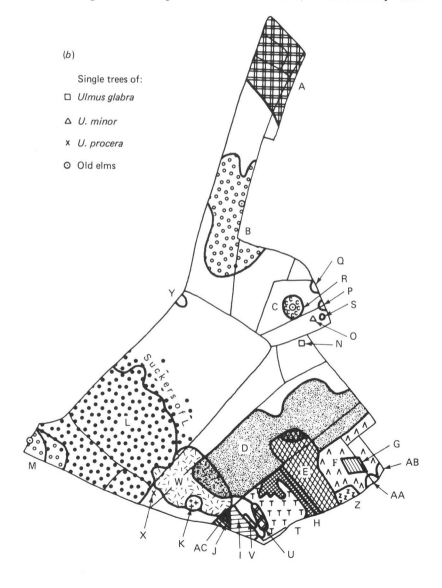

(b)

Single trees of:

□ *Ulmus glabra*

△ *U. minor*

× *U. procera*

⊙ Old elms

realised that many of the most fascinating genecological problems are to be found in man-disturbed habitats (Bishop & Cook, 1981; Mansfield & Freer-Smith, 1981). In such areas the botanist often finds the juxtaposition of extreme habitats, areas in which disruptive selection might be expected to occur. For instance, areas with heavy-metal pollution can occur as islands in a sea of pasture land; grasslands managed for grazing may be found next to hayfields, arable land or woodland; lawns are surrounded by flowerbeds; and improved grasslands, subject to fertiliser and pesticide treatments, are located next to untreated grasslands.

Population biology

In the last decade there has been another major development in ecological research which has influenced the thinking of genecologists. Population biology, which concerns itself with the numbers of organisms, has come of age with the publication of Harper's text book (1977). This most impressive work reviews the information on the demography of plant populations, the processes of colonisation, the effects of over-crowding, etc. Ecologists now study populations in detail, recording the life-spans of individuals, the effects of migration and competition, and details of resource allocation and reproductive success (Willson, 1983). The close study of populations in the field, coupled with imaginative experiments, has begun to offer considerable insight into the factors affecting the number of plants and the consequences of variation in numbers. In their work, often on man-disturbed or artificial populations, ecologists study populations as systems, exploring the plant numbers in one area often over a period of years. Many ecologists are now beginning to consider the implications of genetic variation between and within populations, and are interested in genecological issues as ecologists; in contrast many genecologists of the last 50 years came to the subject through an initial interest in genetics or taxonomy.

Examples of recent experiments

Having considered the main interlocking trends in recent genecological studies, we now examine a number of experiments which have been influenced by some or all of these ideas.

Studies of single factors

The results of many genecological studies have been published in the last 20 years. One must be wary of suggesting common influences for such a varied group of experiments but it is possible to make the following generalisations. First, as we have noted above, genecologists have recognised the importance of comparing plants from *extreme* habitats. If natural selection is to be studied, it will be most readily investigated by studying intraspecific variation in grossly dissimilar sites differing in toxicity, exposure, temperature, light, moisture, salinity, etc. Furthermore, a recognition of the value of studying extreme habitats has led to more rigorous experimentation. Instead of cultivating a haphazard collection of specimens, the genecologist, armed with a specific hypothesis, chooses to collect material from several very different habitats. Sites are sometimes distant from each other, sometimes side by side. Plants are collected and cultivated and subjected to a tolerance test. In such tests a major habitat factor, in which experimental sites are presumed to differ, is investigated under garden or laboratory conditions. For instance, in a recent study of the possibility of ecotypic differentiation in response to trampling, Warwick (1980), working with *Poa annua* and *Plantago* from areas subjected to different trampling stress, investigated the effect on growth and reproduction of treatments with a 'mechanical foot' (a weighted device simulating the pressures of human trampling).

Experiments of this type reveal many patterns explicable in terms of selection (see Table 14.1 for a representative sample). In judging tolerance testing it is important to assess the completeness or otherwise of the evidence. The assumption that the habitats differ in critical factors should be investigated fully by quantitative methods. For instance, if it is supposed that two sites differ in copper contamination, measurement of copper in the soil should be made and the availability of the copper *to the plant* should be examined. Total copper content of the soil might not reveal the level of copper in solution round the plant roots. Bearing in mind the difficulties of extrapolation from laboratory to the field situation, tolerance tests should be as 'natural' as possible. In a large number of cases the investigation has reached the point where, for example, plants from sites A and B have been shown to be subject to different levels of factor X in the field. In a tolerance test, manipulation of factor X reveals a difference in sensitivity to X which is consistent with the level of exposure to X each receives in the wild. In relation to a specific factor each group of plants seems to be best able to survive, grow and reproduce in its native conditions. In such circumstances the genecologist often suggests that the pattern is the result of natural selection. Commonly the evidence for this view is circumstantial, but sometimes contiguous contrasted habitats have been studied and perhaps these are within the limits of gene flow. The notion that selection must be at work on the products of gene flow, maintaining a distinctive pattern in adults, then gains some credence.

Table 14.1. *Examples of studies of ecotypic differentiation selected to indicate the variety of tolerance tests devised by genecologists (see also reviews by Clements, Martin & Long, 1950, Heslop-Harrison, 1964, and Antonovics, Bradshaw & Turner, 1971)*

1. *Soil*
     (i) *Use of natural soils*
   Limestone soil; *Teucrium scorodonia* (Hutchinson, 1967)
   Colliery waste; *Agrostis capillaris* (*A. tenuis*) (Chadwick & Salt, 1969)
     (ii) *Natural soil plus additions*
   Mine soil plus garden soil; test used widely by Liverpool group studying heavy-metal tolerance (Bradshaw & McNeilly, 1981)
     (iii) *Water culture experiments*
   Effects of chromium, nickel, magnesium; *Agrostis* spp. from serpentine (Proctor, 1971a, b)
   Various heavy metals; e.g. arsenic (Pollard, 1980), cadmium (Coughtrey & Martin, 1978)
   Effect of aluminium (Chadwick & Salt, 1969; Davies & Snaydon, 1973b)
     (iv) *Soil water stress*
   Pots allowed to dry out; *Pseudotsuga taxifolia* (Pharis & Ferrell, 1966)
     (v) *Flooding*
   Pots with different water regimes; *Veronica peregrina* (Linhart & Baker, 1973)
2. *Light*
   Variation in light regimes using shade tubes; *Teucrium scorodonia* (Hutchinson, 1967)
3. *Exposure*
   Plants grown in pots in exposed sites; *Agrostis stolonifera* (Aston & Bradshaw, 1966)
4. *Effects of loss of foliage*
   Effects of grazing animals; various taxa (see Watson, 1969; Jones, 1978)
   Effects of mowing; *Plantago major* (Warwick & Briggs, 1980b)
5. *Air pollution*
   Gas chambers, etc.; *Geranium carolinianum* (Taylor & Murdy, 1975; Horsman, Roberts & Bradshaw, 1979) and in studies of various grass species (Ayazloo & Bell, 1981)
6. *Herbicides*
   Various tests on many taxa (see Lebaron & Gressel, 1982)

Studies of several interacting factors: cyanogenesis polymorphism

In many cases where a genetic polymorphism has been discovered, the effect of a single factor has been examined, but it is clear that the relatively simple picture which such studies reveal is part of a larger canvas. This important point is beautifully illustrated by studies of cyanogenesis in plants, in which we now know that many factors interact. It is worthwhile to describe this case in some detail.

In the Sudan campaign of 1896–1900 a number of British transport animals were poisoned by eating a local species of *Lotus* (Dunstan & Henry, 1901). Chemists, interested in the losses, discovered that certain species of the genus contain cyanogenic glucosides. If leaves are bruised, the glucoside is broken down and hydrogen cyanide is liberated. Further studies revealed that there were varying amounts of the glucoside in the leaves and stems of some plants of *Lotus corniculatus:* other plants of the same species proved to be acyanogenic (Armstrong, Armstrong & Horton, 1912). Dawson (1941) studied this polymorphism in the south of England. Using sodium picrate papers, which redden in the presence of hydrogen cyanide, he tested samples from different populations; a selection of his results is given in Table 14.2. By crossing cyanogenic and acyanogenic plants, Dawson was able to show that it was likely that the presence of glucoside is dominant to its absence. The genetics of the situation is complicated, however, as *Lotus corniculatus* is a tetraploid.

In *Trifolium repens*, another species polymorphic for cyanide production, the genetical position is simpler. Atwood & Sullivan (1943) demonstrated that, in this species, glucoside presence (allele *A*) is dominant to glucoside absence (allele *a*). Both *Lotus corniculatus* and *Trifolium repens*

Table 14.2. *Cyanogenesis in* Lotus corniculatus *(Dawson, 1941)*

Localities in England	Numbers of plants	
	Positive test for HCN	Negative test for HCN
Studland Heath, Dorset	77	56
Ballard Down, Dorset	95	56
Ranmore, Surrey	145	8
Crumbles, Sussex	150	5

Table 14.3. *The phenotypes of cyanogenic and acyanogenic* Lotus corniculatus *(Jones, 1973)*

Plant contains	Shorthand notation	Gross phenotype	Varietal type
Cyanogenic glucosides and enzyme	G + E	cyanogenic	*amara*
Cyanogenic glucosides but no enzyme	G + no E	acyanogenic	*dulcis*
Enzyme but no cyanogenic glucosides	no G + E	acyanogenic	*dulcis*
Neither cyanogenic glucosides nor enzyme	no G + no E	acyanogenic	*dulcis*

are also polymorphic for the enzyme which hydrolyses glucoside to produce cyanide. (See Fig. 14.3 and 14.4, and Table 14.3.)

The distribution of the variants in *T. repens* was examined by Daday (1954*a*, *b*), who showed that cyanogenic plants were present with high frequency in South-west Europe. In contrast, acyanogenic plants predominate in North-east Europe (Fig. 14.3). In intermediate sample stations different proportions of the two variants were found; there is in fact a 'ratio-cline' across Europe. Interesting observations were made also upon the frequency of cyanogenic plants at different altitudes in the Alps, and again a ratio-cline was discovered, high frequencies of cyanogenic plants being reported from low altitudes. This frequency declined with increasing elevation, until at high altitudes all the plants in the sample proved to be acyanogenic (Fig. 14.4). In interpreting these findings, Daday showed that there is a correlation between cyanogenesis and January mean. temperatures, a decrease in temperature being associated with an increase in frequency of the acyanogenic variant. It appeared likely from this work that winter temperatures played some direct role, through natural selection, upon the frequency of cyanogenic plants of *Trifolium repens*, or that the locus concerned with glucoside production is genetically linked to genes involved with fitness responses at different temperatures (see Daday, 1965).

In more recent studies new light has been shed upon these patterns. Jones (1962, 1966), investigating the frequency of cyanogenic plants of *Lotus corniculatus* in different English localities, noted that while such plants were relatively free from damage by small invertebrates, many acyanogenic plants showed signs of having been grazed by slugs and snails. Following these observations, he carried out some simple experiments in which various species of slugs and snails were confined with cyanogenic and acyanogenic plants of *Lotus*. The experiments were repeated many times and Jones obtained good evidence that two snails, *Arianta arbustorum* and *Helix aspersa*, and two slugs, *Arion ater* and *Agriolimax reticulatus*, showed selective eating of the acyanogenic plants when offered both variants. This experiment proved of great interest to genecologists and many have repeated the tests using both *Lotus corniculatus* and *Trifolium repens*. Not all their results agree with those of Jones (see, for example, those of Bishop & Korn, 1969). It has been argued that the particular conditions used in the test influence the results. Important factors to take into account are the food materials made available to the animals, whether they are hungry or not, and the variability in animals. For, to quote the opinion of Jones (1972), 'in the same way that some men like beer and others do not, individual molluscs have different palates'.

The balance of evidence seems to favour the view that cyanogenic glucosides provide a defence mechanism against certain small invertebrates. This defence is by no means absolute, however. Recently Crawford-Sidebotham (1971) has examined the food preference of 13 species of slugs and snails. The results of his work (Table 14.4) revealed that while some species showed differential feeding on *Lotus corniculatus* and/or *Trifolium repens*, others did not. Furthermore, Lane (1962) has shown that the larvae of the Common Blue Butterfly (*Polyommatus icarus*) show no preference for acyanogenic plants of *Lotus*; in fact, they produce an enzyme, rhodanese, which converts cyanide into harmless thiocyanate.

Fig. 14.3. Distribution and frequency of the cyanogenic variant in European and near eastern wild populations of *Trifolium repens*. Black section: frequency of the cyanogenic variant. White section: frequency of the acyanogenic variant. ——: January isotherms. (From Jones, 1973, after Daday, 1954*a*.)

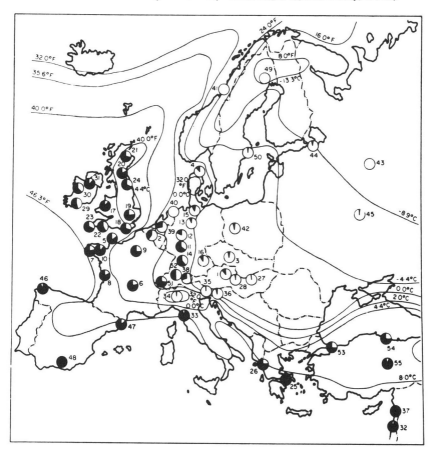

How can one explain the patterns of distribution of acyanogenic and cyanogenic plants discovered by Daday in the light of Jones' findings? It seems most likely that the distributions of animals which selectively eat acyanogenic plants are correlated with climatic factors. In Atlantic regions of Europe with mild winter temperatures and at low altitude, a great number of small invertebrates likely to eat plants may be found. It is clear that one can postulate a selective advantage of the cyanogenic plants in western Europe. Wherein, then, lies the advantage of the acyanogenic condition? The pattern of high frequency of the acyanogenic variant at high altitude and in continental conditions may be explained by the action of frost. Conditions of extreme cold will freeze the cells of plants,

Fig. 14.4. Phenotypic and genotypic frequencies in wild populations of *Trifolium repens* from different altitudes. (From Jones, 1973, after Daday, 1954*b*.)

Phenotypes (left):
AcLi – glucosides and enzyme
Acli – glucosides only
acLi – enzyme only
acli – neither glucosides nor enzyme

Estimated genotypes (right):
Black section = dominant homozygotes
Lined section = heterozygotes
White section = recessive homozygotes

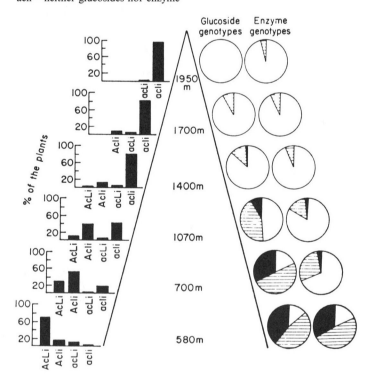

releasing the enzymes which break down any glucosides present. The production of cyanide through its inhibitory effect upon plant metabolism could place the plant at a strong selective disadvantage relative to the acyanogenic variant.

So far our account concentrates on the interaction of climatic and biotic factors on the large scale. While the study of local variation suggests that grazing by small invertebrates is important, other factors may also be involved. It is not possible to summarise all the results discovered in experiments but they generally suggest that a number of interacting factors may influence patterns of cyanogenesis, for there is evidence that water stress, mammal grazing, trampling, insect damage and salt spray may all be important in influencing small-scale pattern (Keymer & Ellis, 1978). For an insight into the way this study of cyanogenesis is leading to a more complex picture, the reviews of Jones (1972, 1973) and associates (Jones, Keymer & Ellis, 1978) should be consulted, together with recent literature on *Trifolium repens* which reports on the grazing behaviour of sheep (Cahn & Harper, 1976*a*, *b*), insect damage (Dritschilo, Krummel, Nafus & Pimentel, 1979) and factors influencing the genetic composition of swards (Turkington & Harper, 1979; Harper, 1983). Clearly, interactions between many factors must be considered in order to obtain a balanced picture of microevolution. In an excellent review of studies of heavy-metal tolerance in various species, Bradshaw & McNeilly (1981) came to the same conclusion.

Table 14.4. *Summary of differential eating experiments with 13 species of slug and snail (from Crawford-Sidebotham, 1971, in Jones, 1972)*

	Lotus corniculatus	*Trifolium repens*
No. of species[a] showing preference for the acyanogenic form	7[b]	7[b]
No. of species[a] showing no selection or no eating of the plant	6	6

[a]These numbers do not contain exactly the same species.
[b]Slugs and snails feeding on both plants: *Agriolimax reticulatus* Müll., *Arion ater* (L.), *A. hortensis* Fér., *Cepaea hortensis* (Müll.), *C. nemoralis* (L.), *Helix aspersa* Müll. and *Monacha cartusiana* (Müll.); feeding only on *T. repens*: *Arianta arbustorum* (L.) and *Helicella virgata* (da Costa); feeding only on *L. corniculatus*: *Arion subfuscus* (Drap.); no grazing on either plant: *Agriolimax caruanae* Poll; *Milax budapestensis* (Hazay) and *Theba pisana* (Müll.).

Reciprocal transplant experiments

By employing reciprocal transplant methods additional evidence about patterns of variation may be obtained. Suppose that two extremely different habitats (a and b) occur in the same region or side by side. Preliminary evidence of tolerance tests suggests that for the species under study each of two morphs (A and B) is best suited to its own native habitat (A in a and B in b). Clearly such a hypothesis might be put to the test by growing plants A and B side by side in each habitat, using either cloned or genetically uniform material for the purpose.

As we have already seen, this type of experiment has had a long history. The early experiments by Bonnier and Clements were technical failures; they were not able to relocate their experimental material with sufficient precision and in all probability their original plants were replaced by plants of local origin. Recently there has been a revival of interest in reciprocal transplant studies. These investigations are characterised by the great care taken in the design of the experiment. Davies & Snaydon (1976) marked their experimental plants with wire, and in the studies by Warwick & Briggs (1980*a, b*) experimental plants were grown in random array at spaced intervals in a carefully mapped area. Details of a number of such experiments are set out in Table 14.5. In almost every case the evidence suggests that 'native' stocks perform best in their own habitats, and alien stocks do less well.

Reciprocal transplant experiments have very clear attractions for the genecologist; they permit examination of responses to the *totality* of the environment, including competition with the native flora at each site. Such investigations subject the plants to more natural conditions than those employed in many tolerance tests. However, while such tests may be of increasing importance in studying patterns of variation in adult plants, they are not designed to reveal all selection processes at work, because they are set up with adult (or near adult) material, and the seedling stages,

Table 14.5. *Examples of reciprocal and other transplant experiments*

1.	Nine grassland taxa: clone transplants into two gardens (McMillan, 1957)
2.	Many taxa: clone transplants (Clausen *et al.* 1940)
3.	Transplant experiments with dwarf and tall subspecies of *Nigella arvensis* in a study of insect visits (Eisikowitch, 1978)
4.	Transplants of *Plantago major* from different grassland types (mown and tall turf) into lawn and tall grass communities (Warwick & Briggs, 1980*b*)
5.	Reciprocal clone transplant experiments on *Anthoxanthum odoratum* from different grassland plots at Rothamsted (Davies & Snaydon, 1976)
6.	Transplant experiments with *Epipactis helleborine* (Weijer, 1952)

on which strong selective forces might act, are not studied. Experiments in which diverse stocks are sown into natural habitats have rarely been attempted by the genecologist because of the extreme difficulty of locating seeds and seedlings and recording their survival and performance. However, with careful mapping, and using stocks bearing appropriately distinctive genetic markers, such experiments could be possible.

Experimental evidence for disruptive selection

Tolerance tests and reciprocal transplant experiments with a number of species suggest that genetic polymorphism occurs and that different morphs may be at a selective advantage in different parts of a mosaic or patchy environment. Developing the ideas on disruptive selection outlined earlier, we may envisage a model system as follows. Across the line of contact of two dissimilar habitats gene flow may occur between the polymorphic variants of an outcrossing species, 'well-adapted' adults on each side of the divide being cross-pollinated by pollen from the other morph. As a result of gene flow at the seed dispersal stage, seed of different genotypes will be scattered round the site. The variation in young plants as they germinate and develop will be subjected to disruptive selection and only the appropriate morph will survive in each area of the patchy environment. The 'potential' population represented by the young plants will be put through the sieve of selection and only well-adapted plants will survive to reproductive maturity.

A number of workers have attempted to compare young and adult generations, and the essential details of one example are shown in Fig. 14.5. In most cases the 'young' generation was represented by collecting mature seed developed in different parts of the patchy environment. Recognising that this type of study neglects the seed dispersal aspect of gene flow, in experiments on *Poa annua* Warwick & Briggs (1978b) investigated the seed bank flora in both lawn and adjacent flower beds. The results of these comparative experiments show that the spectrum of variation in the 'young' is different from that of the adults in the population. This evidence is consistent with the view that disruptive selection is a potent force in certain situations.

These experiments should not, however, be accepted as providing a complete picture of natural situations, but rather should be seen as first attempts to come to terms with the complex patterns and processes found in natural populations. The following complications should be considered in the interpretation of past and future experiments.

The model system predicts that in a particular generation the spectrum of variation in the 'young' will exceed that of the adults. Most of the comparisons so far attempted have compared adults with 'young' of the *next* generation. This comparison raises a number of problems. First, it is not clear whether the habitats under study are at equilibrium and whether the forces of selection are operating at the same intensity year by year. Furthermore, it is difficult to assess the importance of various demographic factors. Many plants are potentially long-lived, and in some of the study areas it is not known how frequently individuals succeed in establishing themselves. Secondly, the model system envisages two entirely different habitats separated by an abrupt boundary. Realistically,

Fig. 14.5. Map of old copper mine workings at Drws y Coed, Caernarvon, Wales, showing positions of transects sampled by McNeilly (1968). In his studies of *Agrostis capillaris* (*A. tenuis*), using a water culture technique, an index of copper tolerance for adult plants and seed produced by different adults was determined for material from two transects: (i) sites 1–6; (ii) sites A–E. Adults from the mines proved to be more copper-tolerant than plants from the non-contaminated pasture adjacent to the mine. Studies of the seedlings, produced from wild collected seeds, revealed a wider spectrum of variation than in the adult plants. This pattern was particularly clear in transect A–E, where evidence for considerable gene flow of copper-tolerance genes downwind of the mine was discovered in progeny of copper-sensitive plants. This experiment is consistent with the view that strong natural selection occurs on the variable products of sexual reproduction. The only seedlings to survive to adulthood are likely to be copper-tolerant on the contaminated areas and non-tolerant variants (which have been shown to be better competitors than copper-tolerant plants) on pasture areas.

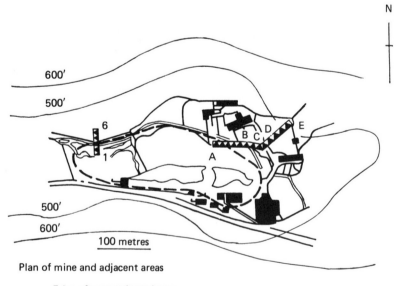

Plan of mine and adjacent areas

— — Edge of contaminated area

(i)

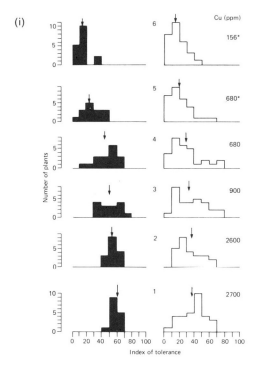

Index of tolerance for adults (black histogram) and seed (open histogram) for transect sites 1–6. (Mean values noted by arrows), with copper contents of soils. * indicates that vegetation cover suggests soils non-toxic.

(ii)

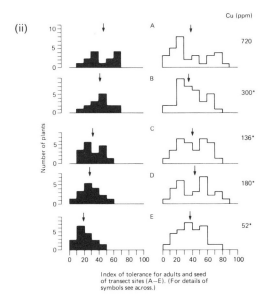

Index of tolerance for adults and seed of transect sites (A–E). (For details of symbols see across.)

models should take account of the evident heterogeneity within habitats in nature and the existence of transitional conditions – the so-called ecotone – between habitats. In a remarkable study of a mine/pasture transition, Antonovics & Bradshaw (1970) discovered very interesting clinal patterns in *Anthoxanthum odoratum*. Thirdly, the complications raised by the existence of a seed bank in the soil have yet to be properly assimilated in experiments on disruptive selection. A huge source of potential variation may be constantly eroding away by loss of seed viability and by predation, whilst at the same time 'variation' is being added by new seed production and dispersal (Harper, 1977).

Given the problems of studying natural and man-disturbed populations in the wild, it is instructive to turn our attention to examples of agricultural experiments on selection. Some of the most convincing examples of the power and speed of selection come from such studies,

Fig. 14.6. Directional selection for high and low oil content in Maize (*Zea*) after 76 generations. (From Solbrig & Solbrig, 1979, after Dudley.)

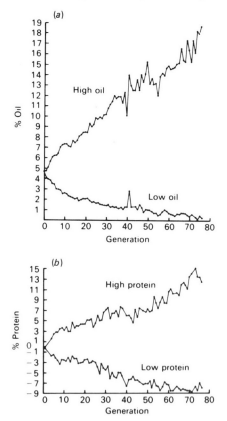

revealing once again how important it is for botanists to be familiar with applied aspects of the subject.

Natural selection in an agricultural experiment

Artificial directional selection has yielded spectacular results in agricultural crops (e.g. Fig. 14.6). Other experiments have studied natural selection, principally to discover which varieties yield best at a given site. For instance, in an investigation by Harlan & Martini (1938), 11 varieties of Barley (*Hordeum*) were mixed together in such proportions that an equal number of plants of each variety might be expected to grow. Seed

Fig. 14.7. Diagrammatic representation of the results of an experiment with Barley (*Hordeum*) in the USA, showing rapid selection of the variety suitable to the particular site. (From Harlan & Martini, 1938)

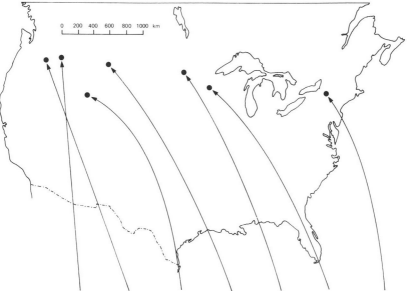

Varieties	Pullman Washington	Moro Oregon	Aberdeen Idaho	Moccasin Montana	Fargo North Dakota	St. Paul Minnisota	Ithaca New York
Coast & Trebi	150	125	159	102	156	121	75
Gatami	1	3	20	73	20	16	46
Smooth Awn	5	10	6	54	23	37	47
Lion	3	3	21	44	14	34	44
Meloy	6	3	9	12	0	5	0
White Smyrna	276	276	119	89	17	5	1
Hannchen	30	48	109	55	152	215	17
Svanhals	23	26	33	31	80	57	8
Deficiens	5	0	7	2	1	0	0
Manchuria	1	6	17	38	37	10	262

Number of plants of each variety in a sample of 500—figures for 1930 samples were sent out to 10 experimental stations in different parts of the USA; all stations received the same spectrum of 'potential variation'. At each site the Barley was harvested and seed saved to sow a plot in the following season. A sample of seed was sent each year (1925–1936) to Washington and the proportional representations of the 11 varieties were determined in a field trial. The results of the annual census for 1930 are shown in Fig. 14.7. It is interesting that different varieties predominate in different areas and there was rapid reduction (or even elimination) of the less well-adapted varieties. In all areas the variety which would eventually dominate the plot in a particular site was quickly evident.

Experiments of this type, where a 'standard' seed mixture is planted out in different sites, have been attempted with a number of crop plants and have yielded broadly similar results (Snaydon, 1978). Such experiments provide clear evidence that selection acts quickly and may be of great intensity. We shall now examine these ideas in more detail.

The speed of microevolutionary change

A number of experimental studies suggest that in certain circumstances rapid change is possible. As this area is one of the most interesting in ecological genetics, we will discuss a number of examples in some detail, beginning with the classical Park Grass Experiment at Rothamsted, England.

The effect of artificial fertilisers on crop plants was an important issue in the mid-nineteenth century, and in 1856 Lawes & Gilbert laid out the famous Park Grass Experiment at Rothamsted which was designed to study the effect of various fertiliser treatments on hay production. For our purposes it is not necessary to give details of treatment and control plots (see Brenchley & Warington, 1969; Thurston, Dyke & Williams, 1976; and Snaydon, 1970, for details). It quickly became apparent that grassland productivity could be increased by various treatments and fortunately the experiment has been continued with an early summer hay crop to the present day. One effect of fertiliser treatments, especially in plots treated with ammonium sulphate, is the lowering of the pH. Gradually plots became acid and in 1903 (after some tentative applications of lime in the 1880s and 1890s) regular liming of the southern half of each plot was undertaken. In 1965 further subdivision of some plots was made to give four subplots, each with a different lime treatment. Looking at the experiment today one is struck by the crispness of the boundaries between plots. Clearly there is little or no sideways movement of nutrients on this more or less level site, and plots differ markedly in vegetation height and productivity. The plots were laid out on a pre-existing grassland said to have been grassland for several centuries (although there are faint traces of plough marks in part of the area). From what was likely to have been a uniform grassland, the experimental regime has produced such marked differences in adjacent plots that the patterns can be seen from the air. Dr Snaydon of the University of Reading realised that, because several species occurred in many of the plots, the Park Grass Experiment offered a unique opportunity to test Darwinian views on selection. Were differences due to fertiliser and lime treatment detectable within one of these common species? How quickly could changes take place?

Studying the grass *Anthoxanthum odoratum*, he discovered that when material was collected from various plots and cultivated in a garden trial, the plants from tall vegetation plots were generally taller and heavier than plants from short vegetation plots (Fig. 14.8). Furthermore, plants showed ecotypic differentiation in relation to edaphic factors. For

Fig. 14.8. Graphs summarising some results of studies of *Anthoxanthum odoratum* collected from the Rothamsted Park Grass Experiment and grown in a special trial. (*a*) Relation between plant height and height of vegetation in source plots. (*b*) Relation between plant weight and yield of herbage in source plots. (*c*) Effect of calcium on dry weight of plants from limed and unlimed source plots. (Snaydon, 1970, 1976; Snaydon & Davies, 1972; Davies & Snaydon, 1973a. See also Davies, 1975 – Responses to potassium and magnesium; Davies & Snaydon, 1974 – Responses to phosphate.)

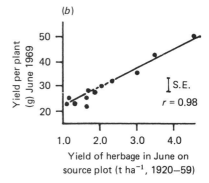

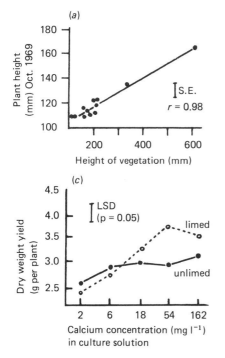

example, plants from acid unlimed plots, where the soil was deficient in calcium, needed much less calcium than plants from limed plots. Studies of gene flow were made and reciprocal transplant experiments carried out from acid to limed plots and *vice versa*, revealing that plants grew best on their native plots. The evidence from all these experiments suggests that population differentiation has taken place in response to fertiliser and lime treatments, plausibly through natural selection acting over the last 120 years. Indications that important changes might be possible in an even shorter time are provided by studies of vegetation following the alteration of the lime addition regime in 1965. In experiments in 1972 it was discovered that after only 7 years changes could be detected in the *Anthoxanthum* populations. The highly imaginative use of this classical experiment has provided one of the most impressive studies of pattern and process in a patchy environment.

Studies of copper tolerance provide another fascinating glimpse into the speed of microevolutionary change. About 1900 a factory refining copper was opened at Prescot, South-west Lancashire. Dust rich in copper was produced by the processing of the metal and the area around the factory became contaminated. The establishment of a number of lawns around the works was dependent upon the removal of contaminated soil and its replacement with copper-free soil. Nevertheless, at the time of the study by Wu, Bradshaw & Thurman (1975), total copper levels in lawn soils were as high as 10 800 parts per million. Investigations revealed that *Agrostis stolonifera* plants in lawns were copper-tolerant to varying degrees and that, while the vegetation cover was complete on the older lawns, it was patchy on a new (8-year-old) lawn. Further, in the establishment of the new lawn, repeated sowing with commercial seed had failed to achieve complete cover. When commercial seed stocks are tested for metal tolerance on contaminated soil, most seedlings die, but a few reveal their metal tolerance by growing normally (Bradshaw & McNeilly, 1981). From this study it was deduced that selection was taking place each time seed was sown on the contaminated new lawn area. Only copper-tolerant varieties survived and such was the toxicity of the ground that only a few survivors were likely from any seed batch. Not all plants on the new lawn need have arisen from direct sown seed; some may be the natural progeny of plants surviving from earlier sowings. This splendid example of the detailed study of a man-disturbed site suggests that, if the appropriate variation is present in a population, then selection will act rapidly.

Other studies of sites of known history provide circumstantial evidence for rapid change in populations under natural selection. Thus, a factory producing smokeless fuel from coal was opened in a country district in West Yorkshire in 1926, and a good deal of sulphur dioxide pollution was produced from the works. Ayazloo & Bell (1981) recently discovered that plants of the grasses *Dactylis glomerata*, *Festuca rubra*, *Holcus lanatus* and *Lolium perenne*, growing in the vicinity of the works, were significantly more tolerant of sulphur dioxide than either samples of ordinary commercial cultivars of these species or samples of the species from districts with clean air.

More direct evidence of the time taken for change comes from agricultural experiments. For instance, an investigation by Brougham & Harris (1967) is particularly revealing. Two varieties of Rye-grass (*Lolium*) were sown (together with White Clover) in an experimental plot. After establishment the sward was subjected to lax grazing for 6 months. The plot was then divided into a number of subplots which were given either lax, moderate or continuous grazing regimes. Here we are not concerned with the detailed results of this experiment; it is sufficient for our purposes to note that major changes in population composition were detected within 4 months of the application of the grazing regimes. The implications of this and many other experiments on agricultural stocks is that strong selective forces may act very quickly to change the composition of populations.

So far we have discussed microevolution in non-mathematical terms. It is, however, important to define more completely what is meant by 'selective advantage' by introducing the concept of 'fitness'. Fitness is relative and is defined by the *number* of descendants left by an individual compared with those of other individuals. What is clearly important in microevolutionary change is the *relative* contribution of offspring made to the next generations.

In some cases where extreme habitats have been studied the calculations of fitness are simplicity itself. For instance, only heavy-metal-tolerant variants survive on highly contaminated soil, and all plants unable to tolerate heavy metals in the rooting medium may die on germination, making no contribution to the next generation. In most habitats, however, conditions are less harsh and individuals of different tolerance survive and reproduce with varying degrees of success. In order to measure fitness in such circumstances it is necessary to have a means of identifying not only the seeds produced by particular individuals but also the location and numbers of individuals establishing themselves from this seed, proper account being taken of overlapping generations and vegetative reproduction.

It is clearly impossible with present techniques to collect the information necessary for calculation of fitness, and genecologists must be content with estimates based on the comparison of the survival and growth of plants in various experiments. Such calculations give measures of 'relative fitness', often called coefficients of selection. For the reader who wishes to examine the formulae used in calculations and some of the results obtained, the following papers provide a useful introduction: Jain & Bradshaw, 1966; Cook, Lefèbvre & McNeilly, 1972; Hickey & McNeilly, 1975; Davies & Snaydon, 1976. 'Relative fitnesses' are likely to be very crude estimates indeed of 'fitness' as defined above. In situations where several estimates of 'relative fitness' have been made from different measurements of vegetative and reproductive characters of the same material, very different results have been obtained (see Warwick & Briggs, 1980*a*, *b*, *c*). Wherever possible, measures of reproductive success should be obtained for the calculation of relative fitness.

Finally, a critical appraisal of this difficult area should not blind us to an important general conclusion which has emerged from genecological studies of the last 20 years. In his famous mathematical studies of selection Fisher (1929) showed that selection would be effective even if there was only a very small selective advantage between individuals in populations. It is now clear from the study of extreme habitats that very large selective advantages must sometimes be operating in nature.

Results of disruptive selection in polymorphic populations

So far in this chapter we have been discussing and illustrating short-term effects of selection. What happens in the long term? As we showed in Chapter 11, there has been considerable interest in the possibility of sympatric speciation, that is to say the possible evolution of new species within, or at the margins of, the distribution of existing species. The situation where a genetically polymorphic species exists in a patchy or mosaic habitat would seem to offer the possibility of speciation without the necessity of geographical isolation (Thoday, 1972).

Several studies of plants have been made on mine debris containing heavy-metal residues, sites which often exist as 'islands' in a 'sea' of pasture. As we have shown, there is evidence for gene flow between pasture and mine plants, followed by disruptive selection. Furthermore, it is clear that pasture plants cannot grow on the mine debris and mine plants grow less well in the pasture (Bradshaw, 1976; Bradshaw & McNeilly, 1981). It seems that crosses within habitat type (mine × mine and pasture × pasture) will produce fitter progeny than the crosses mine × pasture and pasture × mine. Therefore it has been argued that selection would favour any variant arising in the population which would restrict 'gene flow' by promoting within-habitat crosses. Experiments by Antonovics (1968) and McNeilly & Antonovics (1968) working with *Agrostis capillaris* (*A. tenuis*) and *Anthoxanthum odoratum* showed that, although there was some overlap in flowering times between pasture and 'mine' plants, those at the mine edge flowered about 1 week earlier, a difference maintained in cultivation. Further, plants of both species growing on mine debris, though normally self-incompatible, were found to be capable of a degree of self-fertility.

Both these traits may be seen as barriers to free gene flow and could be interpreted as the first steps in the speciation process. It is unclear, however, how much selfing actually takes place in wild populations. The studies of Lefèbvre (1973) suggest that in *Armeria maritima* growing on mine debris the potential for self-fertilisation exists, but in a study of a particular mine population it was discovered that outbreeding was the rule. Whether speciation is a possible outcome of polymorphic variation remains an open question, one which will continue to fascinate genecologists for decades to come.

Patterns of variation in response to seasonal or irregular extreme habitat factors

In many of the examples we have been studying, habitats remain extreme throughout the year; for instance, in a soil contaminated by copper, metal ions may leach out into the soil solution all the year round. It is clear, however, that some habitats are subject to severe conditions only at certain critical times. What patterns are found at such sites? We will consider two very different cases.

The recent evolution of DDT resistance in insects, (Georghiou, 1972) is one of the most convincing cases of microevolution yet described. The increasing use of herbicides suggests the possibility that herbicide-resistant mutants may arise in plants and that such variants could come to dominate the weed floras in agricultural regions. However, the regular application of herbicide does not necessarily lead to populations of herbicide-resistant plants, as was discovered in researches by Holliday & Putwain (1977, 1980). When seed stocks of *Senecio vulgaris* were screened for simazine resistance by sowing seed on to soil containing the herbicide, almost all the seedlings died, a surprising result considering that some of the sites had been treated with simazine for a number of years. A full analysis of the ecological situation revealed an explanation. First, simazine applications were made once in the year. Any simazine-resistant *S. vulgaris* plants arising in the populations were likely to be killed, however, because other herbicides were also used on the plots to keep down the weeds. Furthermore, there appeared to be a bank of simazine-sensitive seed in the soil, which could germinate when brought to the soil surface, and could grow and reproduce in the autumn when herbicide activity was low or absent.

Thus the population dynamics of sites treated with herbicides may be interestingly complex, and intermittant or seasonal factors will not necessarily lead to directional or disruptive selection. Each situation must be carefully examined in its individuality.

Some habitats are characterised by irregularly fluctuating extreme conditions. One of the most familiar concerns a variable water table, the habitat being dry in some parts of the growing season and wet or flooded at others. Cook & Johnson (1968) studied intraspecific variation in the heterophyllous species *Ranunculus flammula*, which has lanceolate aerial leaves and linear leaves under water. They collected plants from a number of habitats in Oregon and grew them under aquatic and terrestrial conditions. They discovered that certain populations, which were likely to experience the most unpredictable regimes in the wild, could produce the extreme heterophyllous leaf types in the appropriate conditions, whereas, in contrast, plants from habitats which were less frequently flooded showed very little heterophylly in the experiments. This investigation raises a number of questions about the modification of the phenotype.

Phenotypic plasticity

In Chapter 6 we discussed the phenomenon of phenotypic plasticity and showed that the phenotype is the product of the interaction of genotype and environment, and that any given genotype will produce different phenotypes in different environments. Each genotype is likely to have a characteristic breadth of phenotypic plasticity, itself under genetic control. In extreme sites plants with a narrow range of responses may be selected, whereas in less extreme or variable sites plants with a wider spectrum of responses might be at a selective advantage. Let us examine why this might be so. Imagine plants growing on an exposed sea cliff. Plant A is genetically programmed to produce a tall phenotype, while B is of genotype appropriate to a dwarf plant. In exposed conditions 'B' is dwarf, but so too is 'A', and 'A' can be said to be a 'phenocopy' of 'B'. In the unusual event of a period with little wind, plant 'A' may grow tall, only to be severely damaged in the next storms. Plant 'B', however, is constrained genetically within a narrow range of phenotypic possibilities in height, and under unusual conditions it does not grow 'too tall' for the habitat. It is easy to see how selection might operate to favour plants in genotype B in such circumstances. However, given unpredictable or highly heterogeneous habitats, it is apparent that a plant with a wide phenotypic plasticity could be at a selective advantage.

Genecologists have not in general given enough attention to phenotypic plasticity; indeed, approaching the subject from a taxonomic standpoint, botanists have frequently regarded phenotypic plasticity as a nuisance. So too have physiologists and biochemists, for different reasons. Thus, phenotypic plasticity is amongst the most neglected phenomena in plant science and deserves much more careful investigation, especially with a view to elucidating the genetic control of plasticity, the spectrum of variation possible with different genotypes in different environments, and the adaptive nature of the responses. Does phenotypic plasticity tend to maximise fitness in the circumstances in which the plant finds itself?

Physiological plasticity should also be examined. An intriguing example of this phenomenon has recently been published. Some cyanogenic plants of *Lotus corniculatus* are of stable phenotype, whilst others are cyanogenic only at certain times of the year and under some conditions. Ellis, Keymer & Jones (1977*b*) are inclined to see this physiological flexibility as adaptively significant, the cyanogenic mechanisms effective against herbivory being 'switched off' at just those times when grazing pressure is likely to be low, and the risk of damage to the plant by other factors is at its highest.

Population variation

As a consequence of selection, only a proportion of the young survives to make a contribution to the next generation. How big a proportion survives, and how variable are the adults within populations?

It is fascinating to study the history of ideas on population variation in the twentieth century. Amongst the most interesting is Turesson's 'biotype depletion hypothesis' (Turesson, 1922*b*, 1925). He supposed that inland populations were variable, having many different biotypes (a biotype being a collection of individuals which are genotypically all essentially the same). Ecotypes arose from inland populations by a process of biotype depletion, i.e. from the pool of variation in inland sites selection favoured particular biotypes in, say, montane or coastal areas, resulting in ecotypes composed of only one or a few biotypes. His ideas were criticised by Faegri (1937) who pointed out that we did not find plants equivalent to fully formed ecotypes in inland populations. For example, there was no sign of succulent, prostrate or dwarf variants 'waiting in the wings' to become coastal ecotypes. Also, as we saw in Chapters 9 and 10, many botanists have supposed that ecotypic differentiation could be the first stage in the process of gradual speciation. If

Turesson's views on populations were correct, ecotypes would be invariable units likely to be rendered extinct by environmental changes, rather than incipient species.

Many other botanists have speculated about population variation, particularly in relation to breeding systems. For instance, Baker (1953) considered that populations of obligate outbreeding species are likely to be quite variable but that, in contrast, depending upon site history, gene flow and the intensity of selection, populations of apomictic, vegetatively reproducing or habitually self-fertilising species might contain only few biotypes. The validity of such speculation must ultimately be tested by observation and experiment. Three lines of evidence might be sought: each is briefly considered below.

1. The amount of genetic variation might be judged by field samples and garden trials; but how many of the genotypic differences are likely to be manifest in the array of phenotypes obtained? Obviously some variation is overt, but cultivation experiments leave the experimenter in the position of guessing how much hidden variation exists.

2. Breeding and other experiments have yielded interesting information on, for example, predominantly self-fertilising species. An experiment by Allard (1965) is instructive. A 50:50 mixture of two pure lines of Lima beans (*Phaseolus lunatus*) was sown in an experimental plot. As a result of five major gene differences the two lines were phenotypically distinct. Seed was collected at harvest for the establishment of a plot of beans in the following season. Classification of the population generation by generation was possible because of the phenotypic differences between lines. Initially one pure line appeared to have a selective advantage over the other, but within a few generations the two pure lines were swamped by hybrid derivatives (identified by their combinations of phenotypic characteristics). Even though the pure lines were probably predominantly self-fertilising with only about 5% crossing between lines, heterozygous derivatives appeared to be at a selective advantage. The simple 'model system' of two competing lines, in which the selective advantage of one line was apparent, was rapidly transformed into a complex, variable dynamic system (Fig. 14.9).

   This experiment sheds light on a number of important issues. First it provides a most elegant example of heterosis, which we discussed in the context of breeding systems in Chapter 7. It also reveals the intensity of selection pressures and the rapidity of changes due to selection. Furthermore, it suggests the possibility that any population of wild or cultivated plants, which is made up initially of pure lines, may not be a stable system if some outcrossing is possible. In a survey of the literature up to 1968, Allard, Jain & Workman (1968) concluded that populations of predominantly self-fertilising species are often more complex than hitherto appreciated and may be as variable as those of regularly outcrossed species. Recent studies have confirmed this impression.

3. With the development of isozyme techniques (already introduced in Chapter 13) the botanist is now provided with a most valuable tool with which to investigate variations. The potential of the method is obvious, and an increasing number of studies are being made. For example, Wu *et al.* (1975), investigating the population structure of copper-tolerant *Agrostis stolonifera* on the lawns around an industrial site, in the experiment discussed above, discovered evidence that many different genotypes of this out-breeding species were found on the lawns, and dispelled the notion that only one or a very few individuals might have colonised the lawn. Isozyme techniques were also used by War-wick & Briggs (1978*a*), who discovered that prostrate individuals of the predominantly self-fertilising species *Poa annua* inhabiting

Fig. 14.9. Graph showing proportions of parental and non-parental types in an experiment with Lima beans (*Phaseolus lunatus*) set up with a mixture of two pure lines (A and B). (From Allard, 1965)

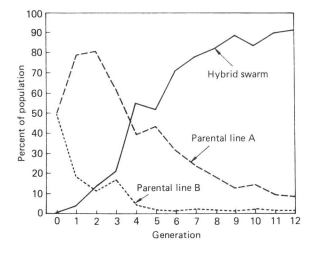

closely mown turf were genetically diverse. Allard and co-workers have investigated populations of *Avena* spp. (Clegg & Allard, 1972; Hamrick & Allard, 1972) and have shown that these largely inbreeding species contain a good deal of variation. An apomictic group has been examined by Solbrig & Simpson (1974), who studied three populations of *Taraxacum* in Michigan (Table 14.6) and Bell & Lester (1978) have used electrophoretic methods in the study of introgression. We have already mentioned (p. 277) the use of isozyme studies in investigating clone structure in *Lárrea tridentata*; and the work by Roose & Gottlieb (1976) on the variation of recently-evolved species of *Tragopogon* (see p. 239) provides another example of the value of the techniques. To sum up, the results available for both plants (Brown, 1979; Gottlieb, 1981; Soltis, 1982) and animals suggest that populations are generally more variable than has been appreciated, and there is intense speculation amongst population geneticists on the extent of variation, its causes and significance in evolution. Much current interest is centred on investigating whether variation in populations, particularly at the molecular biological level, is adaptively significant, or whether some variation is selectively neutral (Clarke, 1979; Solbrig & Solbrig, 1979, and references cited therein).

Table 14.6. *Percentage occurrence of four electrophoretically distinct biotypes of* Taraxacum *in three habitats in the Mathei Botanic Garden at Ann Arbor, University of Michigan, USA*

Habitats	No. of plants sampled	Biotypes			
		A	B	C	D
1. Dry, full sun. Trampled grassy area mown weekly	94	73	13	14	0
2. Dry, shady, mown weekly	96	53	32	14	1
3. Wet semi-shade, less disturbed, mown once yearly	94	17	8	11	64

Solbrig & Simpson (1974, 1977) investigated the resource allocation, fecundity and competitive ability of biotypes A & D. Biotype A, with its greater fecundity, seemed better adapted to disturbed grassy areas. In contrast, biotype D, which was a better competitor in a competition experiment in which A was matched with D, would appear to be at a selective advantage in less disturbed swards.

To understand fully the adaptive significance of variation, we must devise experiments in each particular case which will take into account both the diverse physical features of the environment and the competitive interactions between the different members of the ecosystem. It seems probable that many plant polymorphisms have their explanation, as we saw in the case of cyanogenesis in *Lotus* and *Trifolium*, in terms of the selective advantage of different genotypes in different parts of a complex environmental system. Although many cases of polymorphism have been described (see Turrill, 1948; Valentine, 1975*a*, 1979), few have been experimentally studied and, without experiment, there is little to be gained by guessing at the possible adaptive significance of one morph against another. The case of *Spergula arvensis*, a common weed occurring in two forms differing in seed-coat pattern, illustrates this point very well.

In her main study of *Spergula*, New (1958, 1959) showed that a ratio-cline existed in the British Isles, populations in the north and north-west having significantly higher proportions of the variant with smooth seed-coat than those in the south and east, where papillate seed-coats contributed quite high frequencies (Fig. 14.10). Examination of herbarium material for Europe as a whole confirmed the general tendency for the smooth variant to be characteristic of higher latitudes. Twenty years later, New (1978) was able to show by repeating the sample that the ratio-cline for seed-coat pattern had not significantly changed in Britain (though there *was* evidence for a significant change in proportion of strongly glandular-hairy plants in the populations). Cultivation experiments showed conclusively that the variant with smooth seeds was less tolerant of high temperatures and low humidity than the variant with papillate seeds, and therefore, in her early work, New concluded that the apparently non-adaptive seed-coat character, which she had shown to be determined by a single gene without dominance, must be correlated with genotypic differences in physiology which were themselves clearly of selective importance in different climates. However, the most recent paper (New & Herriott, 1981) shows that papillate seeds germinate more easily than smooth ones under dry conditions, so that an apparently non-adaptive character has after all been shown to have direct adaptive significance. There seem to be two lessons to be drawn from this unusually well-investigated case: first, that assessing characters as 'non-adaptive' may merely reflect ignorance and, secondly, that actual polymorphisms are unlikely to be explained in terms of the operation of any single factor.

Some very familiar cases remain quite obscure. For example, Lords and Ladies (*Arum maculatum*) shows remarkable polymorphism for the spotted *versus* unspotted leaves, and for purple *versus* yellow spadix (Fig. 14.11). Attempts to demonstrate that the spadix colour affects the quantity or quality of small insects trapped by the remarkable pollination mechanism have been unsuccessful; nor has any clear advantage been demonstrated in the presence of anthocyanin spots in the leaves (Prime, 1960).

Recently plant physiologists have studied the properties of plants with leaves of different shapes. A number of species exhibit polymorphism for

Fig. 14.10. Distribution in the British Isles of variants of *Spergula arvensis* with smooth and papillate seed-coats. (From New, 1958)

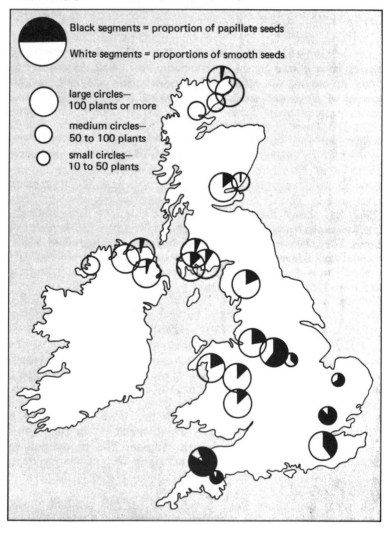

Fig. 14.11. *Arum maculatum*, spotted variant, showing characteristic leaves and inflorescence. (From Ross-Craig, 1973) ($\times$ 1)

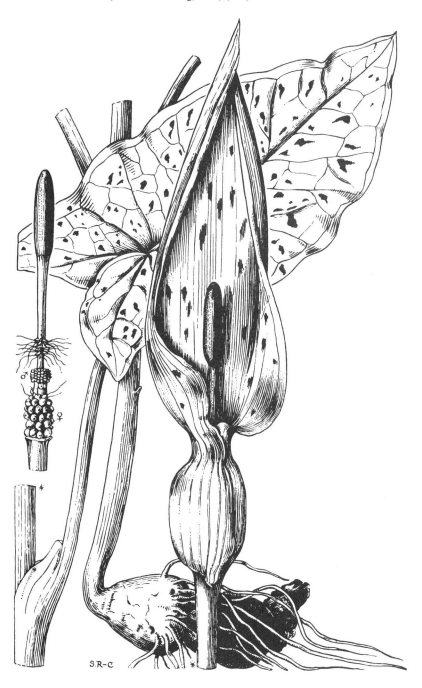

leaf dissection and leaf geometry, and adaptive significance should in these cases perhaps be sought in terms of water relations, gas exchange and photosynthesis, rather than in the leaf-shape *per se* (Givnish & Vermeij, 1976; Givnish, 1979). Many species are polymorphic for hairiness, some individuals being glabrous, others obviously hairy. Again the adaptive significance of such variation might also be 'physiological' (Johnson, 1975), or the indumentum might be important in protecting the plant against herbivore attack (a role suggested by a number of studies; see review by Levin, 1973).

Finally, plants are sometimes polymorphic for 'B' chromosomes. For instance, Fröst (1958) showed with the common perennial *Centaurea scabiosa* a clinal increase in frequency in 'B' chromosomes in plants sampled from west to east in Scandinavia and Finland (Fig. 14.12). The adaptive significance of variation in 'B' chromosomes has puzzled geneticists for many years but it now seems possible that 'B' chromosomes might be important in the control of chromosome pairing at meiosis. (The details of this finding are complicated; the reader is referred for further information to Evans & Macefield, 1973, and Jones & Rees, 1982.)

Fig. 14.12. Distribution of Scandinavian populations of *Centaurea scabiosa* showing the proportion of individuals with B chromosomes. A solid black circle represents an average of two B chromosomes per individual in the particular population. (From Fröst, 1958)

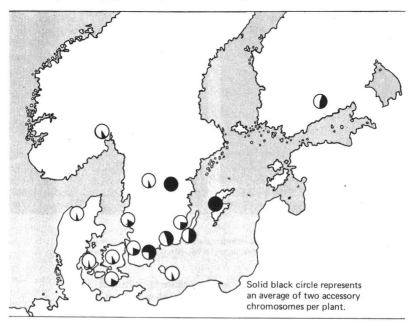

Solid black circle represents
an average of two accessory
chromosomes per plant.

The effects of chance in populations

In our discussions of recent trends in genecology we have been considering patterns in relation to natural selection, and in Chapter 9 we noted that population geneticists have postulated that chance plays a role in population variation. It is exceedingly difficult, however, if not impossible, to separate the effects of chance and selection. It is sometimes claimed that the effects of chance may be revealed in cases where selectively neutral traits are being considered. For example, Rafiński (1979) investigated stigma colour polymorphism in *Crocus scepusiensis* found growing in the Gorce mountains of Poland. Populations differed widely in frequency of the different morphs (Fig. 14.13), and Rafiński considered that the variation could be the result of chance events. In another interesting study, of *Nigella arvensis* growing on Greek islands, Strid (1970) concluded that some of the variation could be caused by founder effects and random drift. Ecological geneticists do not dispute the likely importance of chance but it is easy to see that an ingenious genecologist could devise a selectionist interpretation of any given pattern.

Conclusions

The period since the publication of Heslop-Harrison's review has seen great advances in the study of genecology. Two decades of research have transformed the subject and in concluding our brief survey we wish to make two important points. First, many of the elementary texts on ecological and population genetics give the impression that the subject has an elegant structure very nearly complete in every detail. By a judicious mixture of algebra and experiment drawn from the laboratory and field, the reader may be seduced into thinking that the various aspects of the subject are beginning to come together in a splendid synthesis. The truth is that many of the mathematical models are remote from nature, and field studies are in reality often first essays in the study of complex systems. In our view, the student of evolution must be prepared to face with some humility the complications of both natural and artificial ecosystems, and see the present conceptual framework as a provisional structure which is certain to be modified by future research.

Secondly, it seems likely that genecology will be absorbed wholly, or in part, in the developing subject of population biology (Antonovics, 1976). We applaud this development. No investigation can proceed without a taxonomic framework, however, and it is hoped that many will still approach the subject from a thorough knowledge of the taxonomic background, for the history of the subject reveals how much the

Fig. 14.13. Stigma colour polymorphism in *Crocus scepusiensis* in the Gorce Mountains, East Poland. The boxed region in (*a*) is shown in (*b*). White sections = frequency of plants with white stigma: dark sections = frequency of plants with orange stigma. (From Rafiński, 1979)

(*a*)

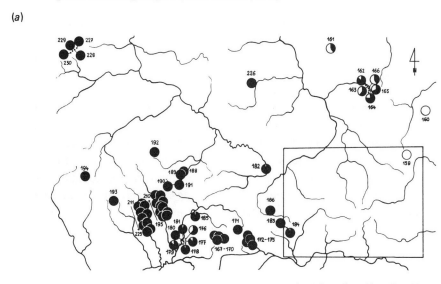

(*b*)

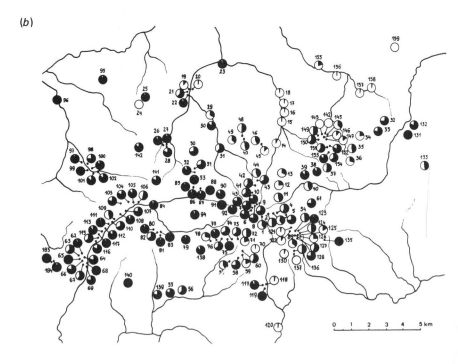

experimenter is indebted to the taxonomist who has an 'eye' for significant patterns of variation.

Finally, we might suggest what could be the most important growth area for the subject in the future. Harper (1977) pointed out that Wallace and Darwin had different views about the forces at work in natural selection, Wallace being concerned to emphasise the effects of inanimate nature – climate, soil, etc. – whilst Darwin stressed the competitive interaction of organisms in his account. Risking a generalisation, it seems that genecologists have, like Wallace, been mainly interested in tolerance of soil and climatic factors. Such a view may give too static a picture of microevolution. Plants are endlessly facing new situations as ecosystems develop, with biotic factors playing a very important part. With some honourable exceptions, genecologists have not yet come to terms with the complex picture of the plant in its ecosystem. An exciting period of research lies ahead as we explore the genetic aspects of the complex interactions between plants in communities, and the intricacies of the relationships between plants and animals.

15

Evolution: some general considerations

In choosing to restrict the material of our book to the study of the nature and causes of variation within and between species, we have largely excluded two areas of enquiry which the reader might expect us to cover. One of these is the traditional field of comparative morphology, which compares and contrasts the structure of a series of representative 'types' of organisms from the algae to the flowering plants. Most elementary textbooks of botany include a survey of this kind and at least an outline knowledge of the most important differences between alga, fungus, moss, fern and flowering plant is required of anyone who would claim to be a botanist. We therefore feel that we are justified in assuming such knowledge. The second excluded area is closely linked, in an important way, to the first – the study of the evolution of the plant kingdom as a whole, and particularly the relevance of our knowledge of the structure of fossil plants to our understanding of evolution. It is quite impracticable to give any detailed account of palaeobotany here, but as certain important questions about the variation of plants can only usefully be considered in relation to the studies of both modern and fossil structures, we will indicate some of these questions in the hope that the interested reader might examine them further.

The variation of plants, even in the relatively restricted sense in which we have interpreted the study, can be approached in two quite different ways. We can focus our attention on the static patterns of variation (this, as we have seen, was the traditional manner); or we can trace the variation in time, which is more difficult. This change of emphasis was one of the main results of Darwinism, and the growth of experimental taxonomy, with its interest in the processes of evolution, is the natural development which we have tried to outline in the main part of this book. We have seen that it is possible to argue, cautiously and tentatively, from the existing, static patterns of plant variation to the dynamics of evolutionary processes and that, within strict limits, an experimental approach is possible to at least some of the key questions in the understanding of evolution. It is now time to summarise briefly the picture we have and to see it in terms of evolution as a whole.

One difficulty is apparent from the outset. By far the greater part of the technical literature on biological evolution is written by zoologists using animal species for their studies and, whilst it is true that the basic principles of genetics apply to plants and animals alike, we have seen that in important respects a typical 'higher plant' differs from a 'higher animal', and the course of botanical evolution is necessarily a special study with its own peculiar complications. In one respect, of course, the study of the evolution of plants is immeasurably simpler than that of animals. The botanist is not involved in those complex and necessarily controversial areas of evolutionary speculation which concern the nature and origin of nervous and mental activity, and which culminate in the study of Man himself as the product of evolution. This does not mean that all botanists agree on philosophical questions concerning evolution, but it does mean that they are more likely to be content with the standard reductionist procedures of scientific enquiry, using an agreed framework of physics and chemistry as a basis for study. If they differ seriously on evolutionary questions, they are likely to differ as laymen interested in the evolution of Man, rather than as scientists with a special knowledge of plants.

The fossil record

We can now turn to look briefly at some aspects of botanical evolution as a whole. The first of these concerns the evidence from fossil remains. Darwin rightly saw that fossil evidence provides overwhelming support for organic evolution as a historical event – indeed it is the only reasonable direct evidence we have, or are ever likely to have. It is, nevertheless, a matter of some difficulty to reconstruct the course of evolution, even in short lengths, from the fossil record. The main reason is the difficulty of interpreting the *absence* of any particular kind of organism from the fossil record, in view of the obviously fragmentary nature of that record. Nowhere is this difficulty greater than in tracing the record of the flowering plants as a whole. While the broad outline of the evolutionary timescale seems to fit the evidence from comparative morphology of living groups of plants, the apparently sudden origin of the angiosperms in the Cretaceous period is still the mystery which it was to Darwin a century ago. Most of the main kinds of floral specialisation, in terms of which our modern flowering plant families are defined, can be found represented in Cretaceous fossils, yet earlier than the Cretaceous the fossil record is extremely scanty. Most writers on angiosperm evolution seem to be unwilling to accept any multiple origin for the group, so that they are forced to one or other of two conclusions: either the main diversification of the flowering plants was extraordinarily rapid in evolutionary time; or, for reasons unknown, the pre-Cretaceous angiosperms were not preserved as fossils. But is the requirement of monophylesis – a single, ancestral angiosperm – a reasonable one? There *are* some writers who challenge this orthodoxy and their point of view deserves to be taken seriously. One of the most consistent of these is the Dutch botanist Meeuse, who stresses the evidence for the gradual acquisition of angiosperm characters in separate lines of gymnosperm evolution, and concludes that modern angiosperms must represent the end-points of several quite different lineages (see Meeuse, 1966; Hill & Crane, 1982).

The reader who would like to know more about these interesting problems is recommended to read Sporne (1971) which is authoritative and well illustrated. More recently Sporne (1974) has written the third text book which completes a series covering the comparative morphology of the main groups of higher plants; these books now provide a very convenient source of the evidence relevant to evolutionary speculation in all the main groups of vascular plants.

Rates of evolution and the origin of angiosperms

This problem of the interpretation of fossil evidence leads naturally to a consideration of rates of evolutionary change. We saw in Chapter 14 that, under certain conditions of strong selection, evolution could proceed so rapidly that we could detect its operation in higher plant populations. If we assume that the main course of evolution has proceeded on the basis of mutation, recombination and selection, we would then expect that in certain situations evolution, as measured by change in the form of the successive generations, would be rapid, while in other circumstances it might be undetectably slow. Broadly speaking, this picture is confirmed by the fossil record, in which certain modern plant genera appear in essentially the same form as ancient fossils and must have survived as virtually unchanged lineages over many millions of years (Fig. 15.1). With these ancient 'conservative' stocks we can contrast the new species, such as the grass *Spartina* discussed in Chapter 11, which has originated as it were in the last second of time on a geological timescale. In the origin of species, then, we must envisage all kinds of situations, from abrupt origin and establishment in a few years to stability over millions of years. The action of selection, as we saw in Chapter 14, can result in stabilising the species or, in other circumstances, it can lead to change.

Stebbins (1950) and others have suggested that these considerations are relevant to the mystery of the origin (or origins) of the flowering plants – the dominant plant group. The striking fact about the diversification of angiosperms in the Cretaceous period is that it coincides with the rise of modern insects, the pollinators of so many angiosperm flowers. Undoubtedly very powerful selection can be exerted by the behaviour of insect pollinators, and in view of the complex relationships of mutual advantage which could arise from the visits of insects foraging for nectar or pollen, evolutionary change could be rapid for both plant and insect visitor (Takhtajan, 1969; Proctor & Yeo, 1973; Faegri & van der Pijl, 1979; Armstrong, Powell & Richards, 1982; see also Fig. 15.2).

In a more recent book, Stebbins (1974) extends his speculative consideration of flowering plant evolution to cover not only the 'classical' flower and fruit structures approached in an ecological framework of thought, but also questions of habit and vegetative form. He asks the question: in what kind of habitats would adaptive radiation proceed most rapidly? His conclusion is that the early angiosperms, which showed rapid diversification in the Cretaceous fossil record, were probably relatively small shrubs

Fig. 15.1. A fossil leaf of *Ginkgo* from Jurassic rock (left) and a leaf of living *Ginkgo biloba* (right). Details of the taxonomy and former distribution of *Ginkgo* and related plants are given in Tralau, 1968. (Fossil: photo: British Museum (Natural History.)

 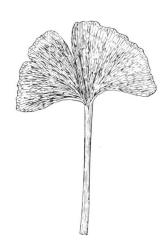

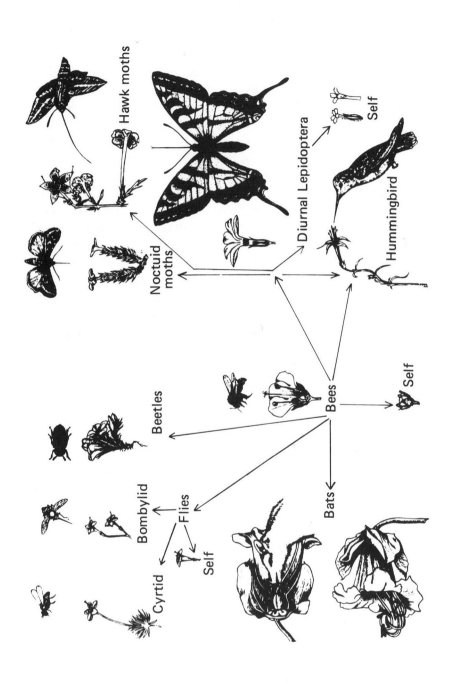

inhabiting somewhat intermediate or marginal communities, not, as often pictured, large trees with single, terminal flowers similar to the modern Magnoliaceae. His explanation of the concentration of 'archaic and apparently primitive Flowering Plants' in Tropical rain forest – which he accepts as an established fact – is that this arises *not* because angiosperms originated in such stable communities, but because the rate of extinction there is so small. The Tropical forest is ancient and therefore contains ancient plants, but it also contains many plants which possess specialised, 'advanced' characters, and its total flora is very rich and diverse.

An entirely novel suggestion to explain the rapid success of angiosperms was made by Whitehouse (1950), who pointed to the widespread occurrence of self-incompatibility mechanisms in many flowering plant families. He proposes that this condition is ancestral and postulates that the significance of the closed carpel 'lies in the protection of the ovules, not from desiccation or the attack of animals, but from fertilization by the individual's own pollen, without appreciably restricting cross-fertilization'. De Nettancourt (1977) supports Whitehouse's view and discusses the implications of the corollary proposition, namely that self-compatibility is a secondary, derived condition in modern angiosperms. He argues that the relative advantages of outbreeding and inbreeding, as we saw in Chapter 7, are likely to be very different under different environmental conditions, and points to the work of Allard *et al.* (1968) which showed remarkably high levels of genetic diversity among self-pollinating populations. It is possible to see both outbreeding and inbreeding as potentially adaptive to particular conditions; applied to the evolution of angiosperms, this would imply that the initial diversification through the Cretaceous period was possible because of the efficiency of the outbreed-

Fig. 15.2. (facing page) It seems probable that, as a consequence of natural selection, the ancestors of a single group may through evolutionary divergence become adapted to a range of environments. A splendid example of presumed adaptive radiation for pollination by different pollen vectors is illustrated by present day members of the Phlox family (Polemoniaceae). The following species of pollinators and flowers are illustrated as representative of each group: bees, *Polemonium reptans* and *Bombus americanorum*; bats, *Cobaea scandens* and *Leptonycteris nivalis*; cyrtid flies, *Linanthus androsaceus croceus* and *Eulonchus smaragdinus*; bombylid flies, *Gilia tenuiflora* and *Bombylius lancifer*; beetles, *Linanthus parryae* and *Trichochrous* sp. (Melyridae); noctuid moths, *Phlox caespitosa* and *Euxoa messoria*; hawk moths (Sphingidae), *Ipomopsis tenuituba* and *Celerio lineata*; diurnal Lepidoptera, *Leptodactylon californicum* and *Papilio philenor*; birds, *Ipomopsis aggregata* and *Stellula calliope*; self (autogamous), *Polemonium micranthum* (bottom), *Gilia splendens*, desert form (left), *Phlox gracilis* (lower right). (From Stebbins, 1974, redrawn from Grant & Grant.)

ing mechanism, but in later periods, as we discuss below, other factors were selectively more important, which favoured other breeding systems.

The coincidence in the fossil record of the rapid rise of angiosperms and insects raises a whole series of questions about co-adaptation. In recent years plant–animal relationships have become popular research fields again, and the classical interpretations of floral structure in terms of insect visitors are being considerably enlarged and modified. Moreover, radically new ideas are appearing, which take into account the predation of animals on plants and the selective importance of defence mechanisms. Thus Grant (1950) argues plausibly for the interpretation of the inferior ovary as such a mechanism, and the whole phenomenon of plant mimicry, curiously neglected in botany, is now being seen as of great adaptive

Fig. 15.3. Examples of host mimicry in Australian Mistletoes (Loranthaceae). (*a*) *Amyema cambagei* parasitising *Casuarina torulosa*. (*a*1) General view: arrow indicates point of attachment of mistletoe. (*a*2) Close-up: mistletoe branch on left, host branch on right. (*b*) *Amyema sanguineum* parasitising *Eucalyptus crebra*. (*b*1) General view: arrows indicate mistletoe plants. (*b*2) Close-up: left arrow indicates host shoot, right arrow the mistletoe. (From Barlow & Wiens, 1977)

significance in reducing predation by animals (see Fig. 15.3 and Wiens, 1978).

Perhaps the most significant development of all is the widespread realisation that many so-called 'secondary chemical compounds' in plants may have adaptive value as defence mechanisms against predators. In extreme cases this has long been recognised, but we are now beginning to see that the chemical defences of plants are at least as complex and varied as are the more traditional structural ones (Harborne, 1978, 1982; Rosenthal & Janzen, 1979; Young & Seigler, 1981; Thompson, 1982).

All these interpretations are orthodox and neo-Darwinian in that they assume that random mutation, recombination, chance and selection are determining the whole evolutionary process. It is important to realise, however, that there are powerful dissentient voices raised against this orthodoxy. In particular, the relative importance of selection and other postulated mechanisms in evolution is still a debated issue and several lines of questioning significant in the history of post-Darwinian thought continue to be heard.

First, Lamarckian ideas of the inheritance of acquired characters are by no means dead. It is usually supposed that 'responses' to environmental factors are not transmitted to subsequent generations. However, Durrant (1958, 1962) discovered that heritable changes could be induced in Flax (*Linum usitatissimum* cultivars 'Stormont Cirrus' and 'Lyrral Prince') in plants given different fertiliser treatments in glasshouse and field. Such changes were stable over a number of generations both in the same environment and in a different environment from that in which they were induced. Similar results have been obtained with *Nicotiana rustica* by Hill (1965). Cullis (1977) has provided a valuable review of the researches on Flax, and offers a model to explain in terms of molecular genetics the stabilisation of environmentally induced changes. Experiments have been carried out on only two species and it remains to be discovered whether the induction of heritable changes by environmental factors is a widespread phenomenon.

Secondly, we have new questioning of the whole selectionist mode of argument. For instance, Gould & Lewontin (1979) have pointed out that evolutionists often 'atomise' the organism into a number of traits, each of which is explained in terms of a structure optimally designed by natural selection for its particular function; any suboptimal functioning is then assumed to be due to the balance between competing demands. Each investigation ends with the telling of a plausible story invoking the force of selection. Plausibility is then often the only test of validity invoked by the evolutionist, and Gould & Lewontin wish to reassert a competing notion ... 'that organisms must be analysed as integrated wholes, with *Baupläne* so constrained by phyletic heritage, pathways of development and general architecture that the constraints themselves become more interesting and more important in delimiting pathways of change than the selective force that may mediate change when it occurs'. Again, botanists might reasonably point to a number of writers on plant evolution who have entered similar *caveats* over the years: an excellent example is provided by Manton (1950) in the final chapter of her now classic work on the Pteridophyta. The difficulty with all such views, which suggest that in some way the direction of evolution is determined by factors internal to the organism, is that they do not allow any single unified explanation; but, as Gould & Lewontin plausibly argue, we should not be looking for simplicity, rather supporting 'Darwin's own pluralistic approach to identifying the agents of evolutionary change'.

Finally, in many ways the most interesting of these continuing controversies is the one concerning the so-called 'saltation' view of evolution, which appears to be directly descended from the classical mutationalist controversies of the late nineteenth century (see, for example, Williamson, 1981). Botanists, to whom the idea of abrupt evolutionary change has never seemed particularly heretical, have always found a place for such views in their speculations on evolution, as the writings of Willis (1922, 1940, 1949), Good (1956) and Lamprecht (1966) bear ample witness. They may therefore be forgiven if they approach with a sense of *déjà vu* some current arguments between palaeontologists and evolutionists, provoked by Gould & Eldredge (1977) who have re-awakened the issue of the relative importance of abrupt change in evolution. These biologists consider that abrupt change dominates the history of life. In their view evolution is concentrated in very rapid events of speciation, most species during their geological history changing little or else fluctuating mildly in morphology with no apparent direction. Clearly the crucial and as yet unresolved question both for the experimentalist and the palaeontologist is the *balance* of gradual and abrupt events.

Polyploid evolution and apomixis

Our attention now focuses on a different topic, namely the interrelation of two phenomena – polyploidy and apomixis – which we have so far discussed separately. As we saw in Chapter 11, the incidence of polyploidy in the three major groups of vascular plants is surprisingly different, being very rare in gymnosperms but widespread in both pteridophytes and angiosperms. Apomixis is similarly distributed, and when the relation between chromosome number and apomixis is investigated in detail, we find a striking correlation, most apomictic species being also polyploid. The converse is, of course, not true; there are very many sexual polyploid species. We might therefore conclude that the occurrence of polyploidy favours the development of apomixis and it was pointed out by Darlington as long ago as 1939 that apomixis could be thought of as 'an escape from sterility'. The argument runs as follows. Hybridisation, whether between diploid or polyploid species, frequently produces more or less sterile clones. Such hybrids, if they are to establish themselves permanently, must be apomictic, relying either on effective vegetative spread or on agamospermy. It is not therefore surprising that there are apomictic taxa of putative hybrid origin, often with very restricted distributions. This neatly explains the correlation of apomixis with hybridisation; it does not, however, fully explain the very high correlation with polyploidy. We can extend the argument to point out, for example, that autopolyploidy and segmental allopolyploidy are both likely to produce meiotic irregularity and partial sterility, and to that extent the view of apomixis as an 'escape' seems reasonable; it hardly seems a complete explanation, however.

The problem may be looked at in quite a different way. The diploid sexual species is generally one in which the individual has definable limits and a more or less predictable life-span. The two main groups of such plants are the annuals and the trees, and it is in these groups that we find least polyploidy and apomixis. Thus the absence of polyploidy and apomixis from the modern conifers is more reasonably connected with their life-form and life-cycle than with any peculiarities they may have as gymnosperms. Conversely, the modern pteridophytes, which are generally rhizomatous and largely herbaceous, show a great deal of polyploidy and some apomixis. We might therefore look upon the polyploid and apomictic lines of evolution more in terms of positive adaptation to particular kinds of habitat than in terms of a negative escape from sterility. We are then asking the same question as in Chapter 7: namely, what are the evolutionary advantages of outbreeding? The argument which we sketched in discussing autogamous species can then be seen to have much wider application. We can postulate certain evolutionary situations and types of habitat in which the advantages of a safe and quick method of reproduction and dispersal outweigh the advantages of outbreeding in maintaining genetic variability. In such situations, auto-gamy (Fig. 15.4) or apomixis could well be favoured.

The most obvious kind of habitat in which quick, safe reproduction is essential is the cultivated or ruderal habitat created by man. The loss of outbreeding mechanisms in many modern weeds is therefore understand-

Fig. 15.4. Diagrams of the flowers of 10 Californian species of *Trichostema*, ranging from small-flowered self-pollinated species to large-flowered species (*T. lanatum*) adapted to pollination by humming-birds.

Species of *Trichostema*	Breeding system (S = self-pollinated; O = outcrossing)	Flower diameter (mm)	Nectar volume (µl)	Pollen:ovule ratio
T. micranthum	S	4.0	0.0	137
T. austromontanum	S	4.0	0.0	233
T. oblongum	S	3.7	0.0	363
T. simulatum	S	4.0	0.0	586
T. rubisepalum	S–O	5.9	0.04	791
T. ovatum	O	7.4	0.78	1035
T. laxum	O	7.5	0.95	1422
T. lanceolatum	O	9.0	0.73	1774
T. parishii	O	12.0	0.47	1826
T. lanatum	O	11.8	5.37	2471

(From Spira, 1980, redrawn from Lewis, 1945.)

This example illustrates a common phenomenon. Autogamous (predominately self-fertilising) species, in comparison with their outcrossed relatives, generally have smaller flowers, little or no nectar and a low pollen:ovule ratio. In outbreeding taxa, much biomass is 'invested' in the uncertain processes of reproduction. In autogamous species, all of which were probably evolved from outbreeding taxa, fertilisation is a near certainty and selection may favour variants in which the 'energy' saved in the reduction of floral parts is 'invested' in additional flowers or fruits. (See Dingle & Hegmann, 1982)

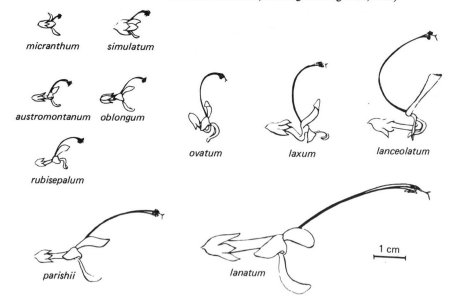

micranthum   simulatum

austromontanum   oblongum

ovatum   laxum   lanceolatum

rubisepalum

parishii   lanatum

1 cm

able in terms of the great selection pressures in favour of quick possession of artificially bared ground, whether arable field or 'waste land'. Moreover, if we look at the natural open habitats in which bare ground is continually made available by agencies of erosion and catastrophic destruction, we can find here in many cases the presumed native habitats of familiar weeds. In Temperate regions these natural open habitats are mostly on coasts or mountains, though river banks, rocky gorges and other local features may also provide them. An excellent example of a weed species which has related native variants in coastal and mountain habitats is the Bladder Campion (*Silene vulgaris*), and many other examples may be found in the review paper by Baker (1974) on the evolution of weeds. The relation between such plants and the crop plants of man is discussed below.

One of the phenomena which has excited interest since the early days of cytogenetic studies on plants is the significance of a general correlation between polyploidy and latitude as seen in the flora of the Northern Hemisphere. The earliest reference to this correlation seems to have been made by the Swedish cytologist Täckholm in his impressive study of the genus *Rosa* published in 1922, followed by Hurst, who published in 1925 a remarkable map of the distribution of diploid and polyploid *Rosa* species (Fig. 15.5). The phenomenon of which the genus *Rosa* is a special case seems to have been first reviewed by Hagerup in 1931. Whilst increasing knowledge has somewhat complicated the picture, the general validity of some correlation of polyploidy with latitude shown by statistics such as those of Morton (1966) (Table 15.1) seems to be widely accepted. We might therefore ask whether an explanation along the lines we have outlined above could plausibly cover this much wider phenomenon. To answer this question we must sketch out our knowledge of the recent evolutionary history of the floras of Northern Temperate regions in order to supply the necessary background.

Table 15.1. *The relation of polyploidy and latitude in angiosperms, summarised from the literature by Morton (1966)*

	Percentage polyploidy in flora
West Africa	26
North Sahara	38
Great Britain	53
Iceland	66
Greenland	71
Arctic Peary Land	86

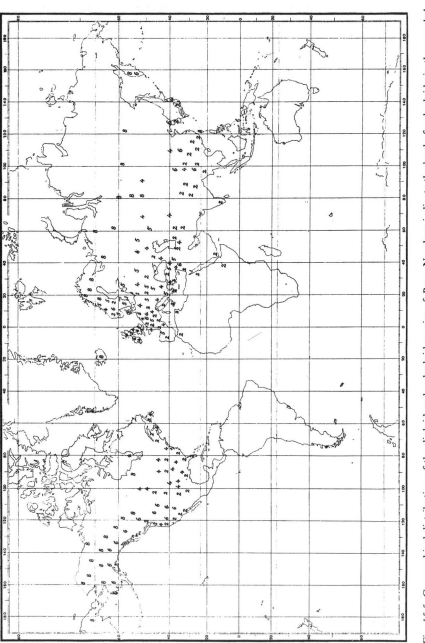

Fig. 15.5. Geographical distribution of the diploid and polyploid species of *Rosa*. Numbers indicate the level of polyploidy in the sampled species. Perhaps the earliest published map illustrating the correlation of polyploidy with latitude. (From Hurst, 1925)

The growth of what are generally known as 'Quaternary studies' is one of the most impressive developments of non-experimental science in the present century. The details obviously lie outside the scope of our book; they can, however, be readily obtained from books such as those by Pennington (1969, 1974) or Iversen (1973) if a wider, European view is preferred. The picture we have is of a series of glaciations interspersed with relatively warm interglacial periods; during the maximum extent of the ice in each glacial period, the previously existing vegetation was more or less completely eliminated from the greater part of the land masses which are present-day Eurasia and North America, and with the subsequent amelioration of the climate the flora re-immigrated northwards to occupy land progressively released from the ice-cover. Such a process of recolonisation can be seen locally at the present day in Scandinavia or the European Alps, where glaciers are known to have retreated in historical time. The most recent glaciation affecting Britain (and indeed Europe as a whole) only came to an end some 15 000 years ago, so that our present flora is a very new re-immigrant flora quite obviously poor in species. This explains why we can in our present climate grow a very large variety of plants most of which are not part of our native flora, and also why some of these garden plants, such as *Rhododendron ponticum*, can escape and become effectively wild in Britain today.

Against this background, what can we say about the roles of polyploidy and apomixis? Obviously the same argument used concerning weeds and open habitats can be applied. The retreat of the ice during the so-called Late Glacial period left behind bare, and often unstable, ground open to rapid colonisation by opportunist species, many of which were polyploids which had adopted apomictic methods of reproduction. It is not necessary to conclude, as earlier workers did, that there is any direct effect of temperature inducing polyploidy; the much more plausible correlation is of apomixis (including vegetative reproduction) with the disturbed and open habit.

If our view of polyploidy is correct, we can postulate periods of relative stability in plant evolution, where the diversification and speciation would be on a diploid level, interspersed with 'catastrophic' destruction of the stable vegetation. The advance of the ice during glaciations would result in radical changes in plant distributions, bringing together diploid stocks which had previously evolved in isolation. Hybrids and polyploid derivatives could be produced, together with apomictic variants. It seems likely that the very recent glaciations of the Northern Hemisphere have had a profound effect on the modern floras of the Northern Temperate and

Arctic zones; can we also find evidence of more ancient polyploidy? If we compare the basic chromosome numbers of vascular plant genera we can certainly find evidence of this kind, admittedly not direct, but nevertheless pertinent. For instance, the whole of the subfamily Pomoideae of the Rosaceae is characterised by the basic number $x = 17$, and diploid sexual species of *Sorbus*, for example, have $2n = 2x = 34$. The other subfamilies of Rosaceae contain the basic numbers $x = 8$ and $x = 9$, and it seems very reasonable to speculate on an ancient allopolyploid origin for the Pomoideae from $x = 8 + 9$. Similar cases can be found in other flowering plant families, and Manton (1950) has shown that the Tropical fern flora contains many cases of 'ancient' polyploidy.

Reverting to the general correlation between polyploidy and latitude, what can we offer in terms of a broader explanation? The first point we should make is that it is by no means an invariable rule that within groups of related taxa in the Northern Hemisphere the diploids are the most southerly in distribution. Indeed, Lewis (1980*a*) gives a number of examples where the reverse is the case, concluding that what characterises most diploid/polyploid groups is not a standard geographic pattern, but the fact that the polyploids, putatively derived by hybridisation from the related diploids, are better adapted to survive under more extreme climatic and edaphic conditions than the parent diploids. The second point to be noted is that our interpretations involve the assumption that, within any group, polyploidy is essentially a one-way process, in which diploid or low-polyploid taxa are the parents of the higher allopolyploids. Whilst there is good reason to think of many of the patterns we see in this light, are we right in assuming that such an interpretation is valid for every case? Until recently, writers on evolution of higher plants have assumed that it was: ancient polyploid species such as make up the genus *Equisetum* certainly look like evolutionary relics that are eventually doomed, and there is no obvious widespread mechanism for descending a polyploid series as there is for ascending it. Raven & Thompson (1964) drew attention, however, to the possible significance of the occasional phenomenon of 'polyhaploidy' – the derivation of functional haploid plants from functional diploids which are themselves polyploid in origin – and, as we saw in Chapter 11, Ornduff (1970) and de Wet (1971) have re-examined the possible evolutionary implications in the light of a few described cases of polyhaploidy. It seems best at present to preserve an open mind to the possibility that polyhaploidy may have evolutionary significance. Stebbins (1980), reviewing the evidence, adopts this attitude.

Before we leave the topics of polyploidy and apomixis, we should note that geneticists have become increasingly interested in apomixis as an evolutionary phenomenon. In a recent paper, Marshall & Brown (1981) argue that, because apomixis is usually under complex genetic control, it remains relatively infrequent, but it (like autogamy) would theoretically eventually become fixed in plant populations unless it radically reduced the fitness of its carriers. They suggest that the strong correlation of apomixis with polyploid hybridisation is due to the fact that such hybrids often exhibit a degree of sterility, which removes the need for one of the simultaneous mutants required for the development of apomixis. Stebbins (1980) seems to be looking for a similar genetical explanation of the correlation of polyploidy and apomixis. These views are all very tentative and many questions remain unanswered.

The domestication of plants

One of the less fortunate effects of the separate development of botany, with its mediaeval roots in medicine, from that of the applied plant sciences of agriculture, horticulture and forestry, is that the rich source of material relevant to our study and to be found in these applied sciences is often neglected in botanical writings. This neglect is all the more remarkable when we recall that, as we saw in Chapter 4, Darwin himself was greatly impressed by the phenomena of domestication in both plants and animals, and indeed saw these phenomena as providing very strong evidence for the power of selection in shaping the organisms we have around us today. Happily, the period when 'pure' botanists largely ignored man's influence, preferring instead to study what they thought, often quite erroneously, to be entirely 'natural' communities of plants, seems to have come to an end. (See Anderson, 1954, for an entertaining and pioneer critique of the now outdated prejudice.) As we have seen in Chapter 13, modern studies of plant populations increasingly take into account historical evidence about man's role in shaping present-day vegetation patterns, and the over-rigid distinction between a postulated 'natural' world and the artificial habitats obviously shaped by man is being progressively abandoned. Having said this, it is of course still true that much relevant research of a highly specialised nature proceeds independently in research institutions, with quite precise applied aims to improve existing crop plants for man's use, and that any detailed discussion of such research would be out of place in our book. In what follows we attempt to indicate a few directions of research which we feel are of special relevance to our general theme, acknowledging our indebtedness to two works in particular, namely the book by Schwanitz (1966) entitled *The origin of cultivated plants* and the review paper by Pickersgill & Heiser (1976) on the genetic and evolutionary aspects of the subject.

Pickersgill & Heiser consider the evidence now available for many different crop plants concerning the nature of genetic difference between the modern crop and its putative wild ancestor. They point out that important qualitative characters of a morphological, physiological or biochemical nature 'are usually determined by one or a few major genes and show relatively simple patterns of inheritance'. Examples cited include the change from brittle to non-brittle rachis in the evolution of Old World cereals, from dehiscent to indehiscent pods in cultivated legumes, and from bitter to mild flesh in cultivated Cucurbitaceae (gourds, cucumbers, etc.). They further consider the role of the breeding system in determining the ease and speed with which such mutants might replace the wild type in domestication. Discussing quantitative characters, such as the increased size of edible seeds and fruits in crop plants, they point out that here the genetic control is usually much more complex, involving a number of genes and only rarely susceptible of detailed analysis. Selection for such characters under cultivation is necessarily a slower and more gradual process. Summing up this part of their survey, they stress that most cultivated plants, however superficially large the difference from their wild ancestors might be, have not in genetic terms diverged far from those ancestors.

Turning to the process of selection in more detail, Pickersgill & Heiser suggest that earlier writers in this subject had not sufficiently taken into account the undoubted fact that conscious or unconscious human selection has not always acted in the direction we might be inclined to postulate from our present-day view. Indeed, for the same crop plant, selection at different periods and by different human cultures could have gone in different, even opposite, directions. In a contribution to the discussion of Pickersgill & Heiser's paper, Harris (1976) suggests that one reason why there is such a striking difference between crop evolution in South-west Asia, where small-seeded grasses, legumes and other mainly annual crop plants were selected, and Central America where relatively large-seeded crop plants such as *Zea* and *Phaseolus* were developed, lies in the different traditions of seed-sowing in the two areas. Diversification of human cultures, on this view, has played a large part in determining the kinds of variants selected as crop plants by man.

Another area of great interest where views are changing with increased knowledge concerns the relationship between crop plants and weeds. Weedy relatives of crop plants have until recent times been thought to be more primitive and generally ancestral to our modern crops; for example, wild Rye (*Secale*) occurs widely in South-west Asia, and annual grasses related to Wheat (*Triticum*) are widespread in South-east Europe and Asia. To some extent we must retain this picture, as in the case of modern Wheats, which are hexaploid with a complicated allopolyploid derivation involving wild diploid grasses; but in general we must revise our view, and look to divergent selection as responsible for the evolution of weeds and related cultivars together (see, for example, Harlan, 1965, and McNeill, 1976). While man was selecting, consciously or unconsciously, his crop plants, weeds were presumably evolving by disruptive natural selection from the same material. Thus we find today weeds related to all the main cereals and to most other important crop plants such as *Beta* (Beet) and *Helianthus* (Sunflower). Furthermore, a new picture, increasingly supported by archaeological evidence about the origins of agriculture, suggests the co-evolution of crops and weeds, involving the possibility of significant interbreeding which has perhaps increased the genetic variability of the evolving crop.

A very different kind of relationship between crops and weeds occurs in a number of instances where the weed, especially in the form of seed, may mimic the systematically unrelated crop with which it is normally associated. The case of *Camelina sativa*, which mimics *Linum* (Flax) in size and weight of seed, is the most commonly quoted example in the literature (Stebbins, 1950) but there are other cases, such as the parasitic weed *Cuscuta* (Dodder) (Wickler, 1968).

The role of polyploidy in the continued evolution of cultivated plants has also come in for some recent reassessment. In view of the frequency of polyploidy amongst garden and crop plants, which has been traditionally explained in terms of the familiar 'gigas' effect of polyploidy (see Chapter 11), it is understandable that artificially induced polyploidy has been very widely used by plant breeders in attempts to produce improved cultivars. A recent assessment by Dewey (1980), however, suggests that the results have not been particularly impressive; indeed only the indirect use of induced polyploids to bring about genetic transfer of desirable characters into crop plants seems to have lived up to expectation. Summarising, Dewey points out that the greatest potential for polyploidy in plant breeding lies with those crop plants such as Potato, *Solanum tuberosum*, which have relatively low chromosome numbers and are harvested primarily for their vegetative parts.

These considerations lead naturally to a question which is being increasingly asked in recent years, namely whether the genetically uniform 'monocultures' favoured by modern agricultural practices are producing in the long-term harmful effects not foreseen in the early days of the agricultural revolution. Barrett (1981) addresses himself to this question in a paper on 'The evolutionary consequences of monoculture'. He outlines the history of several crop plants and their diseases: in each case the increasing reliance on a few cultivars with little genetic variability seems to have been responsible for disastrous crop failure in recent history. It is, of course, easy to see the problem, but more difficult to reverse the general progress towards genetic uniformity in agricultural crops. This type of concern is linked to the general problem of conservation, to which we devote part of our final chapter.

Variety of patterns of evolution

The important correlations between habitat, life-form and breeding system, which we have discussed, are reflected in the varied patterns of relationship shown within genera of flowering plants. Detailed experimental study of several medium-sized or large genera has now been carried far enough to enable us to compare the patterns. We can take the results of a study of five angiosperm genera, and compare them. In the annual genus *Layia*, evolution has been very largely on the diploid level, and the taxonomic species fit the experimentally defined ones reasonably well (Clausen, Keck & Hiesey, 1941). In the large genus *Silene*, most North American species are perennials and polyploid, while in Eurasia there are many annual species in which all the evolution has remained at the diploid level (Kruckeberg, 1955, 1964). On grounds of comparative morphology the annual weed species of *Silene* in Eurasia might be presumed to be specialised and recent products of evolution from more ancient perennial stock. Babcock (1947) advanced a similar argument for the genus *Crepis*, deriving annual sections of the genus from more primitive perennial ones. Polyploidy and apomixis are rare in *Crepis*, being almost confined to a single, North American section (Sect. *Psilochaenia*). In *Viola* there is much polyploidy (Valentine, 1962); the subsection *Rostratae* consists of a number of diploid species together with polyploid derivatives, and most hybrids are sterile. Finally, in this range of examples, there is the perennial genus *Geum*, in which, as we saw in Chapter 10, polyploidy is widespread and species-hybrids show a good deal of fertility.

It is difficult to feel from these examples that the balance of advantage in evolution lies with any one pattern of variation. The impressive feature of the evolution of plants is the variety and complexity of adaptations, including adaptations of the breeding systems themselves; even the most complex of these phenomena, such as we described for the genus *Potentilla* in Chapter 8, must be viewed as 'experiments' of adaptive significance in evolution.

16

Plant variation, nature conservation and the future

The keen interest many people find in the wild plants around the home or in the places they visit on holiday often goes no deeper than identification. Attaching names to plants can be an enjoyable hobby, but we feel confident that many of our readers will reasonably ask whether they might not do more interesting things with their leisure time in pursuing the study of the variation in plants. The last 10 years have seen a great surge of interest in, and indeed concern for, the natural environment of which the wild plants and animals are such an important part, and it is probably true to say that we have never had so many would-be 'naturalists' as we have today. Certainly there has never been a generation so well endowed with accurate, authoritative and attractively presented information on the variety of the world of natural history.

The nature conservation movement

This great outburst of 'field studies' has, however, taken place against a sombre background: we know and value the plant and animal resources of the world apparently when we have become aware, for the first time, of the gravity and urgency of the threat to biological richness and variety posed by the spread of human civilisation. International recognition of these problems led to the United Nations Conference in Stockholm in 1970, where Governments discussed in open, public sessions the 'crisis of the environment'. Most countries now possess, at least in embryo, national organisations devoted to the conservation of nature, and in our own country, both the state Nature Conservancy Council and the voluntary organisations have grown (and are still growing) rapidly in size and influence. We now have a new generation of young naturalists, professional and amateur, who have been brought up in an atmosphere of concern for conservation, to whom the idea of a nature reserve or a system of legally protected rare species is fully acceptable. Such naturalists have had to come to terms with the fact that nature is not inexhaustible and infinitely self-renewing but that, on the contrary, we need to treat our natural heritage with care and respect.

The areas covered by this new professional activity in nature conservation can be briefly summarised under three heads: the creation and management of nature reserves and protected sites (*in situ* conservation); the establishment of reserve stocks of endangered species in Botanic Gardens and elsewhere (*ex situ* conservation); and the investigation of ecological problems relevant to the survival of threatened species or vegetation types, much of which is carried out within existing University Departments using research grants from public bodies. Both *in situ* or *ex situ* conservation efforts pose interesting and often highly complex problems relevant to our study of the variation of plants. Thus, the conditions under which rare species, for which a particular reserve may have been designated in the first place, continue to survive within that reserve are often quite precise and restricted, and an understanding of the problem requires, for example, knowledge of the population size and structure, its variation from year to year, its reproductive capacity, the significance of dormant seed, and many other factors. As an example of such studies, the work of Bradshaw & Doody (1978) on rare species in Teesdale, northern England, can be cited.

Again, similar areas of research are needed when stocks are held, either as clones or as seed, in gene banks; for example, the seed bank operated by the Royal Botanic Gardens, Kew, at Wakehurst Place has undertaken a programme of research on seed viability under controlled conditions which is of considerable theoretical as well as practical interest. Roberts (1975) provides a useful review of the problem of holding stocks of plants as stored seed; the majority of seeds so far tested can be successfully preserved in this way at low temperatures and with low humidity, although there remain 'recalcitrant' species, among them many important Tropical crops such as Cocoa (*Theobroma*), which cannot be so preserved, and the precise conditions for retaining longevity in seeds is obviously a matter which needs individual research. A further problem discussed by Roberts is the genetic effect of seed storage; there is a general correlation between the age of the stored seed and the amount of recessive mutation revealed when the surviving seed is eventually germinated. The importance of this complication in the use of seed banks to preserve particular genotypes is obvious enough; it is also, of course, a significant complication in the conservation of endangered species by this means, although as yet we have no case histories from which we can argue.

What can the keen amateur contribute?

Against this background, what can we recommend to our readers for their study? The first thing we can say is that the new interest in the conservation of nature and the responsible use of the environment means that printed educational and interpretive material is available now as it never was before, enabling anyone to learn more about wild plants and animals, their interrelationships and their places in nature. Most nature reserves and many other places of natural interest and beauty open to the public are now supplied with booklets, nature trail leaflets and other literature. The message of nature conservation has to be that education in the right use and appreciation of nature is one of the most necessary and rewarding of all educational activities, whether at the level of child or adult.

The first step, then, is to use the facilities provided by local and national societies and Conservation Trusts, all of whom owe their existence and success to the interest and ability of their voluntary members. Learn from them what you can and offer them your services in time, expertise or money – all can be used. Above all, do not be afraid to ask questions. In the study of natural history as in other pursuits, it is those enthusiasts who ask the right questions who make really important contributions. This was the case with the great biologists like Darwin, and it is still true today.

What, more specifically, are the areas to which keen amateur botanists interested in the themes of this book can really make important contributions? We would like to suggest three which seem particularly suitable, without in any way implying that there are not many other suitable and interesting topics. The first one concerns the variation of *common* plants. There are several reasons why such studies commend themselves. Most obviously, they cause no problem in the collecting of wild material for, providing the law (embodied in Britain in The Wildlife and Countryside Act, 1981) is not broken by digging up specimens without permission, samples of different populations can be established, preferably from seed or cuttings, as appropriate, and comparative cultivation can reveal how far phenotypic variation detected in the field is or is not hereditary.

A second kind of study which is particularly timely concerns the nature and frequency of wild species-hybrids. This subject, which is full of interest, was comparatively neglected in Britain until the publication in 1975 of *Hybridization and the flora of the British Isles*, an outstanding work edited and compiled by Dr Clive Stace from material collected largely through the membership of the Botanical Society of the British Isles, which sponsored the study. The appearance of this excellent reference book has greatly stimulated the search for hybrids in the British flora, some of which are proving to be much more widespread than realized. The theoretical interest in hybridisation, which we have discussed in our book, is combined with the attractive possibility that, where hybrids can be detected, some kind of microevolutionary change can often be detected also, a change which, in the most favourable circumstances, can be very rapid indeed. Common examples of hybridisation accompanied by rapid microevolutionary change are found at present, for example, in Britain in the weed genera *Epilobium* and *Senecio*.

A third area of study especially attractive at the present time concerns the interrelationships of insects and flowers. For many years this branch of natural history, which for obvious reasons fascinated Charles Darwin among many other nineteenth-century naturalists, has been comparatively neglected. Yet it is, as Darwin saw, central to the theme of the variation of floral structure, and therefore to the whole story of the evolution of flowering plants. It is encouraging to find that many young biologists are returning to this kind of investigation, with its very necessary content of observation in the field, and that, again, an up-to-date text book (Proctor & Yeo, 1973) is available to guide their studies. The papers presented to a joint symposium of the Botanical Society of the British Isles and the Linnean Society held in 1977 (Richards, 1978) give some idea of this burst of new interest in a largely 'observational' field of study.

None of these suggested general areas for field study requires either elaborate equipment or expert knowledge, merely interest and time. For those who may find that they have at their command greater facilities or greater expertise (in botanical identification, for example), of course other possibilities present themselves. We can mention one of these.

Rare species present a particular fascination and a special challenge. In Britain we now know with some accuracy which are our rarest vascular plants and where they are (or were) to be found: and the *British Red Data Book* (Perring & Farrell, 1982) provides us with these data in a condensed form.* What we do *not* know in most cases are the size, and fluctuation with time, of each remaining rare plant population. This information is by its very nature difficult to obtain and requires a long-term commitment; yet it is essential for the organisations, Governmental and voluntary, who

---

* Similar 'Red Data Books' and lists are now available for the rare and threatened plants of many other countries (see Synge, 1981, Appendix 2, for a very useful list) and their further production is continuing at an encouraging rate. The office of the IUCN Threatened Plants Committee at the Royal Botanic Gardens, Kew, will be pleased to supply up-to-date information on this literature to any *bona fide* enquirer.

have the problem of safeguarding the surviving rare species. Amateur botanists are already helping in individual species studies of this type, especially where they have local information and specialised knowledge often accumulated over many years.

Prospects for the future

We close with a speculation about the future. What is the likely effect of the increasingly powerful impact of Man on wild plant species and the communities they live in? As we have explained in this book, rapid environmental change encourages rapid microevolution, and to that extent it is arguable that Man's activities have enriched rather than impoverished the world's flora over the past millenia. It is certainly true that the evolution of specialised weeds and ruderals, adapted to life in the many new habitats opened up by the spread of human civilisation, is an important part of the story of the origin of our modern American and European floras. This is particularly the case in the Mediterranean area, the cradle of our civilization, where much of the richness and beauty of the wild flowers is associated, not with the natural vegetation, but with secondary plant communities which, under the influence of grazing, burning and primitive agriculture, have very largely taken their place. Nature conservation organisations are increasingly aware of a fundamental difficulty in conserving the floras and faunas of any 'developed' region of the world, namely that in most cases the *rare* species, on which for obvious reasons most attention has so far been directed, are precisely those which do not have an obvious niche in the dominant and stable vegetation-types which exist more or less independently of Man. Increasingly we realise that Man is, and always has been since *Homo sapiens* evolved, an important 'biotic' factor.

We may note, however, that, on a world scale, the position is very different. Using modern technology, Man is now able to destroy the natural climax vegetation swiftly and ruthlessly, and vast areas of Tropical forest, together with the whole of their complex, interrelated plant and animal communities, are disappearing very rapidly. Once extinct, they cannot be re-created. Of course, some other vegetation more capable of resisting Man's onslaught will survive and new species will evolve; but, since the evolutionary timescale is so different, we destroy in a decade what took many millions of years to arise and the inevitable result is a drastic impoverishment of the world's flora. Much of our effort in nature conservation is necessarily a rescue operation, at the eleventh hour, of plant and animal species and communities which cannot possibly survive the onslaught of modern Man. There are several reasons why we should support such a rescue operation, most of which would apply to the preservation of a historic building or a work of art; but one argument which is clearly relevant to the title of this book is that we now know that we are losing at an appallingly rapid rate much of the fantastic variation of plants which is our world heritage, and on which in a very literal sense our civilisation has always depended and still depends for food, shelter, energy, medicine and in other ways. The most important international bodies which are working in this field are the International Union for the Conservation of Nature and Natural Resources and the World Wildlife Fund (Office address: 1196 Gland, Switzerland); the role of the Fund is to raise money for urgent conservation projects throughout the world, and all voluntary help is welcomed. Questions about the nature conservation organisations in different countries can be answered by the Office.

Is our nature conservation activity too negative and sentimental? Ought we to be so concerned with 'rescue operations' of species and communities which cannot look after themselves and adapt to Man's destructive power? These are reasonable questions, and it is not easy to give a completely satisfactory answer. The study of plant variation and micro-evolution can certainly proceed, as we have shown, in communities affected or even 'created' by Man's activities, and ironically some of our most telling examples of evolution in action are necessarily of this sort (Chapter 14). Yet there is an understandable desire amongst most field biologists that the 'field' should at least include habitats which look after themselves if left to themselves – the so-called 'natural habitats'. It is true that the more we know of such 'natural habitats', at least in the more familiar lowland areas of the world, the more we realise that the structure of the vegetation is subtly affected by the activities of Man whether direct or indirect, and that, in any case, plant and animal communities are by their nature in dynamic equilibrium when they are apparently stable and persistent. Change, which we call evolution, must inevitably proceed, whatever Man as agent of change or recorder of change might do. At the level of the plant species, this kind of change was the subject matter of our book.

Further reading

Here are a few suggestions for further reading on this topic of world importance. There is an enormous and rapidly expanding litera-ture, from which we select three volumes of special interest. The first, *Conservation and evolution* by Frankel & Soulé (1981), is an outstandingly good review of the whole subject which brings together for the first time a large body of practical knowledge with up-to-date theoretical treatments.

It covers both plant and animal conservation, and presents a very balanced account. From the standpoint of the material in our book, perhaps its most interesting treatment is that of the genetic diversity of plants used by Man and the problems of their conservation. We realise from the treatment of these themes given in this book how the applied botanists, foremost amongst whom has been Sir Otto Frankel himself, have been pioneers in the conservation of genetic resources. It was they who first set up seed banks, for example, to conserve genetic variability of importance for plant breeding programmes especially of the main cereal crops of the world. The second book (Synge, 1981) presents the proceedings of a conference on the biological aspects of rare plant conservation held in Cambridge in 1980; the many papers deal with the important themes of scientific research directed to the problems of protecting rare plants and the genetic variation present in their surviving populations. Finally, we select *Conservation biology* edited by Soulé & Wilcox (1980); although mainly zoological in its examples, it discusses very important issues, especially of Tropical conservation, in an ecological framework. All three books have very extensive bibliographies.

Appendix

Tables of variance ratio are taken from Table V of Fisher & Yates (1963) *Statistical tables for biological, agricultural and medical research*, published by Longman Group Ltd., London (Previously published by Oliver & Boyd Ltd., Edinburgh) and by permission of the authors and publishers.

(i) *0.05 Probability point*

N_2	N_1	1	2	3	4	5	6	12	24	∞
1		161.4	199.5	215.7	224.6	230.2	234.0	243.9	249.0	254.3
2		18.5	19.0	19.2	19.3	19.3	19.3	19.4	19.5	19.5
3		10.1	9.6	9.3	9.1	9.0	8.9	8.7	8.6	8.5
4		7.7	6.9	6.6	6.4	6.3	6.2	5.9	5.8	5.6
5		6.6	5.8	5.4	5.2	5.1	5.0	4.7	4.5	4.4
6		6.0	5.1	4.8	4.5	4.4	4.3	4.0	3.8	3.7
7		5.6	4.7	4.4	4.1	4.0	3.9	3.6	3.4	3.2
8		5.3	4.5	4.1	3.8	3.7	3.6	3.3	3.1	2.9
9		5.1	4.3	3.9	3.6	3.5	3.4	3.1	2.9	2.7
10		5.0	4.1	3.7	3.5	3.3	3.2	2.9	2.7	2.5
11		4.8	4.0	3.6	3.4	3.2	3.1	2.8	2.6	2.4
12		4.8	3.9	3.5	3.3	3.1	3.0	2.7	2.5	2.3
13		4.7	3.8	3.4	3.2	3.0	2.9	2.6	2.4	2.2
14		4.6	3.7	3.3	3.1	3.0	2.9	2.5	2.3	2.1
15		4.5	3.7	3.3	3.1	2.9	2.8	2.5	2.3	2.1
16		4.5	3.6	3.2	3.0	2.9	2.7	2.4	2.2	2.0
17		4.5	3.6	3.2	3.0	2.8	2.7	2.4	2.2	2.0
18		4.4	3.6	3.2	2.9	2.8	2.7	2.3	2.1	1.9
19		4.4	3.5	3.1	2.9	2.7	2.6	2.3	2.1	1.9
20		4.4	3.5	3.1	2.9	2.7	2.6	2.3	2.1	1.8
22		4.3	3.4	3.1	2.8	2.7	2.6	2.2	2.0	1.8
24		4.3	3.4	3.0	2.8	2.6	2.5	2.2	2.0	1.7
26		4.2	3.4	3.0	2.7	2.6	2.5	2.2	2.0	1.7
28		4.2	3.3	3.0	2.7	2.6	2.4	2.1	1.9	1.7
30		4.2	3.3	2.9	2.7	2.5	2.4	2.1	1.9	1.6
60		4.0	3.2	2.8	2.5	2.5	2.3	1.9	1.7	1.4
120		3.9	3.1	2.7	2.5	2.3	2.2	1.8	1.6	1.3
∞		3.8	3.0	2.6	2.4	2.2	2.1	1.8	1.5	1.0

(ii) *0.01 Probability point*

N_2	N_1	1	2	3	4	5	6	12	24	∞
1		4052	4999	5403	5625	5764	5859	6106	6234	6366
2		98.5	99.0	99.2	99.3	99.3	99.3	99.4	99.5	99.5
3		34.1	30.8	29.5	28.7	28.2	27.9	27.1	26.6	26.1
4		21.2	18.0	16.7	16.0	15.5	15.2	14.4	13.9	13.5
5		16.3	13.3	12.1	11.4	11.0	10.7	9.9	9.5	9.0
6		13.7	10.9	9.8	9.2	8.8	8.5	7.7	7.3	6.9
7		12.3	9.6	8.5	7.9	7.5	7.2	6.5	6.1	5.7
8		11.3	8.7	7.6	7.0	6.6	6.4	5.7	5.3	4.9
9		10.6	8.0	7.0	6.4	6.1	5.8	5.1	4.7	4.3
10		10.0	7.6	6.6	6.0	5.6	5.4	4.7	4.3	3.9
11		9.7	7.2	6.2	5.7	5.3	5.1	4.4	4.0	3.6
12		9.3	6.9	6.0	5.4	5.1	4.8	4.2	3.8	3.4
13		9.1	6.7	5.7	5.2	4.9	4.6	4.0	3.6	3.2
14		8.9	6.5	5.6	5.0	4.7	4.5	3.8	3.4	3.0
15		8.7	6.4	5.4	4.9	4.6	4.3	3.7	3.3	2.9
16		8.5	6.2	5.3	4.8	4.4	4.2	3.6	3.2	2.8
17		8.4	6.1	5.2	4.7	4.3	4.1	3.5	3.1	2.7
18		8.3	6.0	5.1	4.6	4.3	4.0	3.4	3.0	2.6
19		8.2	5.9	5.0	4.5	4.2	3.9	3.3	2.9	2.5
20		8.1	5.9	4.9	4.4	4.1	3.9	3.2	2.9	2.4
22		7.9	5.7	4.8	4.3	4.0	3.8	3.1	2.8	2.3
24		7.8	5.6	4.7	4.2	3.9	3.7	3.0	2.7	2.2
26		7.7	5.5	4.6	4.1	3.8	3.6	3.0	2.6	2.1
28		7.6	5.5	4.6	4.1	3.8	3.5	2.9	2.5	2.1
30		7.6	5.4	4.5	4.0	3.7	3.5	2.8	2.5	2.0
60		7.1	5.0	4.1	3.7	3.3	3.1	2.5	2.1	1.6
120		6.9	4.8	4.0	3.5	3.2	3.0	2.3	2.0	1.4
∞		6.6	4.6	3.8	3.3	3.0	2.8	2.2	1.8	1.0

(iii) *0.001 Probability point*

N_2	N_1	1	2	3	4	5	6	12	24	∞
1		405 284	500 000	540 379	562 500	576 405	585 937	610 667	623 497	636 619
2		998.5	999.0	999.2	999.2	999.3	999.3	999.4	999.5	999.5
3		167.5	148.5	141.1	137.1	134.6	132.8	128.3	125.9	123.5
4		74.1	61.3	56.2	53.4	51.7	50.5	47.4	45.8	44.1
5		47.0	36.6	33.2	31.1	29.8	28.8	26.4	25.1	23.8
6		35.5	27.0	23.7	21.9	20.8	20.0	18.0	16.9	15.8
7		29.2	21.7	18.8	17.2	16.2	15.5	13.7	12.7	11.7
8		25.4	18.5	15.8	14.4	13.5	12.9	11.2	10.3	9.3
9		22.9	16.4	13.9	12.6	11.7	11.1	9.6	8.7	7.8
10		21.0	14.9	12.6	11.3	10.5	9.9	8.5	7.6	6.8
11		19.7	13.8	11.6	10.4	9.6	9.1	7.6	6.9	6.0
12		18.6	13.0	10.8	9.6	8.9	8.4	7.0	6.3	5.4
13		17.8	12.3	10.2	9.1	8.4	7.9	6.5	5.8	5.0
14		17.1	11.8	9.7	8.6	7.9	7.4	6.1	5.4	4.6
15		16.6	11.3	9.3	8.3	7.6	7.1	5.8	5.1	4.3
16		16.1	11.0	9.0	7.9	7.3	6.8	5.6	4.9	4.1
17		15.7	10.7	8.7	7.7	7.0	6.6	5.3	4.6	3.9
18		15.4	10.4	8.5	7.5	6.8	6.4	5.1	4.5	3.7
19		15.1	10.2	8.3	7.3	6.6	6.2	5.0	4.3	3.5
20		14.8	10.0	8.1	7.1	6.5	6.0	4.8	4.2	3.4
22		14.4	9.6	7.8	6.8	6.2	5.8	4.6	3.9	3.2
24		14.0	9.3	7.6	6.6	6.0	5.6	4.4	3.7	3.0
26		13.7	9.1	7.4	6.4	5.8	5.4	4.2	3.6	2.8
28		13.5	8.9	7.2	6.3	5.7	5.2	4.1	3.5	2.7
30		13.3	8.8	7.1	6.1	5.5	5.1	4.0	3.4	2.6
60		12.0	7.8	6.2	5.3	4.8	4.4	3.3	2.7	1.9
120		11.4	7.3	5.8	5.0	4.4	4.0	3.0	2.4	1.6
∞		10.8	6.9	5.4	4.6	4.1	3.7	2.7	2.1	1.0

The values for N_1 at the heads of columns give the degrees of freedom for the larger variance. The values of N_2 at the left hand of rows give the degrees of freedom for the smaller variance.

As an example of the use of these tables, we may consider the analysis of variance set out in Table 13.5. The calculated variance ratio is 49.895. This value is compared with those in the tables of variance ratios for 1 and 18 degrees of freedom. As our calculated value exceeds that given for all three probability levels, we may conclude that the probability is less than 1 in 1000, i.e. there is a highly significant difference between the two variances. The null hypothesis of a zero difference between the variances (between and within groups) may safely be rejected.

Glossary

Our list is restricted to scientific expressions used repeatedly throughout the text. Many other technical terms, which are used only once or twice, are defined on first use; definitions of such terms may be sought via the index.

Definitions, some of which have been simplified, have been taken from Heslop-Harrison (1953), Davis & Heywood (1963), Whitehouse (1973), Rieger, Michaelis & Green (1976), Heywood (1976) and Stace (1980).

Agamospermy The production of seeds by asexual means.

Alleles (allelomorphs) Alternative forms of a gene which, on account of their corresponding positions on homologous chromosomes, are subject to Mendelian inheritance.

Allopatry Of species or populations originating in or occurring in different geographical regions.

Allopolyploid A polyploid originating through the addition of unlike chromosome sets, usually following hybridisation between two species.

Aneuploid Individuals having the different chromosomes of the set present in different numbers.

Apogamy The phenomenon shown by some higher plants in which a gametophyte cell gives rise directly to a sporophyte, without the production of a zygote derived by fusion of gametes.

Apomixis Reproduction, including vegetative propagation, which does not involve sexual processes.

Apospory The phenomenon shown by some higher plants in which a diploid embryo-sac is formed directly from a somatic cell of the nucellus or chalaza; an embryo is then formed without fertilisation.

Artificial classification A system of ordering based upon one or a few characters, which gives a convenient arrangement of plants for some specific purpose. In such systems closely related taxa may be placed in different groupings.

Autogamy Self-fertilisation; persistent autogamy may result in an increase in homozygosity and division of a population into a number of 'pure lines'.

Autopolyploid A polyploid originating through the multiplication of the same chromosome set.

B chromosomes Small chromosomes (frequently, but not always, heterochromatic), which are additional to the normal complement of A chromosomes.

Biological species Groups of actually or potentially interbreeding natural populations, which are reproductively isolated from other such groups.

Bivalent The associated pair of homologous chromosomes observed at prophase I in meiosis.

Chiasma (pl. chiasmata) An interchange occurring only at meiosis between chromatids derived from homologous chromosomes. Chiasmata are the visible evidence of genetic crossing-over.

Classification The ordering of plants into a hierarchy of classes to produce an arrangement which serves both to express the interrelationships of plants and to act as an information retrieval system.

Cleistogamy Self-fertilisation within closed flowers.

Cline A variational trend in space found in a population or series of populations of a species.

Clone A group of independent individuals derived vegetatively from a single plant, and therefore of the same genotype.

Crossing-over The occurrence of new combinations of linked characters following the process of exchange between homologous chromosomes at meiosis.

Cytodeme A group of individuals of a taxon differing from others cytologically (most commonly in chromosome number).

'-deme' terminology Devised to bring clarity to a confusion of terms for units below, at and about the species level. (For a discussion of terms, including Turesson's experimental categories, see Stace, 1980). In cases where precise usage is necessary, the term species, and its prefixed derivatives, is to be reserved for taxonomic categories. Experimentalists could use the '-deme' terminology based on the neutral suffix – deme, which denotes a group of individuals of a specified taxon: '-deme' should not stand by itself. Terms are made by adding the appropriate prefixes. The most commonly used terms are:

Groups of individuals of a specified taxon

in a particular area (topodeme);

in a particular habitat (ecodeme);

with a particular chromosome condition (cytodeme);

within which free exchange of genes is possible [in a local area] (gamodeme: equivalent to the term Mendelian population);

which are believed to interbreed with a high level of freedom under a specified set of conditions (hologamodeme: approximately equivalent to the term biological species).

Despite the clear advantages of the '-deme' terminology it has been little used by biologists (Briggs & Block, 1981); indeed the most frequent use of the terminology is the incorrect employment of 'deme', without prefix, in the sense of gamodeme, by many zoologists.

Diploid With two sets of chromosomes: the condition arising at fertilisation.

Diplospory The phenomenon shown by some higher plants in which a diploid embryo-sac is formed directly from a megaspore mother-cell; an embryo is then formed without fertilisation.

Directional selection Selection occurring when the environment is changing in a systematic fashion, leading to a regular change, in a particular direction, of the adaptive characteristics of a gamodeme.

Disruptive selection Selection which breaks up a homogeneous gamodeme into a number of differently-adapted gamodemes.

Ecocline A variational trend correlated with an ecological gradient.

Euchromatin Parts of chromosomes showing the normal cycle of condensation and staining at nuclear divisions.

Euploid A polyploid possessing a chromosome number which is an exact multiple of the basic number of the series.

Gametophyte The haploid gamete-producing phase of the life-cycle of plants.

Gamodeme A group of individuals of a specified taxon which are so situated spatially and temporally that, within the limits of the breeding system, all can interbreed.

Gene flow The dispersal of genes in both gametes and zygotes, within and between breeding populations.

Genecology The study of intraspecific variation in plants in relation to environment.

Genome A single complete set of chromosomes. One such set is present in the gametes of diploid species; two genomes are found in the somatic cells. Polyploid cells contain more than two genomes.

Genotype The totality of the genetic constitution of an individual.

Germ line Cells ancestral to the gametes and set aside from the somatic cells.

Haploid With a single set of chromosomes (one genome), such as occurs at gamete formation.

Heteroblastic change The transition from a juvenile to an adult form accompanied by a more or less abrupt change in morphology.

Heterochromatin Parts of chromosomes, or whole chromosomes, which exhibit an abnormal degree of staining and/or contraction at nuclear divisions.

Heterozygote A zygote or individual carrying two different alleles of a gene (e.g. *Aa*).

Homozygote A zygote or individual formed from the fusion of gametes carrying the same allele of a gene (e.g. *AA* or *aa*).

Introgressive hybridisation (introgression) Genetic modification of one species by another through the intermediacy of hybrids.

Isolating mechanisms Factors which restrict the extent of gene flow between populations by preventing or reducing interbreeding.

Karyotype The appearance and characteristics (shape, size etc.) of the somatic chromosomes at mitotic metaphase.

Matroclinous (maternal) inheritance Condition found where a hybrid is closely similar to its seed parent.

Meiosis A special nuclear division in which the chromosome number is halved.

Meristic variation Variation in numbers of parts or of organs.

Mitosis The nuclear division typical of somatic plant tissues in which a nucleus divides to produce two identical complements of chromosomes (and hence genes).

Multivalent Association of more than two homologous chromosomes at meiosis, e.g. 3 = trivalent; 4 = quadrivalent.

Natural classification A classification based upon overall resemblance, and serving a variety of purposes.

Phenetic classification A classification based upon present-day resemblances and differences between plants.

Phenotype The totality of characteristics of an individual; its appearance as a result of the interaction between genotype and environment.

Phenotypic plasticity The ability of certain genotypes to respond to different environments by producing different phenotypes.

Phylogenetic classification A classification showing the supposed evolution of groups.

Pleiotropism The phenomenon shown by a gene which simultaneously influences more than one characteristic of the phenotype.

Polygenes Genes of small individual effect which act jointly to produce quantitative genetical variation.

Polyhaploid An organism with the gametic chromosome number arising by parthenogenesis in a polyploid, e.g. a diploid ($2x$) plant arising from the parthenogenetic development of an embryo of a tetraploid ($4x$).

Polymorphism The occurrence of two or more distinct genetic variants of a species in a single habitat.

Polyploid Having three or more sets of homologous chromosomes.

Pseudogamy The phenomenon found in some apomictic plants, whereby pollination is necessary for seed development, even though no fertilisation of the egg-cell takes place.

Pure line A lineage of individuals originating from a single homozygous ancestor.

Ramet An individual belonging to a clone.

Ratio-cline Clinal variation occurring in polymorphic species, in which successive populations show progressive change in the proportion of the variants.

Sporophyte The diploid spore-producing phase of the life-cycle of plants arising from the fertilisation of haploid gametes.

Stabilising selection Selection favouring the average individuals of a gamodeme and eliminating extreme variants.

Sympatry Of species or populations, originating in or occupying the same geographical area.

Taxon (pl. taxa) A classificatory unit of any rank: e.g. Daisy: *Bellis perennis* (species); *Bellis* (genus); Compositae (family).

Topocline A geographical variational trend which is not necessarily correlated with an ecological gradient.

Univalent An unpaired chromosome at meiosis.

Variant Any definable individual or group of individuals. A valuable neutral term.

Vivipary The production of small plants or bulbils in place of flowers and seed.

References

Allard, R.W. (1965). Genetic systems associated with colonizing ability in predominantly self-pollinated species. In *The genetics of colonizing species*, ed. H.G. Baker & G.L. Stebbins, pp. 49–75. New York & London: Academic Press.

Allard, R.W., Jain, S.K. & Workman, P.L. (1968). The genetics of inbreeding populations. *Advances in Genetics*, **14**, 55–131.

Alston, R.E. & Turner, B.L. (1963*a*). *Biochemical systematics*. London & New York: Prentice Hall.

Alston, R.E. & Turner, B.L. (1963*b*). Natural hybridization among four species of *Baptisia* (Leguminosae). *American Journal of Botany*, **50**, 159–73.

Amann, J. (1896). Application du calcul des probabilités à l'étude de la variation d'un type végétal. *Bulletin de l'Herbier Boissier*, **4**, 578–90.

Anderson, E. (1949). *Introgressive hybridisation*. London: Chapman & Hall; and New York: Wiley.

Anderson, E. (1953). Introgressive hybridization. *Biological Reviews*, **28**, 280–307.

Anderson, E. (1954). *Plants, man and life*. London: Andrew Melrose.

Anderson, E. & Abbe, L.B. (1933). A comparative anatomical study of a mutant Aquilegia. *American Naturalist*, **67**, 380–4.

Antonovics, J. (1968). Evolution in closely adjacent plant populations. V. Evolution of self-fertility. *Heredity*, **23**, 219–38.

Antonovics, J. (1976). The input from population genetics: 'The new ecological genetics'. *Systematic Botany*, **1**, 233–45.

Antonovics, J. & Bradshaw, A.D. (1970). Evolution in closely adjacent plant populations. VIII. Clinal patterns at a mine boundary. *Heredity*, **25**, 349–62.

Antonovics, J., Bradshaw, A.D. & Turner, R.G. (1971). Heavy metal tolerance in plants. *Advances in ecological Research*, **7**, 1–85.

Armstrong, H. E., Armstrong, F. & Horton, E. (1912). Herbage studies 1. *Lotus corniculatus*, a cyanophoric plant. *Proceedings of the Royal Society, Series B*, **84**, 471–84.

Armstrong, J.A., Powell, J.M. & Richards, A.J. (1982). *Pollination and evolution*. Sydney: Royal Botanic Gardens.

Astié, M. (1962). Tératologie spontanée et expérimentale. *Annales des Sciences naturelles, (Botanique)*, **3**, 619–844.

Aston, J.L. & Bradshaw, A.D. (1966). Evolution in closely adjacent plant populations. II. *Agrostis stolonifera* in maritime habitats. *Heredity*, **21**, 649–64.

Atwood, S.S. & Sullivan, J.T. (1943). Inheritance of a cyanogenic glucoside and its hydrolysing enzyme in *Trifolium repens*. *Journal of Heredity*, **34**, 311–20.

366

Auerbach, C. (1976). *Mutation research: problems, results & perspectives*. London: Chapman & Hall.

Ayazloo, M. & Bell, J.N.B. (1981). Studies on the tolerance to sulphur dioxide of grass populations in polluted areas. I. Identification of tolerant populations. *New Phytologist*, **88**, 203–22.

Babcock, E.B. (1947). *The genus* Crepis. London: Cambridge University Press; and Berkeley: University of California Press.

Baker, H.G. (1947). Criteria of hybridity. *Nature*, **159**, 1–5.

Baker, H.G. (1951). Hybridization and natural gene-flow between higher plants. *Biological Reviews*, **26**, 302–37.

Baker, H.G. (1953). Race formation and reproductive method in flowering plants. In Symposia of the Society for Experimental Biology. No. VII. *Evolution*, pp. 114–145. Cambridge: Cambridge University Press.

Baker, H.G. (1954). Report of meeting of British Ecological Society, April 1953, *Journal of Ecology*, **42**, 570–2.

Baker, H.G. (1955). Self-compatibility and establishment after 'long-distance' dispersal. *Evolution*, **9**, 347–9.

Baker, H.G. (1967). Support for Baker's Law – as a rule. *Evolution*, **21**, 853–6.

Baker, H.G. (1974). The evolution of weeds. *Annual Review of Ecology and Systematics*, **5**, 1–24.

Bannister, M.H. (1965). Variation in the breeding system of *Pinus radiata*. In *The genetics of colonizing species*, ed. H.G. Baker & G.L. Stebbins, pp. 353–72. New York & London: Academic Press.

Bannister, P. (1976). *Introduction to physiological plant ecology*. Oxford: Blackwells.

Barber, H.N. & Jackson, W.D. (1957). Natural selection in action in *Eucalyptus*. *Nature*, **179**, 1267–9.

Barkley, T.M. (1966). *A review of the origin and development of the florists'* Cineraria, *Senecio cruentus*. *Economic Botany*, **20**, 386–95.

Barling, D.M. (1955). Some population studies in *Ranunculus bulbosus* L. *Journal of Ecology*, **43**, 207–18.

Barlow, B.A. & Wiens, D. (1977). Host–parasite resemblance in Australian Mistletoes: the case for cryptic mimicry. *Evolution*, **31**, 69–84.

Barlow, P.W. & Nevin, D. (1976). Quantitative karyology of some species of *Luzula*. *Plant Systematics and Evolution*, **125**, 77–86.

Barrett, J.A. (1981). The evolutionary consequences of monoculture. In *Genetic consequences of man made change*, ed. J.A. Bishop & L.M. Cook, pp. 209–48. London & New York: Academic Press.

Barrett, S.C.H. (1980a). Sexual reproduction in *Eichhornia crassipes* (Water Hyacinth). I. Fertility of clones from diverse regions. *Journal of applied Ecology*, **17**, 101–12.

Barrett, S.C.H. (1980b). Sexual reproduction in *Eichhornia crassipes* (Water Hyacinth). II. Seed production in natural populations. *Journal of applied Ecology*, **17**, 113–24.

Bateson, W. (1895a). The origin of the cultivated *Cineraria*. *Nature*, **51**, 605–7.

Bateson, W. (1895b). The origin of the cultivated *Cineraria*. *Nature*, **52**, 29, 103–4.

Bateson, W. (1897). Notes on hybrid Cinerarias produced by Mr Lynch and Miss Pertz. *Proceedings of the Cambridge Philosophical Society*, **9**, 308–9.

Bateson, W. (1909). *Mendel's principles of heredity*. London: Cambridge University Press; and New York: Macmillan.

Bateson, W. (1913). *Problems of genetics*. London: Oxford University Press; and New Haven, Conn.: Yale University Press.

Bateson, W. & Punnett, R.C. (1911). On gametic series involving reduplication of certain terms. *Journal of Genetics*, **1**, 293–302.

Bateson, W. & Saunders, E.R. (1902). Experimental studies in the physiology of heredity. *Report to the Evolution Committee of the Royal Society*, **1**, 1–160.

Bateson, W., Saunders, E.R. & Punnett, R.C. (1905). Experimental studies in the physiology of heredity. *Report to the Evolution Committee of the Royal Society*, **2**, 1–55, 80–99.

Battaglia, E. (1963). Apomixis. In *Recent advances in the embryology of Angiosperms*, ed. P. Maheshwari, Chapter 8, pp. 221–64. Delhi: University of Delhi Press.

Beale, G. & Knowles, J. (1978). *Extranuclear genetics*. London: Arnold.

Beattie, A. (1978). Plant–animal interactions affecting gene flow in *Viola*. In *The pollination of flowers by insects*, ed. A.J. Richards, pp. 151–64, Linnean Society Symposium Series 6. London: Academic Press.

Beddall, B.G. (1957). Historical notes on avian classification. *Systematic Zoology*, **6**, 129–36.

Bell, N.B. & Lester, L.J. (1978). Genetic and morphological detection of introgression in a clinal population of *Sabatia* section *Campestria* (Gentianaceae). *Systematic Botany*, **3**, 87–104.

Belzer, N.F. & Ownbey, M. (1971). Chromatographic comparison of *Tragopogon* species and hybrids. *American Journal of Botany*, **58**, 791–802.

Bennett, J.H. (1965). (ed.) *Experiments in plant hybridisation*. [Mendel's original paper in English translation with commentary and assessment by R.A. Fisher together with W. Bateson's Biographical Notice of Mendel.] Edinburgh & London: Oliver & Boyd.

Benson, L. (1962). *Plant taxonomy*. New York: Ronald Press.

Benson, M. & Borrill, M. (1969). The significance of clinal variation in *Dactylis marina* Borrill. *New Phytologist*, **68**, 1159–73.

Bergman, B. (1935). Zytologische Studien über sexuelles und asexuelles *Hieracium umbellatum*. *Hereditas*, **20**, 47–64.

Bergman, B. (1941). Studies on the embryo sac mother cell and its development in *Hieracium* subg. *Archieracium*. *Svensk botanisk Tidskrift*, **35**, 1–42.

Berlin, B., Breedlove, D.E. & Raven, P.H. (1974). *Principles of Tzeltal plant classification*. London & New York: Academic Press.

Bernström, P. (1953). Cytogenetic infraspecific studies in *Lamium*. II. *Hereditas*, **39**, 381–437.

Berry, R.J. (1977). *Inheritance and natural history*. London: Collins.

Bishop, J.A. & Cook, L.M. (1981). (eds) *Genetic consequences of man made change*. London: Academic Press.

Bishop, J.A. & Korn, M.E. (1969). Natural selection and cyanogenesis in White Clover. *Trifolium repens*. *Heredity*, **24**, 423–30.

Bishop, O. (1971). *Statistics for biology. A practical guide for the experimental biologist*, 2nd edn. London: Longmans.

Blakeslee, A.F. & Avery, A.G. (1937). Methods of inducing chromosome doubling in plants. *Journal of Heredity*, **28**, 393–411.

Bøcher, T.W. (1944). The leaf size of *Veronica officinalis* in relation to genetic and environmental factors. *Dansk botanisk Arkiv*, **11**(7), 1–20.

Bøcher, T.W. (1949). Racial divergences in *Prunella vulgaris* in relation to habitat and climate. *New Phytologist*, **48**, 285–314.

Bøcher, T.W. (1963). The study of ecotypical variation in relation to experimental morphology. *Regnum Vegetabile*, **27**, 10–16.

Bøcher, T.W. & Larsen, K. (1958). Geographical distribution of initiation of flowering, growth habit and other factors in *Holcus lanatus*. *Botaniska Notiser*, **3**, 289–300.

Bøcher, T.W. & Lewis, M.C. (1962). Experimental and cytological studies on plant species. 7, *Geranium sanguineum*. *Biologiske Skrifter*, **11**, 1–25.

Böhm, W. (1979). *Methods of studying root systems*. Berlin, Heidelberg, New York: Springer-Verlag.

Bolkhovskikh, Z., Grif, V., Matvejeva, T. & Zakharyeva, O. (1969). *Chromosome numbers of flowering plants*. Leningrad: Academy of Sciences of the USSR.

Bonnier, G. (1890). Cultures expérimentales dans les Alpes et les Pyrénées. *Revue générale de Botanique*, **2**, 513–46.

Bonnier, G. (1895). Recherches expérimentales sur l'adaptation des plantes au climat Alpin. *Annales des Sciences naturelles (Botanique)*, **20**, 217–360.

Bonnier, G. (1920). Nouvelles observations sur les cultures expérimentales à diverses altitudes et cultures par semis. *Revue générale de Botanique*, **32**, 305–26.

Borg, S.J. Ter (1972). *Variability of* Rhinanthus serotinus *(Schönh.)* Oborny *in relation to environment*. Thesis, Rijksuniversiteit te Groningen, 158 pp.

Borgström, G. (1939). Formation of cleistogamic and chasmogamic flowers in Wild Violets as a photoperiodic response. *Nature*, **144**, 514–5.

Bose, S. & Fröst, S. (1967). An investigation on the variation of phenolic compounds in *Galeopsis* using thin layer chromatography. *Hereditas*, **58**, 145–64.

Bosemark, N.O. (1954). On accessory chromosomes in *Festuca pratensis*. I. Cytological investigations. *Hereditas*, **40**, 346–76.

Boswell Syme, J.T. (1866). (ed.) *English botany; or Coloured figures of British plants*, 3rd edn., vol. 6. London: Hardwicke.

Bothmer, R. von, Engstrand, L., Gustafsson, M., Persson, J., Snogerup, S. & Bentzer, B. (1971). Clonal variation in populations of *Anemone nemorosa* L. *Botaniska Notiser*, **124**, 505–19.

Box, J.F. (1978). *R.A. Fisher. The life of a scientist*. New York, Chichester, Brisbane, Toronto: Wiley.

Boyer, S.H. (1972). Extraordinary incidence of electrophoretically silent genetic polymorphisms. *Nature*, **239**, 453–4.

Bradshaw, A.D. (1953). Human influence on hybridisation in *Crataegus*. In *The changing flora of Britain*, ed. J.E. Lousley, pp. 181–3. Arbroath: T. Buncle & Co. Ltd for Botanical Society of the British Isles.

Bradshaw, A.D. (1959*a*). Population differentiation in *Agrostis tenuis* Sibth. I. Morphological differentiation. *New Phytologist*, **58**, 208–27.

Bradshaw, A.D. (1959*b*). Population differentiation in *Agrostis tenuis* Sibth. II. The incidence and significance of infection by *Epichloë typhina*. *New Phytologist*, **58**, 310–15.

Bradshaw, A.D. (1959*c*). Studies of variation in bent grass species. II. Variation within *Agrostis tenuis*. *Journal of the Sports Turf Research Institute*, **10**, 1–7.

Bradshaw, A.D. (1960). Population differentiation in *Agrostis tenuis* Sibth. 3. Populations in varied environments. *New Phytologist*, **59**, 92–103.

Bradshaw, A.D. (1965). Evolutionary significance of phenotypic plasticity in Plants. *Advances in Genetics*, **13**, 115–155.

Bradshaw, A.D. (1971). The significance of Hawthorns. In *Hedges and local history, Standing Conference for Local History*. London: National Council of Social Service.

Bradshaw, A.D. (1975). *Crataegus L. In Hybridization and the Flora of the British Isles*, ed. C.A. Stace, pp. 230–231. London & New York: Academic Press.

Bradshaw, A.D. (1976). Pollution and evolution. In *Effects of air pollution on plants*, ed.

T.A. Mansfield, pp. 135–59. London, New York, Melbourne: Cambridge University Press.

Bradshaw, A.D. & McNeilly, T. (1981). *Evolution and pollution*. London: Arnold.

Bradshaw, M.E. (1963*a*). Studies on *Alchemilla filicaulis* Bus., sensu lato and *A. minima* Walters. Introduction and I. Morphological variation in *A. filicaulis*, sensu lato. *Watsonia*, **5**, 304–320.

Bradshaw, M.E. (1963*b*). Studies on *Alchemilla filicaulis* Bus., sensu lato, and *A. minima* Walters. II. Cytology of *A. filicaulis*, sensu lato. *Watsonia*, **5**, 321–6.

Bradshaw, M.E. (1964). Studies on *Alchemilla filicaulis* Bus., sensu lato and *A. minima* Walters. III. *Alchemilla minima*. *Watsonia*, **6**, 76–81.

Bradshaw, M.E. & Doody, J.P. (1978). Plant population studies and their relevance to nature conservation. *Biological Conservation*, **14**, 223–42.

Brand, C.J. & Waldron, L.R. (1910). Cold resistance of Alfalfa and some factors influencing it. *U.S. Department of Agriculture, Bureau of Plant Industry. Bulletin*, No. 185, 1–80.

Brehm, B.G. & Ownbey, M. (1965). Variation in chromatographic patterns in the *Tragopogon dubius – pratensis – porrifolius* complex (Compositae). *American Journal of Botany*, **52**, 811–18.

Brenchley, W.E. & Warington, K. (1969). *The park grass plots at Rothamsted, 1856–1949*. Harpenden: Rothamsted Experimental Station.

Briggs, D. & Block, M. (1981). An investigation into the use of the '-deme' terminology. *New Phytologist*, **89**, 729–35.

Brink, R.A. (1962). Phase changes in higher plants and somatic cell heredity. *Quarterly Review of Biology*, **37**, 1–22.

Brougham, R.W. & Harris, W. (1967). Rapidity and extent of changes in genotypic structure induced by grazing in a Ryegrass population. *New Zealand Journal of agricultural Research*, **10**, 56–65.

Brown, A.H.D. (1979). Enzyme polymorphism in plant populations. *Theoretical Population Biology*, **15**, 1–42.

Brown, R.C. & Schaack, C.G. (1972). Two new species of *Tragopogon* for Arizona. *Madroño*, **21**, 304.

Bruhin, A. (1950). Beiträge zur Zytologie und Genetik schweizerischer *Crepis-Arten*. *Arbeiten aus dem Institut für Allgemeine Botanik der Universität Zurich, Serie B*, **1**, 1–101.

Bruhin, A. (1951). Auslösung von Mutationen in ruhenden Samen durch hohe Temperaturen. *Naturwissenschaften*, **38**, 565–6.

Bulmer, M.G. (1967). *Principles of statistics*, 2nd edn. London & Edinburgh: Oliver & Boyd.

Burchfield, J.D. (1975). *Lord Kelvin and the age of the Earth*. London: Macmillan.

Burkill, I.H. (1895). On the variations in number of stamens and carpels, *Journal of the Linnean Society (Botany)*, **31**, 216–45.

Cahn, M.A. & Harper, J.L. (1976*a*). The biology of the leaf mark polymorphism in *Trifolium repens*. 1. Distribution of phenotypes at a local scale. *Heredity*, **37**, 309–25.

Cahn, M.A. & Harper, J.L. (1976*b*). The biology of the leaf mark polymorphism in *Trifolium repens*. 2. Evidence for the selection of marks by rumen fistulated sheep. *Heredity*, **37**, 327–33.

Cain, A.J. (1958). Logic and memory in Linnaeus's system of taxonomy. *Proceedings of the Linnean Society of London*, **169**, 144–63.

Camp, W.H. & Gilly, C.L. (1943). The structure and origin of species. *Brittonia*, **4**, 323–85.

Campbell, R.C. (1967). *Statistics for biologists*. London & New York: Cambridge University Press.

Campbell, R.C. (1974). *Statistics for biologists*, 2nd edn. London & New York: Cambridge University Press.

Catcheside, D.G. (1939). A position effect in *Oenothera*. *Journal of Genetics*, **38**, 345–52.

Catcheside, D.G. (1947). The *P*-locus position effect in *Oenothera*. *Journal of Genetics*, **48**, 31–42.

Chadwick, M.J. & Salt, J.K. (1969). Population differentiation within *Agrostis tenuis* L. in response to colliery spoil substrate factors. *Nature*, **224**, 186.

Charlesworth, D. & Charlesworth, B. (1979). The evolutionary genetics of sexual systems in flowering plants. *Proceedings of the Royal Society London, B*, **205**, 513–30.

Chew, K., Gray, J.C. & Wildman, S.G. (1975). Fraction I protein and the origin of polyploid wheats. *Science*, **190**, 1304–6.

Chetverikov, S.S. (1926). On certain aspects of the evolutionary process from the standpoint of modern genetics. Translated by Barker, M. Edited by Lerner, I.M. (1961). *Proceedings of the American Philosophical Society*, **105**, 167–195.

Clapham, A.R., Tutin, T.G. & Warburg, E.F. (1981). *Excursion flora of the British Isles*. London: Cambridge University Press.

Clark, B.F.C. (1977). *The genetic code*. London: Arnold.

Clarke, B.C. (1979). The evolution of genetic diversity. *Proceedings of the Royal Society of London, B*, **205**, 453–74.

Clausen, J. (1922). Studies on the collective species *Viola tricolor* L. II. *Botanisk Tidsskrift*, **37**, 363–416.

Clausen, J. (1926). Genetical and cytological investigations on *Viola tricolor* L and *V. arvensis* Murr. *Hereditas*, **8**, 1–156.

Clausen, J. (1951). *Stages in the evolution of plant species*. London: Oxford University Press; and Ithaca, New York: Cornell University Press.

Clausen, J.(1966). Stability of genetic characters in *Tragopogon* species through 200 years. *Transactions and Proceedings of the Botanical Society of Edinburgh*, **40**, 148–58.

Clausen, J. & Hiesey, W.M. (1958). *Experimental studies on the nature of species. IV. Genetic structure of ecological races*. Carnegie Institution of Washington Publication No. 615, Washington D.C.

Clausen, J., Keck, D.D. & Hiesey, W.M. (1939). The concept of species based on experiment. *American Journal of Botany*, **26**, 103–6.

Clausen, J., Keck, D.D. & Hiesey, W.M. (1940). *Experimental studies on the nature of species. I. The effect of varied environments on Western North American plants*. Carnegie Institution of Washington Publication No. 520, pp. 1–452, Washington D.C.

Clausen, J., Keck, D.D. & Hiesey, W.M. (1941). Experimental taxonomy. *Carnegie Institution of Wasington Year Book*, **40**, 160–70.

Clegg, M.T. & Allard, R.W. (1972). Patterns of genetic differentiation in the slender Wild Oat species *Avena barbata*. *Proceedings of the National Academy of Science USA*, **69**, 1820–4.

Clements, F.E., Martin, E.V. & Long, F.L. (1950). *Adaptation and origin in the plant world. The role of environment in evolution*. Waltham, Mass.: Chronica Britanica Co.

Cochran, W.G. (1963). *Sampling techniques*, 2nd edn. New York: Wiley.

Collins, J.L. (1927). A low temperature type of albinism in Barley. *Journal of Heredity*, **33**, 82–6.

Cook, C.D.K. (1974). *Waterplants of the World*. The Hague: Junk.

Cook, S.A. (1962). Genetic system, variation and adaptation in *Eschscholzia californica*. *Evolution*, **16**, 278–99.

Cook, S.A. & Johnson, M.P. (1968). Adaptation to heterogeneous environments. I. Variation in Heterophylly in *Ranunculus flammula* L. *Evolution*, **22**, 496–516.

Cook, S.A., Lefèbvre, C. & McNeilly, T. (1972). Competition between metal tolerant and normal plant populations in normal soil. *Evolution*, **26**, 366–72.

Correns, C. (1909). Vererbungsversuche mit blass (gelb) grunen und buntblättrigen Sippen bei *Mirabilis jalapa, Urtica pilulifera,* und *Lunularia annua. Zeitschrift für Vererbungslehre*, **1**, 291–329.

Correns, C. (1913). Selbsterilität und Individualstoffe. *Biologisches Zentralblatt*, **33**, 389–443.

Cott, H.B. (1940). *Adaptive coloration in animals*. London: Methuen.

Coughtrey, P.J. & Martin, M.H. (1978). Tolerance of *Holcus lanatus* to lead, zinc and cadmium in factorial combination. *New Phytologist*, **81**, 147–54.

Crawford-Sidebotham, T.J. (1971). *Studies of aspects of slug behaviour and the relation between molluscs and cyanogenic plants*. PhD. Thesis, University of Birmingham.

Crew, F.A.E. (1966). Mendelism comes to England. In *G. Mendel Memorial Symposium, 1865–1965*, ed. M. Sosna, pp. 15–30. Prague: Academia Publishing House of the Czechoslovak Academy of Sciences.

Cullis, C.A. (1977). Molecular aspects of the environmental induction of heritable changes in flax. *Heredity*, **38**, 129–54.

Cummings, D.J., Borst, P., Dawid, I.B., Weissman, S.M. & Fox, C.F. (1979). (eds.) *Extrachromosomal DNA*. New York & London: Academic Press.

Curtis, O.F. & Clark D.G. (1950). *An introduction to plant physiology*. London, New York & Toronto: McGraw-Hill.

Daday, H. (1954a). Gene frequencies in wild populations of *Trifolium repens*. 1. Distribution by latitude. *Heredity*, **8**, 61–78.

Daday, H. (1954b). Gene frequencies in wild populations of *Trifolium repens*. 2. Distribution by altitude. *Heredity*, **8**, 377–84.

Daday, H. (1965). Gene frequencies in wild populations of *Trifolium repens* L. IV. Mechanism of natural selection. *Heredity*, **20**, 355–66.

Dahlgren, K.V.O. (1922). Selbststerilität interhalb Klonen von *Lysimachia nummularia. Hereditas*, **3**, 200–10.

D'Amato, F. & Hoffmann-Ostenhof, O. (1956). Metabolism and spontaneous mutations in plants. *Advances in Genetics*, **8**, 1–28.

Darlington, C.D. (1937). *Recent advances in cytology*, 2nd edn. London: Churchill.

Darlington, C.D. (1939). *The evolution of genetic systems*. London: Cambridge University Press.

Darlington, C.D. (1956). *Chromosome botany*. London: Allen & Unwin.

Darlington, C.D. (1963). *Chromosome botany and the origins of cultivated plants*. London: Allen & Unwin.

Darlington, C.D. & Wylie, A.P. (1955). *Chromosome atlas of flowering plants*, 2nd edn. London: Allen & Unwin.

Darrow, G.M. & Camp, W.H. (1945). *Vaccinium* hybrids and the development of new horticultural material. *Bulletin of the Torrey Botanical Club*, **72**, 1–21.

Darwin, C. (1859). *On the origin of species by means of natural selection*, 1st edn. London: Murray.

Darwin, C. (1862). *On the various contrivancies by which British and foreign orchids are fertilised by insects and on the good effects of crossing*. London: Murray.

Darwin, C. (1868). *The variation of plants and animals under domestication.* London: Murray.

Darwin, C. (1871). Pangenesis. *Nature*, **3**, 502–3.

Darwin, C. (1872). *The origin of species by means of natural selection*, 6th edn. London: Murray.

Darwin, C. (1876). *The effects of cross- and self-fertilisation in the vegetable kingdom.* London: Murray.

Darwin, C. (1877*a*). *The different forms of flowers of the same species.* London: Murray.

Darwin, C. (1877*b*). *The various contrivances by which orchids are fertilised by insects.* 2nd edn. London: Murray.

Darwin, C. & Wallace, A. (1859). On the tendency of species to form varieties; and on the perpetuation of varieties and species by natural means of selection. *Proceedings of the Linnean Society of London*, **3**, 45–62.

Darwin, F. (1887–1888). (ed.) *The life and letters of Charles Darwin*, 3 vols. London: Murray.

Darwin, F. (1909*a*). (ed.) *The foundations of the origin of species. A sketch written in 1842 by Charles Darwin.* Cambridge: Cambridge University Press.

Darwin, F. (1909*b*). (ed.) *The foundations of the origin of species. Two essays written in 1842 and 1844 by Charles Darwin.* London: Cambridge University Press.

Davenport, C.B. (1904). *Statistical methods with special reference to biological variation*, 2nd edn. London: Chapman & Hall; and New York: Wiley.

David, F.N. (1971). *A first course in statistics*, 2nd edn. London: Griffin.

Davidson, J.F. (1947). The polygonal graph for simultaneous portrayal of several variables in population analysis. *Madroño*, **9**, 105–10.

Davies, E. (1956). Cytology, evolution, and origin of the aneuploid series in the genus *Carex*. *Hereditas*, **42**, 349–65.

Davies, M.S. (1975). Physiological differences among populations of *Anthoxanthum odoratum* collected from the Park Grass experiment. IV. Response to potassium and magnesium. *Journal of applied Ecology*, **12**, 953–64.

Davies, M.S. & Snaydon, R.W. (1973*a*). Physiological differences among populations of *Anthoxanthum odoratum* collected from the Park Grass experiment. I. Response to calcium. *Journal of applied Ecology*, **10**, 33–45.

Davies, M.S. & Snaydon, R.W. (1973*b*). Physiological differences among populations of *Anthoxanthum odoratum* collected from the Park Grass experiment. II. Response to aluminium. *Journal of applied Ecology*, **10**, 47–55.

Davies, M.S. & Snaydon, R.W. (1974). Physiological differences among populations of *Anthoxanthum odoratum* collected from the Park Grass experiment. III. Response to phosphate. *Journal of applied Ecology*, **11**, 699–707.

Davies, M.S. & Snaydon, R.W. (1976). Rapid population differentiation in a mosaic environment. III. Measures of selection pressures. *Heredity*, **36**, 59–66.

Davies, W.E. (1963). Leaf markings in *Trifolium repens*. In *Teaching genetics in school and university*, ed. C.D. Darlington & A.D. Bradshaw, pp. 94–8. Edinburgh: Oliver & Boyd.

Davis, P.H. & Heywood, V.H. (1963). *Principles of angiosperm taxonomy.* Edinburgh: Oliver & Boyd; and New York: Van Nostrand.

Dawson, C.D.R. (1941). Tetrasomic inheritance in *Lotus corniculatus* L. *Journal of Genetics*, **42**, 49–72.

De Beer, G. (1960–61). (ed.) *'Darwin's notebooks on transmutation of species' I–IV.* Bulletin of British Museum (Natural History). Historical Series 2, Nos. 2–6.

De Beer, G. (1963). *Charles Darwin*. London: Nelson.

De Beer, G. (1964). *Atlas of evolution*. London: Nelson.

De Nettancourt, D. (1977). *Incompatibility in angiosperms*. Berlin, Heidelberg & New York: Springer Verlag.

de Vilmorin, P. (1910). Recherches sur l'hérédité Mendélienne. *Compte Rendu Hebdomadaire des Séances de l'Académie des Sciences, Paris*, **151**, 548–51.

de Vilmorin, P. (1911). Etude sur la caractère adhérence des grains entre eux chez 'le Pois Chenille'. *4th International Conference on Genetics, Paris*, 368–72.

de Vilmorin, P. & Bateson, W. (1911). A case of gametic coupling in *Pisum*. *Proceedings of the Royal Society, B*, **84**, 9–11.

De Vries, H. (1894). Uber halbe Galton-Kurven als Zeichnen diskontinurlichen Variation. *Bericht der Deutschen Botanischen Gesellschaft*, **12**, 197–207.

De Vries, H. (1897). Monstruosités héréditaires offertes en échange aux jardins botaniques. *Botanisch Jaarboek*, **9**, 80–93.

De Vries, H. (1905). *Species and varieties, their origin by mutation*. Chicago: Open Court Publishing Co.

de Wet, J.M.J. (1971). Reversible tetraploidy as an evolutionary mechanism. *Evolution*, **25**, 545–8.

de Wet, J.M.J. (1980). Origins of polyploids. In *Polyploidy*, ed. W.H. Lewis, pp. 3–15. New York & London: Plenum Press.

de Wet, J.M.J. & Harlan, J.R. (1972). Chromosome pairing and phylogenetic affinities. *Taxon*, **21**, 67–70.

Denffer, D. von, Schumacher, W., Mägdefrau, K. & Ehrendorfer, F. (1971). *Strasburger's textbook of botany*. English translation of thirtieth German edition by Bell, P. & Coombe, D.E. (1976). London & New York; Longmans.

Dewey, D.R. (1980). Some applications and misapplications of induced polyploidy to plant breeding. In *Polyploidy*, ed. W.H. Lewis, pp. 445–70. New York & London: Plenum Press.

Di Cesnola, A.P. (1904). Preliminary note on the protective value of colour in *Mantis religiosa*. *Biometrika*, **3**, 58–9.

Digby, L. (1912). The cytology of *Primula kewensis* and of other related *Primula* hybrids. *Annals of Botany*, **26**, 357–88.

Dingle, H. & Hegmann, J.P. (1982). *Evolution and genetics of life histories*. New York, Heidelberg & Berlin: Springer-Verlag.

Dobzhansky, T. (1937). *Genetics and the origin of species*. New York: Colombia University Press.

Dommée, B., Assouad, M.W. & Valdeyron, G. (1978). Natural selection and gynodioecy in *Thymus vulgaris* L. *Botanical Journal of the Linnean Society*, **77**, 17–28.

Doorenbos, J. (1965). Juvenile and adult phases in woody plants. *Handbuch der Pflanzenphysiologie*, **15**(1), 1222–35.

Dover, G.A. & Flavell, R.B. (1982). (eds) *Genome evolution*. London & New York: Academic Press.

Dritschilo, W., Krummel, J., Nafus, D. & Pimentel, D. (1979). Herbivorous insects colonising cyanogenic and acyanogenic *Trifolium repens*. *Heredity*, **42**, 49–56.

Dunstan, W.R. & Henry, T.A. (1901). The nature and origin of the poison of *Lotus arabicus*. *Proceedings of the Royal Society of London*, **68**, 374–8.

Durrant, A. (1958). Environmental conditioning of flax. *Nature*, **181**, 928–9.

Durrant, A. (1962). The environmental induction of heritable changes in *Linum*. *Heredity*, **17**, 27–61.

East, E.M. (1913). Inheritance of flower size in crosses between species of *Nicotiana*. *Botanical Gazette*, **55**, 177–88.

East, E.M. & Mangelsdorf, A.J. (1925). A new interpretation of the hereditary behaviour of self-sterile plants. *Proceedings of the National Academy of Science, Washington,* **11,** 166–83.

Edmonds, J.M. (1978). Numerical taxonomic studies on *Solanum* L. Section *Solanum* (Maurella). *Botanical Journal of the Linnean Society,* **76,** 27–51.

Edwards, K.J.R. (1977). *Evolution in modern biology.* Studies in biology, No. 87. London: Arnold.

Ehrendorfer, F. (1959). Differentiation–hybridization cycles and polyploidy in *Achillea. Cold Spring Harbour Symposium of quantitative Biology,* **24,** 141–52.

Ehrlich, P.R. & Raven, P.H. (1969). Differentiation of populations. *Science,* **165,** 1228–32.

Eigsti, C.J. & Dustin, P. (1955). *Colchicine in agriculture, medicine, biology and chemistry.* Ames, Iowa: Iowa State College Press.

Eisikowitch, D. (1978). Insect visiting of two subspecies of *Nigella arvensis* under adverse seaside conditions. In *The pollination of flowers by insects,* ed. A.J. Richards, pp. 125–32. Linnean Society Symposium Series No. 6. London: Academic Press.

Eldredge, N. & Cracraft, J. (1980). *Phylogenetic patterns and the evolutionary process: method and theory in comparative biology.* New York: Columbia University Press.

Elkington, T.T. (1968). Introgressive hybridization between *Betula nana* L. and *B. pubescens* Ehrh. in North-west Iceland. *New Phytologist,* **67,** 109–18.

Elliot, E. (1914), see Lamarck.

Ellis, W.M., Keymer, R.J. & Jones, D.A. (1977a). The effect of temperature on the polymorphism of cyanogenesis in *Lotus corniculatus* L. *Heredity,* **38,** 339–47.

Ellis, W.M., Keymer, R.J. & Jones, D.A. (1977b). On the polymorphism of cyanogenesis in *Lotus corniculatus* L. VIII. Ecological studies in Anglesey. *Heredity,* **39,** 45–65.

Evans, G.M. & Macefield, A.J. (1973). The effect of B chromosomes on homoeologous pairing in species hybrids. I. *Lolium temulentum* × *Lolium perenne. Chromosoma,* **41,** 63–73.

Faegri, K. (1937). Some fundamental problems of taxonomy and phylogenetics. *Botanical Review,* **3,** 400–23.

Faegri, K. & van der Pijl, L. (1979). *The principles of pollination ecology,* 3rd revised edn. Oxford & New York: Pergamon Press.

Fagerlind, F. (1937). Embryologische, zytologische und bestäubungs-experimentelle Studien in der Familie Rubiaceae nebst Bemerkungen über einige Polyploiditätsprobleme. *Acta Horti Bergiani,* **11,** 195–470.

Favarger, C. & Villard, M. (1965). *Nouvelles récherches cytotaxinomiques sur Chrysanthemum leucanthemum* L. sens. lat. *Bericht der Schweizerischen Botanischen Gesellschaft,* **75,** 57–79.

Ferguson, A. (1980). *Biochemical systematics and evolution.* Glasgow & London: Blackie.

Fisher, R.A. (1929). *The genetical theory of natural selection,* 2nd edn, reprinted 1958. London: Constable; and New York: Dover Books.

Fisher, R.A. (1935). *The design of experiments.* Edinburgh & London: Oliver & Boyd.

Fisher, R.A. (1936). Has Mendel's work been rediscovered? *Annals of Science,* **1,** 115–37. [Reprinted in Bennett (1965), pp. 59–87, see above.]

Fisher, R.A. & Yates, F. (1963). *Statistical tables for biological, agricultural and medical research,* 6th edn. Edinburgh: Longman (Oliver & Boyd).

Flake, R.H., Urbatsch, L. & Turner, B.L. (1978). Chemical documentation of allopatric introgression in *Juniperus. Systematic Botany,* **3,** 129–44.

Flake, R.H., von Rudloff, E. & Turner, B.L. (1969). Quantitative study of clinal variation in *Juniperus virginiana* using terpenoid data. *Proceedings of the National Academy of Sciences, USA*, **64**, 487–94.

Frankel, O.H. & Soulé, M.E. (1981). *Conservation and evolution*. London: Cambridge University Press.

Frost, H.B. (1938). Nucellar embryony and juvenile characters in clonal varieties of *Citrus. Journal of Heredity*, **29**, 423–32.

Fröst, S. (1958). The geographical distribution of accessory chromosomes in *Centaurea scabiosa. Hereditas*, **44**, 75–111.

Fröst, S. & Ising, G. (1968). An investigation into the phenolic compounds in *Vaccinium myrtillus* L. (Bilberries), *Vaccinium vitis-idaea* L. (Cowberries), and the hybrid between them *V. intermedium* Ruthe employing thin layer chromatography. *Hereditas*, **60**, 72–6.

Gajewski, W. (1957). A cytogenetic study on the genus *Geum. Monographiae Botanicae*, No. 4.

Galton, F. (1871). Experiments in pangenesis, by breeding from rabbits of a pure variety, into whose circulation blood taken from other varieties had previously been largely transfused. *Proceedings of the Royal Society of London*, **19**, 393–410.

Galton, F. (1876). The history of twins, as a criterion of the relative powers of nature and nurture. *Journal of the Royal Anthropological Institute*, **5**, 391–406.

Galton, F. (1889). *Natural inheritance*. London & New York: Macmillan.

Garrick, M.D., Bricker, J. & Garrick, L.M. (1974). An electrophoretically silent polymorphism for the beta chains of rabbit hemoglobin and associated polyribosome patterns. *Genetics*, **76**, 99–108.

Gates, R.R. (1909). The stature and chromosomes of *Oenothera gigas* de Vries. *Archiv für Zellforschung*, **3**, 525–52.

Gay, P.A. (1960). A new method for the comparison of populations that contain hybrids. *New Phytologist*, **59**, 219–26.

Geiger, R. (1965). *The climate near the ground*. Cambridge, Mass.: Harvard University Press.

Georghiou, G.P. (1972). The evolution of resistance to pesticides. *Annual Review of Ecology and Systematics*, **3**, 133–68.

Gerstel, D.U. (1950). Self-incompatibility studies in Guayule. II. Inheritance. *Genetics*, **35**, 482–506.

Gibson, P.B. & Schultz, E.F., Jr. (1953). The effect of environment on the number of vascular bundles per petiole in White Clover. *Agronomy Journal*, **45**, 123–4.

Gill, B.S. & Kimber, G. (1974). Giemsa C-banding and the evolution of Wheat. *Proceedings of the National Academy of Sciences, USA*, **71**, 4086–90.

Gillham, N.W. (1978). *Organelle heredity*. New York: Raven Press.

Gilmour, J.S.L. & Gregor, J.W. (1939). Demes: a suggested new terminology. *Nature*, **144**, 333–4.

Gilmour, J.S.L. & Heslop-Harrison, J. (1954). The deme terminology and the units of micro-evolutionary change. *Genetica*, **27**, 147–61.

Givnish, T. (1979). On the adaptive significance of leaf form. In *Tropics in plant population biology*, ed. O.T. Solbrig, S. Jain, G.B. Johnson & P.H. Raven, pp. 375–407. London: Columbia University Press.

Givnish, T.J. & Vermeij, G.J. (1976). Sizes and shapes of Liane leaves. *American Naturalist*, **110**, 743–76.

Glass, B. (1959). Heredity and variation in the eighteenth century concept of the species.

In *Forerunners of Darwin 1745–1859*, ed. B. Glass, O. Temkin & W.I. Straus, pp. 144–72. London: Oxford University Press.

Goebel, K. (1897). *Uber Jugendformen von Pflanzen und deren künstliche Wiederhervorrufung*. Sitzungsbericht der Mathematisch-Physikalischen Class der Königlich-Bayerischen Akademie der Wissenschaften, München, 26.

Goldblatt, P. (1980). Polyploidy in angiosperms: monocotyledons. In *Polyploidy*, ed. W.H. Lewis, pp. 219–39. New York & London: Plenum Press.

Goldschmidt, R.B. (1940). *The material basis of evolution*. New Haven: Yale University Press.

Goldschmidt, R.B. (1955). *Theoretical genetics*. Berkeley & Los Angeles: University of California Press.

Good, R. d'O. (1956). *Features of evolution in the flowering plants*. London & New York: Longmans.

Gottlieb, L.D. (1981). Electrophoretic evidence and plant populations. *Progress in Phytochemistry*, **7**, 1–45.

Gould, S.J. (1979). Species Are Not Specious. *New Scientist*, **83**, 374–6.

Gould, S.J. & Eldredge, N. (1977). Punctuated equilibria: the tempo and mode of evolution reconsidered. *Paleobiology*, **3**, 115–51.

Gould, S.J. & Lewontin, R.G. (1979). The spandrels of San Marco and the panglossian paradigm: a critique of the adaptionist programme. *Proceedings of the Royal Society of London, B*, **205**, 581–8.

Graham, B.F., Jr. & Bormann, F.H. (1966). Natural root grafts. *The botanical Review*, **32**, 255–92.

Grant, V. (1950). The protection of the ovules in flowering plants. *Evolution*, **4**, 179–201.

Grant, V. (1957). The plant species in theory and practice. In *The species problem*, ed. E. Mayr, pp. 39–80. Washington: American Association for the Advancement of Science, Publication 50.

Grant, V. (1966). The selective origin of incompatibility barriers in the plant genus *Gilia*. *The American Naturalist*, **100**, 99–118.

Grant, V. (1971). *Plant speciation*. New York & London: Columbia University Press.

Grant, V. (1975). *Genetics of flowering plants*. New York & London: Columbia University Press.

Grant, W.F. (1965). A chromosome atlas and interspecific hybridisation index for the genus *Lotus* (Leguminosae). *Canadian Journal of Genetics and Cytology*, **7**, 457–71.

Gray, J.C. (1980). Fraction I protein and plant phylogeny. In *Chemosystematics: principles and practice*, ed. F.A. Bisby, J.G. Vaughan & C.A. Wright, pp. 167–93. London & New York: Academic Press.

Green, R.H. (1979). *Sampling design and statistical methods for environmental biologists*. New York, Chichester, Brisbane, Toronto: Wiley.

Greene, E.L. (1909). Linnaeus as an evolutionist. *Proceedings of the Washington Academy of Sciences*, **11**, 17–26.

Gregor, J.W. (1930). Experiments on the genetics of wild populations, I. *Plantago maritima*. *Journal of Genetics*, **22**, 15–25.

Gregor, J.W. (1931). Experimental delimitation of species. *New Phytologist*, **30**, 204–17.

Gregor, J.W. (1938). Experimental taxonomy. 2. Initial population differentiation in *Plantago maritima* in Britain. *New Phytologist*, **37**, 15–49.

Gregor, J.W. (1939). Experimental taxonomy. 4. Population differentiation in North American and European Sea Plantains allied to *Plantago maritima* L. *New Phytologist*, **38**, 293–322.

Gregor, J.W. (1944). The ecotype. *Biological Reviews*, **19**, 20–30.

Gregor, J.W. (1946). Ecotypic differentiation. *New Phytologist*, **45**, 254–70.

Gregor, J.W., Davey, V.McM. & Lang, J.M.S. (1936). Experimental taxonomy. I. Experimental garden technique in relation to the recognition of small taxonomic units. *New Phytologist*, **35**, 323–50.

Gregor, J.W. & Lang, J.M.S. (1950). Intra-colonial variation in plant size and habit in Sea Plantains. *New Phytologist*, **49**, 135–41.

Greig-Smith, P. (1964). *Quantitative plant ecology*, 2nd edn. London: Butterworth.

Groot, J. & Boschuizen, R. (1970). A preliminary investigation into the genecology of *Plantago major* L. *Journal of experimental Botany*, **21**, 835–41.

Guignard, L. (1891). Nouvelles études sur la fécondation. *Annales des Sciences naturelles (Botanique)*, **14**, 163–296.

Gunther, R.W.T. (1928). *Further correspondence of John Ray*. London & New York: Oxford University Press.

Gustafsson, Å. (1946). Apomixis in higher plants. 1. The mechanism of apomixis. *Acta Universitatis lundensis*, **42**(3), 1–67.

Gustafsson, Å. (1947a). Apomixis in higher plants. 2. The causal aspect of apomixis. *Acta Universitatis lundensis*, **43**(3), 69–179.

Gustafsson, Å. (1947b). Apomixis in higher plants. 3. Biotype and species formation. *Acta Universitatis lundensis*, **43**(12), 183–370.

Habakkuk, H.J. (1960). Thomas Robert Malthus, F.R.S. (1766–1834). *Notes and Records of the Royal Society of London*, **14**, 99–108.

Haeckel, E.H.P.A. (1876). *The history of creation*. (Translation revised by E.R. Lankester.) London: Routledge.

Hagerup, O. (1931). Uber Polyploide in Beziehung zu Klima, Okologie und Phylogenie. *Hereditas*, **16**, 19–40.

Hagerup, O. (1947). The spontaneous formation of haploid, polyploid and aneuploid embryos in some orchids. *Biologiske Meddelelser*, **20**(9), 1–22.

Hair, J.B. (1956). Subsexual reproduction in *Agropyron*. *Heredity*, **10**, 129–60.

Hall, M.T. (1952). A hybrid swarm in *Juniperus*. *Evolution*, **6**, 347–66.

Hamrick, J.L. & Allard, R.W. (1972). Microgeographical variation in allozyme frequencies in *Avena barbata*. *Proceedings of the National Academy of Sciences, USA*, **69**, 2100–4.

Harberd, D.J. (1957). The within population variance in genecological trials. *New Phytologist*, **56**, 269–80.

Harberd, D.J. (1958). Progress and prospects in genecology. *Record of Scottish Plant Breeding Station*, 1958, 52–60.

Harberd, D.J. (1961). Observations on population structure and longevity of *Festuca rubra*. L. *New Phytologist*, **60**, 184–206.

Harberd, D.J. (1962). Some observations on natural clones in *Festuca ovina*. *New Phytologist*, **61**, 85–100.

Harberd, D.J. (1963). Observations on natural clones of *Trifolium repens* L. *New Phytologist*, **62**, 198–204.

Harborne, J.B. (1978). (ed.) Biochemical aspects of plant and animal coevolution. *Proceedings of the Phytochemical Society Symposium No. 15, Reading 1977*. London, New York, San Francisco: Academic Press.

Harborne, J.B. (1982). *Introduction to ecological biochemistry*, 2nd edn. London & New York: Academic Press.

Harborne, J.B., Williams, C.A. & Smith, D.M. (1973). Species-specific Kaempferol derivatives in ferns of the Appalachian *Asplenium* complex. *Biochemical Systematics*, **1**, 51–4.

Hardin, G. (1966). *Biology. Its principles and implications*, 2nd edn. London & San Francisco: Freeman.

Harlan, J.R. (1965). The possible role of weed races in the evolution of cultivated plants. *Euphytica*, **14**, 173–6.

Harlan, J.R. & de Wet, J.M.J. (1975). On Ö Winge and a prayer: The origins of polyploidy. *The botanical Review*, **41**, 361–90.

Harlan, H.V. & Martini, M.L. (1938). The effect of natural selection on a mixture of barley varieties. *Journal of agricultural Research*, **57**, 189–99.

Harper, J.L. (1977). *Population biology of plants*. London & New York: Academic Press.

Harper, J.L. (1978). The demography of plants with clonal growth. In *Structure and functioning of plant populations*, ed. Freyson, A.H.J. & Waldendorp, J.W., pp. 27–48. Amsterdam, Oxford & New York: North Holland Publishing Company.

Harper, J.L. (1983). A Darwinian plant ecology. In *Evolution from molecules to men*. ed. D.S. Bendall, pp. 323–45. Cambridge: Cambridge University Press.

Harris, R. (1976). Discussion, in Pickersgill, B. & Heiser, C.B. (1976).

Harshberger, J.W. (1901). The limits of variation in plants. *Proceedings of the National Academy of Sciences, USA*, **53**, 303–19.

Hartman, P.E. & Suskind, S.R. (1965). *Gene action*. New York: Prentice-Hall.

Harvey, W.H. (1860). Darwin on the origin of species. *The Gardeners' Chronicle and agricultural Gazette*, February 18 1860, 145–6.

Hathaway, W.H. (1962). Weighted hybrid index. *Evolution*, **16**, 1–10.

Hayes, W. (1964). *The genetics of bacteria and their viruses*. Oxford & Edinburgh: Blackwells.

Heath, O.V.S. (1970). *Investigation by experiment*. London: Arnold.

Heiser, C.B., Jr. (1949). Natural hybridization with particular reference to introgression. *The botanical Review*, **15**, 645–87.

Heiser, C.B., Jr. (1973). Introgression re-examined. *The botanical Review*, **39**, 347–66.

Heiser, C.B., Jr. (1979). Hybrid populations of *Helianthus divaricatus* and *H. microcephalus* after 22 years. *Taxon*, **28**, 71–5.

Heslop-Harrison, J. (1952). A reconsideration of plant teratology. *Phyton*, **4**, 19–34.

Heslop-Harrison, J. (1953). *New concepts in flowering-plant taxonomy*. London: Heinemann.

Heslop-Harrison, J. (1955). The conflict of categories. In *Species studies in the British flora*, ed. J.E. Lousley, pp. 160–72. Arbroath: Botanical Society of the British Isles.

Heslop-Harrison, J. (1964). Forty years of genecology. *Advances in ecological Research*, **2**, 159–247.

Heslop-Harrison, J. (1975). Incompatibility and the pollen–stigma interaction. *Annual Review of Plant Physiology*, **26**, 403–25.

Heslop-Harrison, J. (1978). *Cellular recognition systems in plants*. Studies in biology, No. 100. London: Arnold.

Hesselman, H. (1919). Iakttagelser över skogsträdpollens spridningsförmåga. *Meddelanden från Statens Skogsförsoksanstalt*, **16**, 27.

Heywood, V.H. (1976). *Plant taxonomy*, 2nd edn. Studies in biology, No. 5. London: Arnold.

Heywood, V.H. (1980). The impact of Linnaeus on botanical taxonomy – past, present and future. *Veröffentlichungen der Joachim Jungius-Gesellschaft der Wissenschaften, Hamburg*, **43**, 97–115.

Hickey, D.A. & McNeilly, T. (1975). Competition between metal tolerance and normal plant populations; a field experiment on normal soil. *Evolution*, **29**, 458–64.

Hiesey, W.M. & Milner, H.W. (1965). Physiology of ecological races and species. *Annual Review of Plant Physiology*, **16**, 203–16.

Hill, C.R. & Crane, P.R. (1982). Evolutionary cladistics and the origin of angiosperms. In *Problems of phylogenetic reconstruction*. ed. K.A. Joysey & A.E. Friday, pp. 269–361. London & New York: Academic Press.

Hill, J. (1965). Environmental induction of heritable changes in *Nicotiana rustica*. *Nature*, **207**, 732–4.

Hinton, W.F. (1976). The evolution of insect-mediated self-pollination from an outcrossing system in *Calyptridium* (Portulacaceae). *American Journal of Botany*, **63**, 979–86.

Hoffmann, H. (1881). Rückblick auf meine Variations-Versuche von 1855–80. *Botanische Zeitung*, 1881, 345–51, 361–7, 377–83, 393–9, 409–15, 424–31.

Holliday, R.J. & Putwain, P.D. (1977). Evolution of resistance to simazine in *Senecio vulgaris* L. *Weed Research*, **17**, 291–6.

Holliday, R.J. & Putwain, P.D. (1980). Evolution of herbicide resistance in *Senecio vulgaris*: variation in susceptibility to simazine between and within populations. *Journal of applied Ecology*, **17**, 779–91.

Horsman, D.C., Roberts, T.M. & Bradshaw, A.D. (1979). Studies on the effect of sulphur dioxide on perennial ryegrass (*Lolium perenne*, L.). II. Evolution of Sulphur Dioxide Tolerance. *Journal of experimental Botany*, **30**, 495–501.

Hort, A. (1938). *The Critica botanica of Linnaeus*. London: Ray Society, British Museum.

Hsiao, J.-Y. & Li, H.-L. (1973). Chromatographic studies on the Red Horsechestnut (*Aesculus* × *carnea*) and its putative parent species. *Brittonia*, **25**, 57–63.

Hughes, A. (1959). *A history of cytology*. London & New York: Abelard-Schuman.

Hughes, M.B. & Babcock, E.B. (1950). Self-incompatibility in *Crepis foetida* L. subsp. *rhoeadifolia*. *Genetics*, **35**, 570–88.

Hull, P. (1974). Self-fertilisation and the distribution of the radiate form of *Senecio vulgaris* L. in Central Scotland. *Watsonia*, **10**, 69–75.

Hutchinson, A.H. (1936). The polygonal representation of polyphase phenomena. *Transactions of the Royal Society of Canada, Ser. 3, Sect. V*, **30**, 19–26.

Hurst, C.C. (1925). *Experiments in genetics*. Cambridge: Cambridge University Press.

Huskins, C.L. (1930). The origin of *Spartina townsendii*. *Genetica*, **12**, 531–8.

Hutchinson, T.C. (1967). Ecotype differentiation in *Teucrium scorodonia* with respect to susceptibility to lime-induced chlorosis and to shade factors. *New Phytologist*, **66**, 439–53.

Huxley, J. (1938). Clines: an auxiliary taxonomic principle. *Nature*, **142**, 219–20.

Huxley, J.S. (1940). (ed.) *The new systematics*. Oxford: Clarendon Press.

Huxley, J.S. (1942). *Evolution: the modern synthesis*. London: Allen & Unwin.

Iversen, J. (1973). *The development of Denmark's nature since the last glacial*. Copenhagen: Reitzels Forlag.

Jackson, R.C. (1962). Interspecific hybridization in *Haplopappus* and its bearing on chromosome evolution in the Blepharodon section. *American Journal of Botany*, **49**, 119–132.

Jackson, R.C. (1965). A cytogenetic study of a three-paired race of *Haplopappus gracilis*. *American Journal of Botany*, **52**, 946–53.

Jain, S.K. & Bradshaw, A.D. (1966). Evolutionary divergence among adjacent plant populations. I. The evidence and its theoretical analysis. *Heredity*, **21**, 407–41.

Jameson, D.L. (1977). (ed.) *Evolutionary genetics*. Stroudsburg, Pa.: Dowden, Hutchinson & Ross.

Jardine, N. & Sibson, R. (1971). *Mathematical taxonomy*. London & New York: Wiley.

Jauhar, P.P. (1975). Genetic control of diploid-like meiosis in hexaploid Tall Fescue. *Nature*, **254**, 595–7.

Jenkin, F. (1867). Unsigned Review of Darwin's 'On the origin of species'. *The North British Review*, July 1867, 277–318.

Jenkins, M.T. (1924). Heritable characters of Maize 20 Iojap-striping, a chlorophyll defect. *Journal of Heredity*, **15**, 467–72.

Jensen, I. & Bogh, H. (1941). On conditions influencing the danger of crossing in the case of wind-pollinated cultivated plants. *Tidsskrift for Planteavl*, **46**, 238–66.

Jinks, J.L. (1964). *Extrachromosomal inheritance*. Englewood Cliffs, N.J.: Prentice-Hall.

Johannsen, W. (1909). *Elemente der exakten Erblichkeitslehre*. Jena: Fischer.

Johnson, B.L. (1972). Protein electrophoretic profiles and the origin of the B genome of wheat. *Proceedings of the National Academy of Sciences, USA*, **69**, 1398–402.

Johnson, H. (1945). Interspecific hybridization within the genus *Betula*. *Hereditas*, **31**, 163–76.

Johnson, H.B. (1975). Plant pubescence: an ecological perspective. *Botanical Review*, **41**, 233–58.

Jones, D.A. (1962). Selective eating of the acyanogenic form of the plant *Lotus corniculatus* L. by various animals. *Nature*, **193**, 1109–10.

Jones, D.A. (1966). On the polymorphism of cyanogenesis in *Lotus corniculatus*. Selection by animals. *Canadian Journal of Genetics and Cytology*, **8**, 556–67.

Jones, D.A. (1972). Cyanogenic glycosides and their function. In *Phytochemical ecology*, ed. J.B. Harborne, pp. 103–24. London & New York: Academic Press.

Jones, D.A. (1973). Co-evolution and cyanogenesis. In *Taxonomy and ecology*, ed. V.H. Heywood, pp. 213–42. Systematics Association Special Volume No. 5. London & New York: Academic Press.

Jones, D.A., Keymer, R.J. & Ellis, W.M. (1978). Cyanogenesis in plants and animal feeding. In *Biochemical aspects of plant and animal coevolution*, ed. J.B. Harborne, pp. 21–34. London, New York & San Francisco: Academic Press.

Jones, D.F. (1924). The attainment of homozygosity in inbred strains of Maize. *Genetics*, **9**, 405–18.

Jones, K. (1958). Cytotaxonomic studies in *Holcus*. I. The chromosome complex in *Holcus mollis* L. *New Phytologist*, **57**, 191–210.

Jones, K. (1964). Chromosomes and the nature and origin of *Anthoxanthum odoratum* L. *Chromosoma*, **15**, 248–74.

Jones, K. (1978). Aspects of chromosome evolution in higher plants. *Advances in botanical Research*, **6**, 120–94.

Jones, K. & Borrill, M. (1961). Chromosomal status, gene exchange and evolution in *Dactylis*. 3. The role of the inter-ploid hybrids. *Genetica*, **32**, 296–322.

Jones, K. & Carroll, C.P. (1962). Cytotaxonomic studies in *Holcus*. II. Morphological relationships in *Holcus mollis* L. *New Phytologist*, **61**, 63–84.

Jones, M.D. & Brooks, J.S. (1952). *Effect of tree barriers on outcrossing in Corn*. Oklahoma Agricultural Experiment Station Bulletin, No. T-45.

Jones, R.N. (1975). B-chromosome systems in flowering plants and animal species. *International Review of Cytology*, **40**, 1–100.

Jones, R.N. & Rees, H. (1982). *B chromosomes*. London: Academic Press.

Jonsell, B. (1978). Linnaeus's views on plant classification and evolution. *Botaniska Notiser*, **131**, 523–30.

Jordan, A. (1864). *Diagnoses d'espèces nouvelles ou méconnues pour servir de matériaux a une flore réformée de la France et des Contrées voisines.* Paris: Savy.

Kasha, K.J. (1974). (ed.) *Haploids in higher plants. Advances and potential.* Guelph, Canada: The University of Guelph.

Kay, Q.O.N. (1978). The role of preferential and assortative pollination in the maintenance of flower colour polymorphisms. In *The pollination of flowers by insects,* ed. A.J. Richards, pp. 175–90. Linnean Society Symposium Series 6. London: Academic Press.

Kellerman, W.A. (1901). Variation in *Syndesmon thalictroides. Ohio Naturalist,* **1**, 107–11.

Kelvin, Lord. (Sir W. Thompson) (1871). On geological time. *Transactions of the Geological Society of Glasgow,* **3**, 1–28.

Kendall, M.G. & Plackett, R.L. (1977). *Studies in the history of statistics and probability,* vol. II. London: Griffin.

Kerner, A. (1895). The natural history of plants, their forms, growth, reproduction and distribution. Translated and edited by F.W. Oliver. London: Blackie.

Kernick, M.D. (1961). Seed production of specific crops. In *Agricultural and horticultural seeds,* pp. 181–547. FAO Agriculture Studies No. 55.

Kerster, H.W. & Levin, D.A. (1968). Neighborhood size in *Lithospermum caroliniense. Genetics,* **60**, 577–87.

Keymer, R. & Ellis, W.M. (1978). Experimental studies on plants of *Lotus corniculatus* L. from Anglesey polymorphic for cyanogenesis. *Heredity,* **40**, 189–206.

Kihara, H. & Ono, T. (1926). Chromosomenzahlen und systematische Gruppierung der *Rumex*-Arten. *Zeitschrift für Zellforschung und mikroskopische Anatomie, Berlin, Wien,* **4**, 475–81.

Kimber, G. (1961). Basis of the diploid-like meiotic behaviour of polyploid Cotton. *Nature,* **191**, 98–100.

Kimber, G. & Athwal, R.S. (1972). A reassessment of the course of evolution of Wheat. *Proceedings of the National Academy of Sciences, USA,* **69**, 912–15.

Knight, G.R., Robertson, A. & Waddington, C.H. (1956). Selection for sexual isolation within a species. *Evolution,* **10**, 14–22.

Knott, D.R. & Dvořák, J. (1976). Alien germ plasm as a source of resistance to disease. *Annual Review of Phytopathology,* **14**, 211–35.

Knox, R.B. & Heslop-Harrison, J. (1963). Experimental control of aposporous apomixis in a grass of the Andropogoneae. *Botaniska Notiser,* **116**, 127–41.

Koopman, K.F. (1950). Natural selection for reproductive isolation between *Drosophila pseudoobscura* and *Drosophila persimilis. Evolution,* **4**, 135–48.

Koshy, T.K. (1968). Evolutionary origin of *Poa annua* L. in the light of karyotypic studies. *Canadian Journal of Genetics and Cytology,* **10**, 112–18.

Kruckeberg, A.R. (1951). Intraspecific variability in the response of certain native plant species to serpentine soil. *American Journal of Botany,* **38**, 408–19.

Kruckeberg, A.R. (1954). The ecology of serpentine soils. III. Plant species in relation to serpentine soils. *Ecology,* **35**, 267–74.

Kruckeberg, A.R. (1955). Interspecific hybridizations of *Silene. American Journal of Botany,* **42**, 373–8.

Kruckeberg, A.R. (1964). Artificial crosses involving North American *Silenes. Brittonia,* **16**, 95–105.

Kyhos, D.W. (1965). The independent aneuploid origin of two species of *Chaenactis* (Compositae) from a common ancestor. *Evolution,* **19**, 26–43.

Lamarck, J.B. (1809). *Philosophie zoologique.* (English translation, *Zoological philosophy*, translated by H. Elliot, published 1914, London & New York: Macmillan.)

Lamb, H.H. (1970). Our changing climate. In *Flora of a changing Britain*, ed. F.H. Perring, pp. 11–24. Hampton, Middlesex: Botanical Society of the British Isles.

Lamprecht, H. (1966). *Die Entstehung der Arten und höheren Kategorien.* New York & Vienna: Springer.

Lane, C. (1962). Notes on the Common Blue (*Polyommatus icarus*) egg laying and feeding on the cyanogenic strains of the Bird's-foot Trefoil (*Lotus corniculatus*). *Entomologist's Gazette*, **13**, 112–16.

Lang, A. (1965). Physiology of flower initiation. Section 5: Blüten und Fruchtbildung. *Handbuch der Pflanzenphysiologie*, **15**(1), 1380–536.

Langlet, O. (1934). Om variationen hos tallen *Pinus sylvestris* och dess samband med climatet. *Meddelanden från Statens Skogsförsöksanstalt*, **27**, 87–93.

Langlet, O. (1971). Two hundred years genecology. *Taxon*, **20**, 653–722.

Lankester, E. (1848). *The correspondence of John Ray.* London: Ray Society, British Museum.

Larsen, E.C. (1947). Photoperiodic responses of geographical strains of *Andropogon scoparius*. *Botanical Gazette*, **109**, 132–50.

Lawrence, W.E. (1945). Some ecotypic relations of *Deschampsia caespitosa*. *American Journal of Botany*, **32**, 298–314.

Lawrence, W.J.C. (1950). *Science and the glasshouse.* Edinburgh & London: Oliver & Boyd.

Lebaron, H.M. & Gressel, J. (1982). (eds) *Herbicide resistance in plants.* New York:Wiley.

Lee, A. (1902). Dr Ludwig on variation and correlation in plants. *Biometrika*, **1**, 316–19.

Lefèbvre, C. (1973). Outbreeding and inbreeding in a zinc–lead mine population of *Armeria maritima*. *Nature*, **243**, 96–7.

Lehninger, A.L. (1975). *Biochemistry*, 2nd edn. New York: Worth.

Levan, A. (1938). The effect of colchicine on root mitosis in *Allium*. *Hereditas*, **24**,471–86.

Levin, D.A. (1973). The role of trichomes in plant defense. *Quarterly Review of Biology*, **48**, 3–15.

Levin, D.A. (1978*a*). Pollinator behaviour and the breeding structure of plant populations. In *The pollination of flowers by insects*, ed. A.J. Richards, pp. 133–50. Linnean Society Symposium Series 6. London: Academic Press.

Levin, D.A. (1978*b*). The origin of isolating mechanisms in flowering plants. *Evolutionary Biology*, **11**, 185–317.

Levin, D.A. (1979). The nature of plant species. *Science*, **204**, 381–4.

Levin, D.A. & Kerster, H.W. (1967). An analysis of interspecific pollen exchange in *Phlox*. *American Naturalist*, **101**, 387–400.

Levin, D.A. & Kerster, H.W. (1968). Local gene dispersal in *Phlox pilosa*. *Evolution*, **22**, 130–9.

Levin, D.A. & Kerster, H.W. (1969). Density-dependent gene dispersal in *Liatris*. *American Naturalist*, **103**, 61–74.

Levin, D.A. & Kerster, H.W. (1974). *Gene flow in seed plants. Evolutionary Biology*, **7**, 139–220.

Lewis, D. (1979). *Sexual incompatibility in plants.* Studies in biology, No. 110. London: Arnold.

Lewis, D. & Crowe, L.K. (1956). The genetics and evolution of gynodioecy. *Evolution*, **10**, 115–25.

Lewis, H. (1973). The origin of diploid neospecies in *Clarkia*. *The American Naturalist*, **107**, 161–70.

Lewis, W.H. (1976). Temporal adaptation correlated with ploidy in *Claytonia virginia*. *Systematic Botany*, **1**, 340–7.

Lewis, W.H. (1980*a*). (ed.) *Polyploidy*. London & New York: Plenum Press.

Lewis, W.H. (1980*b*). Polyploidy in species populations. In *Polyploidy*, ed. W.H. Lewis, pp. 103–44. New York & London: Plenum Press.

Lewis, W.H. (1980*c*). Polyploidy in angiosperms: dicotyledons. In *Polyploidy*, ed. W.H. Lewis, pp. 241–68. New York & London: Plenum Press.

Lewontin, R.C. (1974). *The genetic basis of evolutionary change*. New York: Columbia University Press.

Lewontin, R.C. & Birch, L.C. (1966). Hybridization as a source of variation for adaptation to new environments. *Evolution*, **20**, 315–36.

Li, H.L. (1956). The story of the cultivated Horse-Chestnuts. *Morris Arboretum Bulletin*, **7**, 35–9.

Linhart, V.B. & Baker, I. (1973). Intra-population differentiation of physiological response to flooding in a population of *Veronica peregrina*. *Nature*, **242**, 275–6.

Linhart, Y.B., Mitton, J.B., Sturgeon, K.B. & Davis, M.L. (1981). Genetic variation in space and time in a population of Ponderosa Pine. *Heredity*, **46**, 407–26.

Linnaeus, C. (Carl von Linné). (1737). *Critica botanica*. (English translation by A. Hort, 1938, London: Ray Society, British Museum.)

Linnaeus, C. (1737, but not distributed until 1738). *Hortus cliffortianus*. Amsterdam.

Linnaeus, C. (1744). Peloria. In *Amoenitates academicae* (1749–90). (See Stearn (1957) for details of the many editions.)

Linnaeus, C. (1749–90). *Amoenitates academicae*. (For details of the many editions see Stearn (1957).)

Linnaeus, C. (1751). *Philosophia botanica*. Stockholm.

Linnaeus, C. (1753). *Species plantarum*. (Facsimile edition 1957, London: Ray Society, British Museum.)

Linnaeus, C. (1762–3). *Species plantarum*, 2nd edn. Stockholm.

Litardière, R. de (1939). Sur les caractères chromosomiques et la systématique des *Poa* du group du *P. annua* L. *Revue de Cytologie et de Cytophysiologie végétales*, **4**, 82–5.

Lloyd, D.G. (1975). The maintenance of gynodioecy and androdioecy in angiosperms. *Genetica*, **45**, 325–39.

Löve, A. (1962). The biosystematic species concept. *Preslia*, **34**, 127–39.

Löve, A. & Löve, D. (1961). Chromosome numbers of Central and Northwest European plant species. *Opera Botanica*, **5**, 1–581.

Lovis, J.D. (1977). Evolutionary patterns and processes in ferns. *Advances in botanical Research*, **4**, 229–415.

Lövkvist, B. (1956). The *Cardamine pratensis* complex. *Symbolae Botanicae Upsalienses*, **14**(2), 1–131.

Lövkvist, B. (1962). Chromosome and differentiation studies in flowering plants of Skåne, South Sweden. 1. General aspects. Type species with coastal differentiation. *Botaniska Notiser*, **115**, 261–87.

Ludwig, F. (1895). Uber Variationskurven und Variationsflächen der Pflanzen. *Botanisches Zentralblatt*, **64**, 1–8, 33–41, 65–72, 97–105.

Ludwig, F. (1901). Variationsstatistische Probleme und Materialen. *Biometrika*, **1**, 11–29.

Lynch, R.I. (1900). Hybrid Cinerarias. *Journal of the Royal Horticultural Society*, **24**, 269–74.

McFadden, E.S. & Sears, E.R. (1946). The origin of *Triticum spelta* and its free-threshing hexaploid relatives. Hybrids of synthetic *T. spelta* with cultivated hexaploids. *Journal of Heredity*, **37**, 81–9, 107–16.

McLean, R.C. & Ivimey-Cook, W.R. (1956). *Textbook of theoretical botany*. London, New York & Toronto: Longmans.

McLeish, J. & Snoad, B. (1962). *Looking at chromosomes*. London & New York: Macmillan. [2nd edn, 1972].

McMillan, C. (1957). Nature of the plant community. III. Flowering Behaviour within two grassland communities under reciprocal transplanting. *American Journal of Botany*, **44**, 144–53.

McMillan, C. (1970). Photoperiod in *Xanthium* populations from Texas and Mexico. *American Journal of Botany*, **57**, 881–8.

McMillan, C. (1971). Photoperiod evidence in the introduction of *Xanthium* (Cocklebur) to Australia. *Science*, **171**, 1029–31.

McNeill, J. (1976). The taxonomy and evolution of weeds. *Weed Research*, **16**, 399–413.

McNeilly, T. (1968). Evolution in closely adjacent plant populations. III. *Agrostis tenuis* on a small copper mine. *Heredity*, **23**, 99–108.

McNeilly, T. & Antonovics, J. (1968). Evolution in closely adjacent plant populations. IV. Barriers to gene flow. *Heredity*, **23**, 205–18.

McVean, D.N. (1953). Regional variation of *Alnus glutinosa* (L.) Gaertn. in Britain. *Watsonia*, **3**, 26–32.

Mansfield, T.A. & Freer-Smith, P.H. (1981). Effects of urban air pollution on plant growth. *Biological Reviews*, **56**, 343–68.

Manton, I. (1950). *Problems of cytology and evolution in the Pteridophyta*. London & New York: Cambridge University Press.

Marchant, C.J. (1963). Corrected chromosome numbers for *Spartina* × *townsendii* and its parent species. *Nature*, **199**, 299.

Marchant, C.J. (1967). Evolution in *Spartina* (Gramineae). 1. The history and morphology of the genus in Britain. *Journal of the Linnean Society (Botany)*, **60**, 1–24.

Marchant, C.J. (1968). Evolution in *Spartina* (Gramineae). 2. Chromosomes, basic relationships and the problem of *S.* × *townsendii* agg. *Journal of the Linnean Society (Botany)*, **60**, 381–409.

Marsden-Jones, E. (1930). The genetics of *Geum intermedium* Willd. haud Ehrh. and its back-crosses. *Journal of Genetics*, **23**, 377–95.

Marsden-Jones, E.M. & Turrill, W.B. (1945). Report of the transplant experiments of the British Ecological Society. *Journal of Ecology*, **33**, 59–81. [See also earlier reports in the *Journal of Ecology*: **18**, 352; **21**, 268; **23**, 443; **25**, 189; **26**, 359 & 380.]

Marshall, D.R. & Brown, A.H.D. (1981). The evolution of apomixis. *Heredity*, **47**, 1–15.

Massart, J. (1902). L'accomodation individuelle chez le *Polygonum amphibium*. *Bulletin de Jardin Botanique de l'Etat à Bruxelles*, **1**, 73–95.

Mather, K. (1966). Breeding systems and response to selection. In *Reproductive biology and taxonomy of vascular plants*, ed. J.G. Hawkes, pp. 13–19. Conference report of Botanical Society of the British Isles. Oxford: Pergamon Press.

Matthew, P. (1831). Ideas on evolution, published in an Appendix to *On naval timber and arboriculture*. London.

Mayr, E. (1940). Speciation phenomena in birds. *The American Naturalist*, **74**, 249–78.

Mayr, E. (1942). *Systematics and the origin of species*. New York: Columbia University Press.

Mayr, E. (1957). *Species concepts and definitions*. In *The species problem*, ed. E. Mayr, pp. 1–22. Washington: American Association for the Advancement of Science Publication 50.

Mayr, E. (1963). *Animal species and evolution.* London: Oxford University Press; and Cambridge, Mass.: Harvard University Press.

Mays, L.L. (1981). *Genetics. A molecular approach.* New York: Macmillan; and London: Collier Macmillan.

Meeuse, A.D.J. (1966). *Fundamentals of phytomorphology.* New York: Ronald Press.

Melville, R. (1944). The British elm flora. *Nature,* **153**, 198–9.

Mendel, G. (1866). Versuche über Planzenhybriden. *Verhandlungen des Naturforschenden Vereins in Brünn,* **4**, 3–44. [English translation in Bateson, W. (1909), *Mendel's principles of heredity.* London: Cambridge University Press; also Bennett, J.H. (1965). (ed.) *Experiments in plant hybridisation.* Edinburgh & London: Oliver & Boyd.]

Mergen, F. (1963). Ecotypic variation in *Pinus strobus. Ecology,* **44**, 716–27.

Merrell, D.J. (1962). *Evolution and genetics: the modern theory of evolution.* New York: Holt, Rinehart & Winston.

Merrell, D.J. (1981). *Ecological genetics.* London: Longman.

Merxmüller, H. (1970). 1. Biosystematics: still alive? Provocation of biosystematics. *Taxon,* **19**, 140–5.

Meyer, V.G. (1966). Flower abnormalities. *Botanical Reviews,* **32**, 165–218.

Michaelis, P. (1954). Cytoplasmic inheritance in *Epilobium* and its theoretical significance. *Advances in Genetics,* **6**, 287–401.

Millener, L.H. (1961). Day length as related to vegetative development in *Ulex europaeus.* 1. The experimental approach. *New Phytologist,* **60**, 339–54.

Mirsky, A.E. (1968). The discovery of DNA. *Scientific American,* **218**, 76–88.

Mitchell, R.S. (1968). Variation in the *Polygonum amphibium* complex and its taxonomic significance. *University of California Publications in Botany,* **45**, 1–54.

Mivart, St.G. (1871). *The genesis of species,* 2nd edn. London: Macmillan.

Mølgaard, P. (1976). *Plantago major* ssp. *major* and ssp. *pleiosperma.* Morphology, biology and ecology in Denmark. *Botanik Tidsskrift,* **71**, 31–56.

Mooney, H.A. & Billings, W.D. (1961). Comparative physiological ecology of Arctic and Alpine Populations of *Oxyria digyna. Ecological Monographs,* **31**, 1–29.

Moore, D.M. (1959). Population studies on *Viola lactea* Sm. and its wild hybrids. *Evolution,* **13**, 318–32.

Moore, D.M. (1976). *Plant cytogenetics.* London: Chapman & Hall; and New York: Wiley & Sons.

Moore, D.M. (1982). *Flora Europaea check-list and chromosome index.* London: Cambridge University Press.

Moore, D.M. & Harvey, M.J. (1961). Cytogenetic relationships of *Viola lactea* Sm. and other West European arosulate violets. *New Phytologist,* **60**, 85–95.

Moore, R.J. & Mulligan, G.A. (1964). Further studies on natural selection among hybrids of *Carduus acanthoides* and *Carduus nutans. Canadian Journal of Botany,* **42**, 1605–13.

Morison, R. (1672). *Plantarum umbelliferarum distributio nova.* Oxford.

Morris, M.G. & Perring, F.H. (1974). (eds) *The British oak: its history and natural history.* Published for The Botanical Society of the British Isles. Classey: Faringdon.

Morton, J.K. (1966). The role of polyploidy in the evolution of a tropical flora. In *Chromosomes today,* vol. 1, ed. C.D. Darlington & K.R. Lewis, pp. 73–6, Edinburgh: Oliver & Boyd.

Müntzing, A. (1929). Cases of partial sterility in crosses within a Linnean species. *Hereditas*, **12**, 297–319.

Müntzing, A. (1930*a*). Uber Chromosomenvermehrung in *Galeopsis*-kreuzungen und ihre phylogenetische Bedeutung. *Hereditas*, **14**, 153–72.

Müntzing, A. (1930*b*). Outlines to a genetic monograph of the genus *Galeopsis* with special reference to the nature and inheritance of partial sterility. *Hereditas*, **13**, 185–341.

Müntzing, A. (1932). Untersuchungen über Periodizität und Saison-Dimorphismus bei einigen Annuellen *Lamium*-arten. *Botaniska Notiser*, 1932, 153–76.

Müntzing, A. (1938). Sterility and chromosome pairing in intraspecific *Galeopsis* hybrids. *Hereditas*, **24**, 117–88.

Müntzing, A. (1961). *Genetic research*. Stockholm: L.T. Førlag.

Müntzing, A., Tedin, O. & Turesson, G. (1931). Field studies and experimental methods in taxonomy. *Hereditas*, **15**, 1–12.

Nägeli, C. von (1865). Die Bastardbindung im Pflanzenreiche. *Sitzungsbericht der Königlich-Bayerischen Akademie der Wissenschaften zu München Botanische Mitteilungen*, **2**, 159–87.

Nägeli, C. von & Peter, A. (1885). *Die Hieracien Mitteleuropas. Vol. 1, Piloselloiden*. Munich: Oldenbourg.

Nägeli, C. von & Peter, A. (1886–1889). *Die Hieracien Mitteleuropas. Vol. 2, Archieracion* (1886–1889). Munich: Oldenbourg.

Nannfeldt, J.A. (1937). The chromosome numbers of *Poa*, Sect. *Ochlopoa* A. and Gr. and their taxonomical significance. *Botaniska Notiser*, 1937, 238–57.

Navashin, M. (1926). Variabilität des Zellkerns bei *Crepis*-Arten in Bezug auf die Artbildung. *Zeitschrift für Zellforschung und mikroskopische Anatomie*, **4**, 171–215.

New, J.K. (1958). A population study of *Spergula arvensis* 1. *Annals of Botany, New Series*, **22**, 457–77.

New, J.K. (1959). A population study of *Spergula arvensis* 2. *Annals of Botany, New Series*, **23**, 23–33.

New, J.K. (1978). Change and stability of clines in *Spergula arvensis* L. (Corn Spurrey) after 20 years. *Watsonia* 12(2), 137–43.

New, J.K. & Herriott, J.C. (1981). Moisture for germination as a factor affecting the distribution of the seedcoat morphs of *Spergula arvensis* L. *Watsonia* 13(4), 323–4.

Newton, W.C.F. & Pellew, C. (1929). *Primula kewensis* and its derivatives. *Journal of Genetics*, **20**, 405–66.

Nieuwhof, M. (1963). Pollination and contamination of *Brassica oleracea* L. *Euphytica*, **12**, 17–26.

Nilsson-Ehle, E. (1909). Kreuzungsuntersuchungen an Hafer und Weizen. *Acta Universitatis lundensis*, Ser. 2, **5**(2), 1–122.

Njoku, E. (1956). Studies on the morphogenesis of leaves. II. The effect of light intensity on leaf shape in *Ipomoea caerulea*. *New Phytologist*, **55**, 91–110.

Nordenskiöld, H. (1949). The somatic chromosomes of some *Luzula* species. *Botaniska Notiser*, 1949, 81–92.

Nordenskiöld, H. (1951). Cyto-taxonomical studies in the genus *Luzula*. I. Somatic chromosomes and chromosome numbers. *Hereditas*, **37**, 325–55.

Nordenskiöld, H. (1956). Cyto-taxonomical studies in the genus *Luzula*. 2. Hybridization experiments in the *campestris–multiflora* complex. *Hereditas*, **42**, 7–73.

Nordenskiöld, H. (1961). Tetrad analysis and the course of meiosis in three hybrids of *Luzula campestris*. *Hereditas*, **47**, 203–38.

Olby, R.C. (1966). *The origins of Mendelism*. London: Constable; and New York: Schocken.

Olby, R. (1974). *The path to the double helix*. London: Macmillan.

Ornduff, R. (1970). Pathways and patterns of evolution – a discussion. *Taxon*, **19**, 202–4.

Osawa, J. (1913). Studies on the cytology of some species of *Taraxacum*. *Archiv für Zellforschung*, **10**, 450–69.

Osborn, H.F. (1894). *From the Greeks to Darwin: An outline of the development of the evolution idea*. London & New York: Macmillan.

Ownbey, M. (1950). Natural hybridisation and amphidiploidy in the genus *Tragopogon*. *American Journal of Botany*, **37**, 487–99.

Ownbey, M. & McCollum, G.D. (1953). Cytoplasmic inheritance and reciprocal amphiploidy in *Tragopogon*. *American Journal of Botany*, **40**, 788–96.

Ownbey, M. & McCollum, G.D. (1954). The chromosome of *Tragopogon*. *Rhodora*, **56**, 7–21.

Pankhurst, R.J. (1975). (ed.) *Biological identification with computers*. London & New York: Academic Press.

Pantin, C.F.A. (1960). Alfred Russel Wallace, F.R.S. and his essays of 1858 and 1855. *Notes and Records of the Royal Society of London*, **14**, 67–84.

Parker, R.E. (1973). *Introductory statistics for biology*. London: Arnold.

Parsons, P.A. (1959). Some problems in inbreeding and random mating in tetrasomics. *Agronomy Journal*, **51**, 465–7.

Paterniani, E. (1969). Selection for reproductive isolation between two populations of Maize, *Zea mays* L. *Evolution*, **23**, 534–47.

Pazy, B. & Zohary, D. (1965). The process of introgression between *Aegilops* polyploids: natural hybridization between *A. variabilis*, *A. ovata* and *A. biuncialis*. *Evolution*, **19**, 385–94.

Pearson, E.S. & Kendall, M.G. (1970). *Studies in the history of statistics and probability*. London: Griffin.

Pearson, K. (1900). *The grammar of science*, 2nd edn. London: Black.

Pearson, K. (1924). *The life, letters and labours of Francis Galton*, Vol. II. Cambridge: Cambridge University Press.

Pearson, K., Lee, A., Warren, E., Fry, A., Fawcett, C.D. *et al*. (1901). Mathematical contributions to the theory of evolution. IX. On the principle of homotyposis and its relation to heredity, to the variability of the individual, and to that of the race. Part I – Homotyposis in the vegetable kingdom. *Philosophical Transactions of the Royal Society*, *A*, **197**, 285–379.

Pearson, K. *et al*. (1903). Cooperative investigation on plants. 2. Variation and correlation in Lesser Celandine from diverse localities. *Biometrika* **2**, 145–64.

Pearson, K. & Yule, G.U. (1902). Variation in ray-flowers of *Chrysanthemum leucanthemum*, 1133 heads gathered at Keswick during July 1895. *Biometrika*, **1**, 319.

Peckham, M. (1959). *The origin of species by Charles Darwin. A variorum text*. London: Oxford University Press; and Philadelphia: University of Pennsylvania Press.

Pellew, C. (1913). Note on gametic reduplication in *Pisum*. *Journal of Genetics*, **3**, 105–6.

Pennington, W. (1969). *The history of British vegetation*. London: English Universities Press.

Pennington, W. (1974). *The history of British vegetation*, 2nd edn. London: English Universities Press.

Perring, F.H. & Farrell, L. (1977). *British Red Data Books. Vol. 1, Vascular plants*. Lincoln: The Society for the Promotion of Nature Conservation with the financial support of the World Wildlife Fund. [2nd edn 1982]

Pharis, R.P. & Ferrell, W.K. (1966). Differences in drought resistance between coastal and inland sources of Douglas Fir. *Canadian Journal of Botany*, **44**, 1651–9.

Pickersgill, B. & Heiser, C.B. (1976). Cytogenetics and evolutionary change under domestication. *Philosophical Transactions of the Royal Society of London, B*, **275**, 55–69.

Pollard, A.J. (1980). Diversity of metal tolerances in *Plantago lanceolata* L. from the southeastern United States. *New Phytologist*, **86**, 109–17.

Pope, O.A., Simpson, D.M., & Duncan, E.N. (1944). Effect of Corn barriers on natural crossing in Cotton. *Journal of Agricultural Research*, **68**, 347–61.

Portugal, F.H. & Cohen, J.S. (1977). *A century of DNA*. Cambridge, Mass. & London: The MIT Press.

Prentice, H.C. (1979). Numerical analysis of infraspecific variation in European *Silene alba* and *S. dioica* (Caryophylaceae). *Botanical Journal of the Linnean Society*, **78**, 181–212.

Prime, C.T. (1955). Vegetative reproduction in *Arum maculatum*. *Watsonia*, **3**, 175–8.

Prime, C.T. (1960). *Lords and ladies*. London & New York: Collins.

Proctor, J. (1971a). The plant ecology of serpentine. II. Plant response to serpentine soils. *Journal of Ecology*, **59**, 397–410.

Proctor, J. (1971b). The plant ecology of serpentine. III. The influence of a high magnesium/calcium ratio and high nickel and chromium levels in some British and Swedish serpentine soils. *Journal of Ecology*, **59**, 827–42.

Proctor, M.C.F. & Yeo, P.F. (1973). *The pollination of flowers*. London: Collins.

Provine, W.B. (1971). *The origins of theoretical population genetics*. Chicago & London: University of Chicago Press.

Quetelet, M.A. (1846). *Lettres à S.A.R. le Duc Régnant de Saxe-Coburg et Gotha, sur la théorie des probabilités, appliquée aux sciences morales et politiques*. Brussels. Translation by O.G. Downes (1849): *Letters addressed to H.R.H. the Grand Duke of Saxe-Coburg and Gotha on the theory of probabilities as applied to the moral and political sciences*. London: Charles & Edwin Layton.

Quinn, J.A. (1978). Plant ecotypes: ecological or evolutionary units. *Bulletin of the Torrey Botanical Club*, **105**, 58–64.

Rackham, O. (1975). *Hayley Wood. Its history and ecology*. Cambridge: Cambridgeshire and Isle of Ely Naturalists Trust.

Rackham, O. (1980). *Ancient woodland: its history, vegetation and uses in England*. London: Arnold.

Rafiński, J.N. (1979). Geographic variability of flower colour in *Crocus scepusiensis* (Iridaceae). *Plant Systematics and Evolution*, **131**, 107–25.

Rajhathy, T. & Thomas, H. (1972). Genetic control of chromosome pairing in hexaploid oats. *Nature*, **239**, 217–19.

Ramsbottom, J. (1938). Linnaeus and the species concept. *Proceedings of the Linnean Society of London*, **150**, 192–219.

Randolph, L.F., Nelson, I.S. & Plaisted, R.L. (1967). Negative evidence of introgression affecting the stability of Louisiana *Iris* species. *Cornell University Agriculture Experimental Station Memoir*, No. 398.

Raven, P.H. (1976). Systematics and plant population biology. *Systematic Botany*, **1**, 284–316.

Raven, P.H. & Thompson, H.J. (1964). Haploidy and angiosperm evolution. *The American Naturalist*, **98**, 251–2.

Rayner, A.A. (1969). *A first course in biometry for agricultural students.* Pietermaritzburg: University of Natal Press.

Rees, H. & Jones, R.N. (1977). *Chromosome genetics.* London: Arnold.

Rhoades, M.M. (1943). Genic induction of an inherited cytoplasmic difference. *Proceedings of the National Academy of Sciences, USA,* **29**, 327–9.

Rice, E.L. (1974). *Allelopathy.* New York & London: Academic Press.

Richards, A.J. (1978). (ed.) *The pollination of flowers by insects.* Linnean Society Symposium 6. London: Academic Press.

Richards, A.J. (1979). Reproduction in flowering plants. *Nature,* **278**, 306.

Richards, A.J. & Ibrahim, H. (1978). Estimation of neighbourhood size in two populations of *Primula veris.* In *The pollination of flowers by insects,* ed. A.J. Richards, pp. 165–74. Linnean Society Symposium Series 6. London: Academic Press.

Ridgman, W.J. (1975). *Experimentation in biology.* Glasgow: Blackie.

Rieger, R., Michaelis, A. & Green, M.M. (1976). *Glossary of genetics and cytogenetics,* 4th edn. Berlin, Heidelberg & New York: Springer-Verlag.

Riley, H.P. (1938). A character analysis of colonies of *Iris fulva* and *Iris hexagona* var. *giganticaerulea* and natural hybrids. *American Journal of Botany,* **25**, 727–38.

Riley, R. (1965). Cytogenetics and the evolution of Wheat. In *Essays on crop plant evoluion,* ed. J. Hutchinson, pp. 103–22. London: Cambridge University Press.

Riley, R. & Chapman, V. (1958). Genetic control of the cytologically diploid behaviour of hexaploid wheat. *Nature,* **183**, 713–5.

Riley, R. & Law, C.N. (1965). Genetic variation in chromosome pairing. *Advances in Genetics,* **13**, 57–114.

Riley, R., Unrau, J. & Chapman, V. (1958). Evidence on the origin of the B genome of Wheat. *Journal of Heredity,* **49**, 91–8.

Ritchie, J.C. (1955a). A natural hybrid in *Vaccinium.* 1. The structure, performance and chorology of the cross *Vaccinium intermedium* Ruthe. *New Phytologist,* **54**, 49–67.

Ritchie, J.C. (1955b). A natural hybrid in *Vaccinium.* 2. Genetic studies in *Vaccinium intermedium* Ruthe. *New Phytologist,* **54**, 320–35.

Robbins, W.W., Weier, T.E. & Stocking, C.R. (1962). *An introduction to plant science,* 2nd edn. London & New York: Wiley.

Roberts, E.H. (1975). Problems of long-term storage of seed and pollen for genetic resources conservation. In *Crop genetic resources for today and tomorrow,* ed. O.H. Frankel & J.G. Hawkes, pp. 269–95, 316. International Biological Programme 2. Cambridge: Cambridge University Press.

Roberts, H.F. (1929). *Plant hybridisation before Mendel.* Princeton: Princeton University Press; and London: Oxford University Press.

Roles, S.J. (1960). Illustrations [Part II] to *Flora of the British Isles,* Clapham, A.R., Tutin, T.G. & Warburg, E.F. Cambridge: Cambridge University Press.

Roose, M.L. & Gottlieb, L.D. (1976). Genetic and biochemical conequences of polyploidy in *Tragopogon. Evolution,* **30**, 818–30.

Rosen, F. (1889). Systematische und biologische Beobachtungen über *Erophila verna. Botanische Zeitung,* **47**, 565–80, 581–91, 597–608, 613–20.

Rosenthal, G.A. & Janzen, D.H. (1979). *Herbivores. Their interaction with secondary plant metabolites.* New York & London: Academic Press.

Ross-Craig, S. (1948–1973). *Drawings of British plants.* London: Bell & Sons Ltd.

Rückert, J. (1892). Zur Entwicklungs Geschichte des Ovarioleies bei Selachiern. *Anatomischer Anzeiger,* **7**, 107.

Rushton, B.S. (1978). *Quercus robur* L. and *Quercus petraea* (Matt.) Liebl: a multivariate approach to the hybrid problem. I. Data acquisition, analysis and interpretation. *Watsonia*, **12**, 81–101.

Rushton, B.S. (1979). *Quercus robur* L. and *Quercus petraea* (Matt.) Liebl.: a multivariate approach to the hybrid problem. 2. The geographical distribution of population types. *Watsonia*, **12**, 209–24.

Russell, B. (1931). *The scientific outlook*. London: Allen & Unwin.

Salmon, S.C. & Hanson, A.A. (1964). *The principles and practice of agricultural research*. London: Leonard Hill.

Sarkar, P. & Stebbins, G.L. (1956). Morphological evidence concerning the origin of the B genome in Wheat. *American Journal of Botany*, **43**, 297–304.

Saunders, E.R. (1897). On discontinuous variation occurring in *Biscutella laevigata*. *Proceedings of the Royal Society, B*, **62**, 11–26.

Schlising, R.A. & Turpin, R.A. (1971). Hummingbird dispersal of *Delphinium cardinale* pollen treated with radioactive iodine. *American Journal of Botany*, **58**, 401–6.

Schmidt, J. (1899). Om ydre faktorers indflydelse paa løvbladets anatomiske bygning hos en af vore strandplanter. *Botanisk Tidsskrift*, **22**, 145–65.

Schrödinger, E. (1944). *What is Life?* London: Cambridge University Press; and New York: Macmillan.

Schwanitz, F. (1966). *The origin of cultivated plants*. Cambridge, Mass.: Harvard University Press.

Schweber, S.S. (1977). The origin of the origin revisited. *Journal of the History of Biology*, **10**, 229–316.

Sherman, M. (1946). Karyotype evolution: a cytogenetic study of seven species and six interspecific hybrids of *Crepis*. *University of California Publications in Botany*, **18**, 369–408.

Shivas, M.G. (1961*a*). Contributions to the cytology and taxonomy of species of *Polypodium* in Europe and America. 1. Cytology. *Journal of the Linnean Society*, **58**, 13–25.

Shivas, M.G. (1961*b*). Contributions to the cytology and taxonomy of species of *Polypodium* in Europe and America. 2. Taxonomy. *Journal of the Linnean Society*, **58**, 27–38.

Simmonds, N.W. (1976). (ed.) *Evolution of Crop Plants*. London: Longmans.

Simpson, D.M. (1954). Natural cross-pollination in Cotton. *U.S. Department of Agriculture Technical Bulletin*, No. 1094.

Simpson, G.G. (1944). *Tempo and mode in evolution*. New York: Columbia University Press.

Sindu, A.S. & Singh, S. (1961). Studies on the agents of cross pollination of Cotton. *Indian Cotton Growing Review*, **15**, 341–53.

Smith, A. (1965). The assessment of patterns of variation in *Festuca rubra* L. in relation to environmental gradients. *Scottish Plant Breeding Station Record*, 1965, 163–95.

Smith, A. (1972). The pattern of distribution of *Agrostis* and *Festuca* plants of *various genotypes in a sward*. *New Phytologist*, **71**, 937–45.

Smith, A.C. (1957). Fifty years of botanical nomenclature. *Brittonia*, **9**, 2–8.

Smith, D.C., Nielsen, E.L. & Ahlgren, H.L. (1946). Variation in ecotypes of *Poa pratensis*. *The botanical Gazette*, **108**, 143–66.

Smith, D.M. & Levin, D.A. (1963). A chromatographic study of reticulate evolution in the Appalachian *Asplenium* complex. *American Journal of Botany*, **50**, 952–8.

Smith, G.L. (1963*a*). Studies in *Potentilla* L. 1. Embryological investigations into the mechanism of agamospermy in British *P. tabernaemontani* Aschers. *New Phytologist*, **62**, 264–82.

Smith, G.L. (1963*b*). Studies in *Potentilla* L. 2. Cytological aspects of apomixis in *P. crantzii* (Cr.) Beck ex Fritsch. *New Phytologist*, **62**, 283–300.

Smith, G.L. (1971). Studies in *Potentilla* L. III. Variation in British *P. tabernaemontani* Aschers. and *P. crantzii* (Cr.) Beck ex Fritsch. *New Phytologist*, **70**, 607–18.

Smith, J. (1841). Notice of a plant which produces perfect seeds without any apparent action of pollen. *Transactions of the Linnean Society of London*, **18**, 509–12.

Smith, K.M. (1977). *Plant viruses*, 6th edn. London: Chapman & Hall.

Smith, P.M. (1976). *The chemotaxonomy of plants*. London: Arnold.

Smith, S. (1960). The origin of the Origin. *Advancement of Science*, **16**, 391–401.

Snaydon, R.W. (1970). Rapid population differentiation in a mosaic environment. 1. The response of *Anthoxanthum odoratum* populations to soils. *Evolution*, **24**, 257–69.

Snaydon, R.W. (1976). Genetic change within species. In *The Park Grass experiment on the effect of fertilisers and liming on the botanical composition of permanent grassland and on the yield of hay*, Thurston, J.M., Dyke, G.V. & Williams, E.D., Appendix. Harpenden: Rothamsted Experimental Station.

Snaydon, R.W. (1978). Genetic changes in pasture populations. In *Plant relations in pastures*, ed. J.R. Wilson, pp. 253–69. Melbourne: CSIRO.

Snaydon, R.W. & Davies, M.S. (1972). Rapid population differentiation in a mosaic environment. II. Morphological variation in *Anthoxanthum odoratum*. *Evolution*, **26**, 390–405.

Sneath, P.H.A. & Sokal, R.R. (1973). *Numerical taxonomy*. San Francisco: Freeman.

Snedecor, G.W. & Cochran, W.G. (1980). *Statistical methods*, 7th edn. Ames, Iowa: Iowa State University Press.

Snyder, L.A. (1950). Morphological variability and hybrid development in *Elymus glaucus*. *American Journal of Botany*, **37**, 628–35.

Snyder, L.A. (1951). Cytology of inter-strain hybrids and the probable origin of variability in *Elymus glaucus*. *American Journal of Botany*, **38**, 195–202.

Sokal, R.R. & Crovello, T.J. (1970). The biological species concept: a critical evaluation. *American Naturalist*, **104**, 127–53.

Sokal, R.R. & Rohlf, F.J. (1969). *Biometry. The principles and practice of statistics in biological research*. San Francisco: Freeman.

Sokal, R.R. & Rohlf, F.J. (1981). *Biometry. The principles and practice of statistics in biological research*, 2nd edn. San Francisco: Freeman.

Sokal, R.R. & Sneath, P.H.A. (1963). *Principles of numerical taxonomy*. London & San Francisco: Freeman.

Solbrig, O.T. & Simpson, B.B. (1974). Components of regulation of a population of Dandelions in Michigan. *Journal of Ecology*, **62**, 473–86.

Solbrig, O.T. & Simpson, B.B. (1977). A garden experiment on competition between biotypes of the Common Dandelion (*Taraxacum officinale*). *Journal of Ecology*, **65**, 427–30.

Solbrig, O.T. & Solbrig, D.J. (1979). *Introduction to population biology and evolution*. London: Addison-Wesley Publishing Company.

Soltis, D.E. (1982). Allozymic variability in *Sullivantia* (Saxifragaceae). *Systematic Botany*, **7**, 26–34.

Somaroo, B.H. & Grant, W.F. (1971). Interspecific hybridization between diploid species of *Lotus* (Leguminosae). *Genetica*, **42**, 353–67.

Sørenson, T. & Gudjónsson, G. (1946). Spontaneous chromosome-aberrants in apomictic *Taraxaca*. *Biologiske Skrifter, K. Danske Videnskabernes Selskab*, **4**, No. 2.

Soulé, M.E. & Wilcox, B.A. (1980). (eds.) *Conservation biology: An evolutionary–ecological perspective*. Sunderland, Mass.: Sinauer Associates Inc.

Spiess, E.B. (1977). *Genes in populations*. New York & London: Wiley.

Spira, T.P. (1980). Floral parameters, breeding system and pollinator type in *Trichostema* (Labiatae). *American Journal of Botany*, **67**, 278–84.

Sporne, K.R. (1971). *The mysterious origin of flowering plants*. London: Oxford University Press.

Sporne, K.R. (1974). *The morphology of angiosperms*. London: Hutchinson.

Sprengel, C.K. (1793). *Das entdeckte Geheimniss der Natur im Bau und in der Befruchtung der Blumen*. Berlin.

Srb, A.M. & Owen, R.D. (1958). *General genetics*. San Francisco: Freeman.

Stace, C.A. (1975). (ed.) *Hybridization and the flora of the British Isles*. London: Academic Press.

Stace, C.A. (1980). *Plant taxonomy and biosystematics*. London: Arnold.

Stadler, L.J. (1942). *Some observations on gene variability and spontaneous mutation*. The Spragg Memorial Lectures, Michigan State University.

Stapledon, R.G. (1928). Cocksfoot grass (*Dactylis glomerata* L.): ecotypes in relation to the biotic factor. *Journal of Ecology*, **16**, 72–104.

Stearn, W.T. (1957). *Introduction to facsimile edition of Linnaeus' Species plantarum*. London: Ray Society, British Museum.

Stebbins, G.L. (1947). *Types of polyploids: their classification and significance*. *Advances in Genetics*, **1**, 403–29.

Stebbins, G.L. (1950). *Variation and evolution in plants*. London: Oxford University Press; and New York: Columbia University Press.

Stebbins, G.L. (1966). *Processes of organic evolution*. Englewood Cliffs, NJ: Prentice-Hall. [2nd edn 1971.]

Stebbins, G.L. (1971). *Chromosomal evolution in higher plants*. London: Arnold.

Stebbins, G.L. (1974). *Flowering plants. Evolution above the species level*. London: Arnold.

Stebbins, G.L. (1980). Polyploidy in plants: unsolved problems and prospects. In *Polyploidy, biological relevance*, ed. W.H. Lewis, pp. 495–520. New York & London: Plenum Press.

Stebbins, G.L. & Daly, K. (1961). Changes in the variation pattern of a hybrid population of *Helianthus* over an eight-year period. *Evolution*, **15**, 60–71.

Stebbins, G.L., Harvey, B.L., Cox, E.L., Rutger, J.N., Jelencovic, G. & Yagil, E. (1963). Identification of the ancestry of an amphiploid *Viola* with the aid of paper chromatography. *American Journal of Botany*, **50**, 830–9.

Stelleman, P. (1978). The possible role of insect visits in pollination of reputedly anemophilous plants, exemplified by *Plantago lanceolata*, and syrphid flies. In *The pollination of flowers by insects*, ed. A.J. Richards, pp. 41–6. Linnean Society Symposium Series 6. London: Academic Press.

Stent, G.S. (1968). 'That was the molecular biology that was'. *Science*, **160**, 390–5.

Stern, C. & Sherwood, E.R. (1966). *The origin of genetics*. A Mendel source book. London & San Francisco: Freeman.

Sternberg, L. (1976). Growth forms of *Larrea tridentata*. *Madroño*, **23**, 408–17.

Stewart, R.N. (1947). The morphology of somatic chromosomes in *Lilium*. *American Journal of Botany*, **34**, 9–26.

Strasburger, E. (1910). Chromosomenzahl. *Flora*, **100**, 398–446.

Strickberger, M.W. (1976). *Genetics*, 2nd edn. New York & London: Macmillan.

Strid, A. (1970). Studies in the Aegean flora. XVI. Biosystematics of the *Nigella arvensis* complex. With special reference to the problem of non-adaptive radiation. *Opera Botanica*, No. 28. Lund: Gleerup.

Sturtevant, A.H. (1965). *A history of genetics*. New York: Harper & Row.

Sunderland, N. (1980). Guidelines in the culture of pollen in vitro. In *Tissue culture methods for plant pathologists*, ed. D.S. Ingram & J.P. Helgeson, pp. 33–40. Oxford: Blackwells.

Sutton, W.S. (1902). On the morphology of the chromosome group in *Brachystola magna*. *Biological Bulletin, Marine Biological Laboratory, Woods Hole, Mass.*, **4**, 24–39.

Sutton, W.S. (1903). The chromosomes in heredity. *Biological Bulletin, Marine Biological Laboratory, Woods Hole, Mass.*, **4**, 231–48.

Synge, H. (1981). (ed.) *The biological aspects of rare plant conservation*. Chichester: Wiley.

Täckholm, G. (1922). Zytologische studien über die Gattung *Rosa*. *Acta Horti Bergiani*, **7**, 97–381.

Takhtajan, A. (1969). *Flowering plants – origin and dispersal*. Authorised translation from the Russian by C. Jeffrey. Edinburgh: Oliver & Boyd.

Taylor, G.E., Jr. & Murdy, W.H. (1975). Population differentiation of an annual plant species, *Geranium carolinianum* in response to sulfur dioxide. *Botanical Gazette*, **136**, 212–15.

Taylor, K. & Markham, B. (1978). *Ranunculus ficaria* L. Biological Flora of the British Isles. *Journal of Ecology*, **66**, 1011–31.

Thiselton-Dyer, W.T. (1895*a*). Variation and specific stability. *Nature*, **51**, 459–61.

Thiselton-Dyer, W.T. (1895*b*). Origin of the cultivated *Cineraria*. *Nature*, **52**, 3–4, 78–9, 128–9.

Thoday, J.M. (1972). Disruptive selection. *Proceedings of the Royal Society of London, B*, **182**, 109–43.

Thomas, D.A. & Barber, H.N. (1974). Studies of leaf characteristics of a cline of *Eucalyptus urnigera* from Mount Wellington, Tasmania. II. Reflection, transmission and absorption of radiation. *Australian Journal of Botany*, **22**, 701–7.

Thomas, H.H. (1947). The rise of geology and its influence on contemporary thought. *Annals of Science*, **5**, 325–41.

Thompson, J.N. (1982). *Interaction and coevolution*. New York: Wiley.

Thurston, J.M., Dyke, G.V. & Williams, E.D. (1976). *The Park Grass experiment on the effect of fertilisers and liming on the botanical composition of permanent grassland and on the yield of hay*. Harpenden: Rothamsted Experimental Station.

Tischler, G. (1950). *Die Chromosomenzahlen der Gefässpflanzen Mitteleuropas*. 'S-Gravenhage: Junk.

Tobgy, H.A. (1943). A cytological study of *Crepis fuliginosa, C. neglecta* and their F_1 hybrid, and its bearing on the mechanism of phylogenetic reduction in chromosome number. *Journal of Genetics*, **45**, 67–111.

Tome, G.A. & Johnson, I.J. (1945). Self- and Cross- Fertility Relationships in *Lotus corniculatus* L. and *Lotus tenuis* Wald. et. Kit. *Journal of the American Society of Agronomy*, **37**, 1011–23.

Tower, W.L. (1902). Variation in the ray-flowers of *Chrysanthemum leucanthemum* L. at Yellow Springs, Green County, O, with remarks upon the determination of the modes. *Biometrika*, **1**, 309–15.

Tralau, H. (1968). Evolutionary trends in the genus *Ginkgo*. *Lethaia*, **1**, 63–101.

Tubeuf, K.F.von. (1923). *Monographie der Mistel*. Munich & Berlin: Oldenbourg.

Turesson, G. (1922*a*). The species and variety as ecological units. *Hereditas*, **3**, 100–13.

Turesson, G. (1922*b*). The genotypical response of the plant species to the habitat. *Hereditas*, **3**, 211–350.

Turesson, G. (1925). The plant species in relation to habitat and climate. *Hereditas*, **6**, 147–236.

Turesson, G. (1927*a*). Erbliche Transpirationsdifferenzen zwischen Ökotypen derselben Pflanzen Art. *Hereditas*, **11**, 193–206.

Turesson, G. (1927*b*). Untersuchungen über Grenzplasmolyse und Saugkraftwerte in verschiedenen Ökotypen derselben Art. *Jahrbücher für wissenschaftliche Botanik*, **66**, 723–47.

Turesson, G. (1930). The selective effect of climate upon the plant species. *Hereditas*, **14**, 99–152.

Turesson, G. (1943). Variation in the apomictic microspecies of *Alchemilla vulgaris* L. *Botaniska Notiser*, 1943, 413–27.

Turesson, G. (1961). Habitat modifications in some widespread plant species. *Botaniska Notiser*, **114**, 435–52.

Turkington, R. & Harper, J.L. (1979). The growth, distribution and neighbour relationships of *Trifolium repens* in permanent pasture. IV. Fine-scale biotic differentiation. *Journal of Ecology*, **67**, 245–54.

Turrill, W.B. (1938). The expansion of taxonomy with special reference to Spermatophyta. *Biological Reviews*, **13**, 342–73.

Turrill, W.B. (1940). Experimental and synthetic plant taxonomy. In *The new systematics*, ed. J.S. Huxley, pp. 47–71. Oxford: Clarendon Press.

Turrill, W.B. (1948). *British plant life*. London: Collins.

Tutin, T.G. (1957). A contribution to the experimental taxonomy of *Poa annua* L. *Watsonia*, **4**, 1–10.

Tutin, T.G., Heywood, V.H., Burges, N.A., Moore, D.M., Valentine, D.H., Walters, S.M. & Webb, D.A. (1964–80). *Flora Europaea*. London: Cambridge University Press.

Upcott, M. (1940). The nature of tetraploidy in *Primula kewensis*. *Journal of Genetics*, **39**, 79–100.

Uphof, J.C.Th. (1938). Cleistogamic flowers. *The botanical Review*, **4**, 21–49.

Valentine, D.H. (1939). The Butterbur. *Discovery (New Series)*, **11**(14), 246–50.

Valentine, D.H. (1941). Variation in *Viola riviniana* Rchb. *New Phytologist*, **40**, 189–209.

Valentine, D.H. (1956). Studies in British Primulas. V. The inheritance of seed incompatibility. *New Phytologist*, **55**, 305–18.

Valentine, D.H. (1962). Variation and evolution in the Genus *Viola*. *Preslia*, **34**, 190–206.

Valentine, D.H. (1966). The experimental taxonomy of some *Primula* species. *Transactions and Proceedings of the Botanical Society of Edinburgh*, **40**, 169–80.

Valentine, D.H. (1975*a*). The taxonomic treatment of polymorphic variation. *Watsonia*, **10**, 385–90.

Valentine, D.H. (1975*b*). *Primula*. In *Hybridization and the flora of the British Isles*, ed. C.A. Stace, pp. 346–8. London & New York: Academic Press.

Valentine, D.H. (1979). Experimental work on the British flora. *Watsonia*, **12**, 201–7.

Vasek, F.C. (1980). Creosote Bush: long-lived clones in the Mojave Desert. *American Journal of Botany*, **67**, 246–55.

Vernon, H.M. (1903). *Variation in animals and plants.* London: Kegan Paul.

Verschaffelt, E. (1899). Galton's regression to mediocrity bij ongeslachtelijke verplanting. In *Livre Jubilaire dédié à Charles van Bambeke*, pp. 1–5. Brussels: Lamerton.

Vickery, R.K. (1964). Barriers to gene exchange between members of the *Mimulus guttatus* complex (Scrophulariaceae). *Evolution*, **18**, 52–69.

Vorzimmer, P.J. (1972). *Charles Darwin: the years of controversy.* London: University of London Press.

Vosa, C.G. (1975). The use of giemsa and other staining techniques in karyotype analysis. *Current Advances in Plant Science*, **6**, 495–510.

Waddington, C.H. (1966). Mendel and evolution. In *G. Mendel Memorial Symposium, 1865–1965*, ed. M. Sosna, pp. 145–50. Prague: Academia Publishing House of the Czechoslovak Academy of Sciences.

Waddington, C.H. (1969). Some European contributions to the prehistory of molecular biology. *Nature*, **221**, 318–21.

Wagner, M. (1868). *Die Darwin'sche Theorie und das Migrationgesetz der Organismen.* Leipzig: Duncker & Humblot.

Waldron, L.R. (1912). Hardiness in successive Alfalfa generations. *The American Naturalist*, **46**, 463–9.

Wallace, B. (1981). *Basic population genetics.* New York: Columbia University Press.

Walters, S.M. (1961). The shaping of angiosperm taxonomy. *New Phytologist*, **60**, 74–84.

Walters, S.M. (1962). Generic and specific concepts in the European flora. *Preslia*, **34**, 207–26.

Walters, S.M. (1970). Dwarf variants of *Alchemilla* L. *Fragmenta Floristica et Geobotanica*, **16**, 91–8.

Walters, S.M. (1972). Endemism in the genus *Alchemilla* in Europe. In *Taxonomy, phytogeography and evolution*, ed. D.H. Valentine, pp. 301–5. London & New York: Academic Press.

Wang, C.W., Perry, T.O. & Johnson, A.G. (1960). *Pollen dispersal of Slash Pine (Pinus elliottii* Engelm.) with special reference to seed orchard management. *Silvae Genetica*, **9**, 78–86.

Ward, D.B. (1974). The 'ignorant man' technique of sampling plant populations. *Taxon*, **23**, 325–30.

Wareing, P.F. & Phillips, I.D.J. (1981). *Growth and differentiation in plants*, 3rd edn. Oxford: Pergamon Press.

Warwick, S.I. (1980). The genecology of lawn weeds. VII. The response of different growth forms of *Plantago major* L. and *Poa annua* L. to simulated trampling. *New Phytologist*, **85**, 461–9.

Warwick, S.I. & Briggs, D. (1978a). The genecology of lawn weeds. I. Population differentiation in *Poa annua* L. in a mosaic environment of bowling green lawns and flower beds. *New Phytologist*, **81**, 711–23.

Warwick, S.I. & Briggs, D. (1978b). The genecology of lawn weeds. II. Evidence for disruptive selection in *Poa annua* L. in a mosaic environment of bowling green lawns and flower beds. *New Phytologist*, **81**, 725–37.

Warwick, S.I. & Briggs, D. (1979). The genecology of lawn weeds. III. Cultivation experiments with *Achillea millefolium* L., *Bellis perennis* L., *Plantago lanceolata* L., *Plantago major* L. and *Prunella vulgaris* L. collected from lawns and contrasting grassland habitats. *New Phytologist*, **83**, 509–36.

Warwick, S.I. & Briggs, D. (1980a). The genecology of lawn weeds. IV. Adaptive significance of variation in *Bellis perennis* L. as revealed in a transplant experiment. *New Phytologist*, **85**, 275–88.

Warwick, S.I. & Briggs, D. (1980*b*). The genecology of lawn weeds. V. The adaptive significance of different growth habit in lawn and roadside populations of *Plantago major* L. *New Phytologist*, **85**, 289–300.

Warwick, S.I. & Briggs, D. (1980*c*). The genecology of lawn weeds. VI. The adaptive significance of variation in *Achillea millefolium* L. as investigated by transplant experiments. *New Phytologist*, **85**, 451–60.

Watson, J.D. (1968). *The double helix*. London: Weidenfeld & Nicolson.

Watson, J.D. (1970). *Molecular biology of the gene*, 2nd edn. New York: Benjamin.

Watson, J.D. & Crick, F.H.C. (1953). A structure of deoxyribose nucleic acid. *Nature*, **171**, 737–8.

Watson, P.J. (1969). Evolution in closely adjacent plant populations. VI. An entomophilous species, *Potentilla erecta*, in two contrasting habitats. *Heredity*, **24**, 407–22.

Webb, D.A. (1954). Notes on four Irish heaths. *Irish Naturalist Journal*, **11**, 187–92, 215–19.

Weijer, J. (1952). The colour-differences in *Epipactis helleborine* (L.) Cr. Wats. & Coult., and the selection of the genetical varieties by environment. *Genetica*, **26**, 1–32.

Weismann, A. (1883). *Uber die Vererbung*. English translation, *On heredity* (1889), translated by A.E. Shipley: Oxford: Clarendon Press.

Weldon, W.F.R. (1895*a*). The origin of the cultivated *Cineraria*. *Nature*, **52**, 54, 104, 129.

Weldon, W.F.R. (1895*b*). Remarks on variation in animals and plants. *Proceedings of the Royal Society of London*, **57**, 379–82.

Weldon, W.F.R. (1898). Presidential address. Section D. Zoology. *Nature*, **58**, 499–506.

Weldon, W.F.R. (1902*a*). On the ambiguity of Mendel's categories. *Biometrika*, **2**, 44–55.

Weldon, W.F.R. (1902*b*). Seasonal changes in the characters of *Aster prenanthoides* Muhl. *Biometrika*, **2**, 113–4.

Wells, W.C. (1818). An account of a White female, part of whose skin resembles that of a Negro. [Paper given at the Royal Society, 1813] In *Two essays upon dew and single vision*. London.

Wettstein, R. von (1895). Der Saison-Dimorphismus als Ausgangpunkt für die Bildung neuer Arten im Planzenreich. *Berichte der Deutschen botanischen Gesellschaft*, **13**,303–13.

White, G. (1789). *The natural history of Selborne*. World Classics Edition (1951): London: Oxford University Press.

White, M.J.D. (1978). *Modes of speciation*. San Francisco: Freeman & Company.

White, O.E. (1917). Inheritance studies in *Pisum*. 2. The present state of knowledge of heredity and variation in peas. *Proceedings of the American Philosophical Society*, **56**, 487–588.

Whitehouse, H.L.K. (1950). Multiple-allelomorph incompatibility of pollen and style in the evolution of the angiosperms. *Annals of Botany*, **14**, 199–216.

Whitehouse, H.L.K. (1959). Cross- and self-fertilisation in plants. In *Darwin's biological work*, ed. P.R. Bell, pp. 207–61. London: Cambridge University Press.

Whitehouse, H.L.K. (1965). *Towards an understanding of the mechanism of heredity*. London: Arnold. [3rd edn, 1973.]

Wickler, W. (1968). *Mimicry in plants and animals*. New York: McGraw-Hill.

Wiens, D. (1978). Mimicry in plants. *Evolutionary Biology*, **11**, 365–403.

Wilkins, D.A. (1959). Sampling for genecology. *Record of the Scottish Plant Breeding Station*, 1959, 92–6.

Wilkins, D.A. (1960). Recognising adaptive variants. *Proceedings of the Linnean Society of London*, **171**, 122–6.

Williams, G.C. (1966). *Adaptation and natural selection. A critique of some current evolutionary thought*. Princeton, NJ: Princeton University Press.

Williamson, P.G. (1981). Morphological stasis and developmental constraint: real problems for Neo-Darwinism. *Nature*, **294**, 214–15.

Willis, J.C. (1922). *Age and area*. London: Cambridge University Press; and New York: Macmillan.

Willis, J.C. (1940). *The course of evolution*. London: Cambridge University Press.

Willis, J.C. (1949). *The birth and spread of plants*. Geneva: Conservatoire et Jardin Botaniques.

Willson, M.F. (1983). *Plant reproductive ecology*. New York: Wiley.

Wilmott, A.J. (1949). Intraspecific categories of variation. In *British flowering plants and modern systematic methods*, ed. A.J. Wilmott, pp. 28–45. London: Botanical Society of the British Isles.

Winge, Ø. (1917). The chromosomes, their numbers and general importance. *Comptes Rendus des Travaux du Laboratoire Carlsberg*, **13**, 131–275.

Winge, Ø. (1940). Taxonomic and evolutionary studies in *Erophila* based on cytogenetic investigations. *Comptes Rendus des Travaux du Laboratoire Carlsberg*, (Ser. Physiol.) **23**, 41–74.

Winkler, H. (1908). Uber Parthenogenesis und Apogamie im Pflanzenreich. *Progressus rei Botanicae*, **2**, 293–454.

Winkler, H. (1916). Uber die experimentelle Erzeugung von Pflanzen mit abweichenden Chromosomenzahlen. *Zeitschrift für Botanik*, **8**, 417–531.

Woodell, S.R.J. (1965). Natural hybridization between the Cowslip (*Primula veris* L.) and the Primrose (*P. vulgaris* Huds.) in Britain. *Watsonia*, **6**, 190–202.

Woodson, R.E., Jr. (1964). The geography of flower color in Butterflyweed. *Evolution*, **18**, 143–63.

Wright, J.W. (1953). Pollen dispersion studies: some practical applications. *Journal of Forestry*, **51**, 114–18.

Wright, S. (1931). Evolution in Mendelian populations. *Genetics*, **16**, 97–159.

Wright, S. (1966). Mendel's ratios. In *The origin of genetics*, ed. C. Stern & E.R. Sherwood, pp. 173–5. London & San Francisco: Freeman.

Wright, S. (1977). *Evolution and the genetics of populations*. Vol. 3, *Experimental results and evolutionary deductions*. Chicago & London: The University of Chicago Press.

Wu, L., Bradshaw, A.D. & Thurman, D.A. (1975). The potential for evolution of heavy metal tolerance in plants. III. The rapid evolution of copper tolerance in *Agrostis stolonifera*. *Heredity*, **34**, 165–87.

Yates, F. (1960). *Sampling methods for censuses and surveys*, 3rd edn. London: Griffin.

Yeo, P.F. (1975). Some aspects of heterostyly. *New Phytologist*, **75**, 147–53.

Young, D.A. & Seigler, D.S. (1981). (eds) *Phytochemistry and angiosperm phylogeny*. New York: Praeger.

Youngner, V.B. (1960). Environmental control of initiation of the inflorescence, reproductive structures and proliferations in *Poa bulbosa*. *American Journal of Botany*, **47**, 753–7.

Yule, G.U. (1902). Mendel's laws and their probable relations to intra-racial heredity. *New Phytologist*, **1**, 193–207, 222–38.

Zirkle, C. (1941). Natural selection before the 'Origin of species'. *Proceedings of the American Philosophical Society*, **84**, 71–123.

Zirkle, C. (1966). Some anomalies in the history of Mendelism. In *G. Mendel Memorial Symposium 1865–1965*, ed. M. Sosna, pp. 31–7. Prague: Academia Publishing House of the Czechoslovak Academy of Sciences.

Zohary, D. & Feldman, M. (1962). Hybridization between amphidiploids and the evolution of polyploids in the wheat (*Aegilops–Triticum*) group. *Evolution*, **16**, 44–61.

Zohary, D. & Nur, V. (1959). Natural triploids in the Orchard Grass *Dactylis glomerata* polyploid complex and their significance for gene flow from diploid to tetraploid levels. *Evolution*, **13**, 311–17.

Index

Plant names If a genus is represented only by a particular species, the full binomial is indexed (e.g. *Acer pseudoplatanus*). Generic names standing alone (e.g. *Achillea*) may include references to the genus only, and to individual species of the genus. In the case of the large genera *Potentilla* and *Ranunculus*, subsidiary references are given to individual species.

Authors cited Only those authors are included whose works are cited in the main text (as opposed to bracketed references). The full references are in the Bibliography.

Page numbers in **bold type** refer to figures.